Prune Production Manual

Technical Editor • Richard P. Buchner

Publication 3507
2012

University *of* California
Agriculture and Natural Resources

To order or obtain ANR publications and other products, visit the ANR Communication Services online catalog at http://anrcatalog.ucanr.edu/ or phone 1-800-994-8849. Direct inquiries to

University of California
Agriculture and Natural Resources
Communication Services
2801 Second Street
Davis, CA 95618
Telephone 1-800-994-8849
E-mail: anrcatalog@ucanr.edu

Publication 3507
ISBN-13: 978-1-60107-702-8; PDF ISBN: 978-1-60107-825-4
Library of Congress Control Number: 2012937267

Photo credits are given in the captions. Front cover: prune trees in bloom; back cover: bloom stages of prune. Cover photos by Michael Poe; design by Celeste Rusconi.

This publication has been anonymously peer reviewed for technical accuracy by University of California scientists and other qualified professionals. This review process was managed by ANR Associate Editor for Pomology, Viticulture, and Subtropical Horticulture Ben Faber.

POD-7/18-SB/CR/SO

Warning on the Use of Chemicals

Pesticides are poisonous. Always read and carefully follow all precautions and safety recommendations given on the container label. Store all chemicals in the original labeled containers in a locked cabinet or shed, away from food or feeds, and out of the reach of children, unauthorized persons, pets, and livestock.

Confine chemicals to the property being treated. Avoid drift onto neighboring properties, especially gardens containing fruits or vegetables ready to be picked.

Do not place containers containing pesticide in the trash nor pour pesticides down sink or toilet. Either use the pesticide according to the label or take unwanted pesticides to a Household Hazardous Waste Collection site. Contact your county agricultural commissioner for additional information on safe container disposal and for the location of the Hazardous Waste Collection site nearest you.

Dispose of empty containers by following label directions. Never reuse or burn the containers or dispose of them in such a manner that they may contaminate water supplies or natural waterways.

Contents

Preface and Acknowledgments

This *Prune Production Manual* updates the *Prune Orchard Management* publication first printed in August 1981 by the University of California Division of Agricultural Sciences. The new manual is a complete rewrite and provides the most current information on all phases of prune production, along with an index and a comprehensive glossary defining technical terminology. The text is written in a nontechnical format designed to be practical and well suited for field application. Most chapters list references so interested readers can research topics in greater detail if they wish. This book is meant to be a stand-alone publication that describes all aspects of prune production. Throughout the manual, tree and crop are described as prune. The term "dried plum" represents a marketing strategy and is described in chapter 2.

Forty-four authors contributed to this publication. Information was drawn from the University of California (UC), the United States Department of Agriculture (USDA), and industry experts. Authors included UC Cooperative Extension farm advisors, specialists, and research assistants; UC faculty from Davis and Riverside; USDA scientists; emeritus advisors; and highly skilled experts working in the prune industry. These authors' efforts are greatly appreciated. Each approached their assignments with a positive, can-do attitude that made this publication possible. Appreciation is also extended to the researchers who donated their time and expertise for the confidential peer review process.

Gary Obenauf, Bill Olson, and Steve Sibbett served on the technical steering committee. Their help was invaluable in planning, implementation, and chapter reviews. Thank you to the California Dried Plum Board for financial support to initiate this project. Appreciation is also extended to the Agriculture and Natural Resources Communication Services staff, especially Publications Manager Ann Senuta and photographer Michael Poe, along with associate editors Ben Faber and Jim Stapleton. Publications project manager Stephen W. Barnett edited and assembled the manual, and it was proofread and indexed by Hazel White. Everyone worked together to make this *Prune Production Manual* a reality.

Richard P. Buchner

Authors

James E. Adaskaveg, Professor, Department of Plant Pathology, UC Riverside
Richard P. Buchner, UC Cooperative Extension Farm Advisor, Tehama County
Joseph H. Connell, UC Cooperative Extension Farm Advisor, Butte County
Kevin R. Day, UC Cooperative Extension Farm Advisor, Tulare County
Carolyn J. DeBuse, UC Cooperative Extension Farm Advisor, Yolo and Solano Counties
Matthew DeCeault, Graduate Student, Department of Plant Sciences, UC Davis
Theodore M. DeJong, Professor, Department of Plant Sciences, UC Davis
James F. Doyle, UC Cooperative Extension Prune Breeding Program Emeritus
John P. Edstrom, UC Cooperative Extension Farm Advisor Emeritus, Colusa County
Patrick Ferreira, Commodity Manager, North Valley Office, California Dried Fruit Association
Allan E. Fulton, UC Cooperative Extension Farm Advisor, Tehama County
Cyndi K. Gilles, UC Cooperative Extension Research Associate, Tehama County
Kurt J. Hembree, UC Cooperative Extension Farm Advisor, Fresno County
Chuck A. Ingels, UC Cooperative Extension Farm Advisor, Sacramento County
Judy Johnson, Research Entomologist, USDA San Joaquin Valley Agricultural Sciences Center, Parlier, CA
Mark Krause, Mariani Packing Co., Vacaville, CA
William H. Krueger, UC Cooperative Extension Farm Advisor, Glenn County
Bruce D. Lampinen, UC Cooperative Extension Integrated Orchard Management Specialist, UC Davis
Rex E. Marsh, UC Cooperative Extension Ecologist Emeritus, UC Davis
Michael V. McKenry, UC Cooperative Extension Nematologist, Kearney Agricultural Center
Themis J. Michailides, UC Cooperative Extension Plant Pathologist, Kearney Agricultural Center
Franz J. A. Niederholzer, UC Cooperative Extension Farm Advisor, Sutter and Yuba Counties
Maxwell V. Norton, UC Cooperative Extension Farm Advisor, Merced County
William H. Olson, UC Cooperative Extension Farm Advisor Emeritus, Butte County
Ken Peterson, KB Packing Co., Marysville, CA
Richard L. Peterson, Executive Director, California Dried Plum Board, Sacramento
Carolyn Pickel, UC Cooperative Extension Integrated Pest Management Advisor Emeritus, Sutter and Yuba Counties
Vito S. Polito, Professor, Department of Plant Sciences, UC Davis
Terry L. Prichard, UC Cooperative Extension Irrigation Water Management Specialist Emeritus, San Joaquin County
Wilbur O. Reil, UC Cooperative Extension Farm Advisor Emeritus, Yolo and Solano Counties
Terrell P. Salmon, UC Cooperative Extension Wildlife Specialist Emeritus, San Diego County
Blake L. Sanden, UC Cooperative Extension Farm Advisor, Kern County
Lawrence J. Schwankl, UC Cooperative Extension Irrigation Specialist, Kearney Agricultural Center
Ken A. Shackel, Professor, Department of Plant Sciences, UC Davis
G. Steven Sibbett, UC Cooperative Extension Farm Advisor Emeritus, Tulare County
George Sousa Sr., Mariani Packing Co., Vacaville, CA
Stephen M. Southwick, Research Director, 0-G Packing Co., Stockton, CA
William E. Steinke, Former UC Cooperative Extension Agricultural Engineering Specialist, UC Davis
Beth L. Teviotdale, UC Cooperative Extension Plant Pathologist Emeritus, Kearney Agricultural Center
Fred Thomas, CERUS Consulting, Chico, CA
Greg Thompson, General Manager, Prune Bargaining Association, Yuba City, CA
James F. Thompson, UC Cooperative Extension Postharvest Operations Specialist Emeritus, UC Davis
Becky B. Westerdahl, UC Cooperative Extension Nematologist, UC Davis
James T. Yeager, Former UC Cooperative Extension Staff Research Associate, UC Davis

Part 1 Prune Industry Overview

1 The Prune Industry in California

• Joseph H. Connell

Prunes are varieties of European plum (*Prunus domestica* L.) that can be satisfactorily dried whole without fermenting at the pit. All prunes are plums, but all plums are not prunes. Plums other than those classified as prunes are not dried whole because they ferment around the pit while drying. High sugar content is one of the factors necessary for successful drying. Like all plums, prunes can be eaten fresh if a very sweet fruit is desired. The California industry grows prune trees and produces prunes. Once prunes are harvested and dried they are marketed domestically as dried plums for better consumer appeal. Many domestic consumers were unaware that prunes were actually dried plums, and dried plums were perceived more positively in market research.

Prune production is an intensive, specialized industry found only in California and a few other areas around the world.

History

The European plum, believed to have originated in western Asia, has been grown in parts of Europe for over 2,000 years. According to Theophrastus, the prune was cultivated in Asia Minor in ancient times. In his *Natural History*, the Roman scholar-politician Pliny the Elder indicated that prunes were already being cultivated at the beginning of the Roman Empire and mentioned 11 varieties that had been introduced into Italy. Growing without cultivation in the Damascus area, the prune was probably introduced into France by the Crusaders. The date plum, or *pruneau d'ente*, was widely known in southwest France by 1529. Through its culture in France, the *petite pruneau d'Agen*—now known as the French prune—had become the leading commercial dried fruit of the world long before its horticultural development in California (see chapter 9, "Varieties").

The *pruneau d'Agen* was introduced to California from France by Louis Pellier (fig. 1.1), a French horticulturist who had come to California seeking gold. Louis had spent his early years working in the orchards and vineyards of France, and in 1850 he established a nursery business in San Jose. Louis's brother Pierre came from France to join the nursery operation in 1853. The two brothers worked the nursery together until the spring of 1856, when Pierre returned to France to marry and bring back a large collection of nursery stock, including fruit scions and cuttings, grape cuttings, and many kinds of seeds.

Pierre Pellier returned to San Jose in December 1856 with the *petite pruneau d'Agen* (procured in the Ville Neuve d'Agen) as part of the collection. Louis Pellier provided French prune scions to John Q. A. Ballou and George W. Tarleton, who top-grafted them on native plum and Damson plum rootstocks previously

Figure 1.1 Louis Pellier, native of France and founder of California's prune industry, established Pellier's City Gardens nursery in San Jose in October 1850. He introduced the French Prune, *la petite pruneau d'Agen*, to California during the winter of 1856–1857. *Photo:* History San Jose.

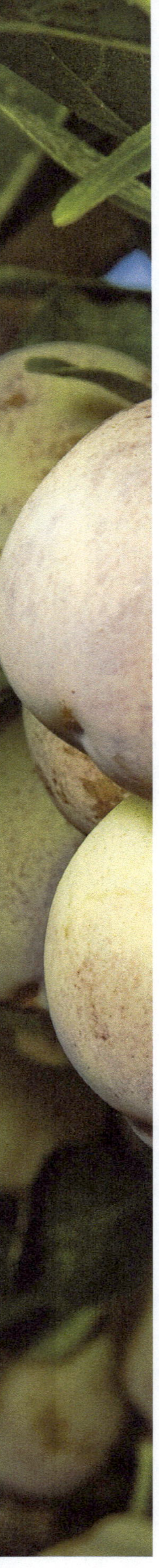

established with material brought from the eastern United States. Their adjoining plots became the state's first prune orchard. Nurseryman B. Kamp also procured grafts from Pellier and worked to increase propagation and distribution of the French prune. He was one of the first to plant prune trees in orchard rows. The French prune grew vigorously and bore heavily, and these early growers were soon producing good dried prunes. In the first record of commercial prune production in California, John Ballou shipped 130 pounds of dried prunes to San Francisco in 1859. In 1868, Ballou sent 11 tons of dried prunes to eastern U.S. markets by ship.

Completion of the transcontinental railroad in 1869 opened a tremendous market for the economical and nutritious California prunes that were available when fresh fruits were out of season. By 1870, there were over 19,000 prune trees in California. The success of the Bradley orchard, planted in 1870 on Stevens Creek Road in San Jose, led others to go into prune growing, and the industry began to grow rapidly. Prune production on a large scale began with the O'Banion and Kent orchard, near Saratoga, planted in 1878; the Dr. Handy orchard of 100 acres, planted in Saratoga in 1880; and the Buxton orchard, planted at Saratoga in 1881. In 1886 California shipped 2 million pounds of dried prunes. Five years later, prune shipments reached 27 million pounds, with over 21 million pounds being shipped from San Jose alone. Pellier's City Gardens nursery was responsible for sparking the beginning of the California prune industry (fig. 1.2) and revolutionizing the Santa Clara Valley. By 1900, more than 35,000 acres of prunes were grown in Santa Clara County, with 90,000 acres statewide. The prune industry in California was firmly established and continued to experience phenomenal growth for the next 30 years.

Figure 1.2 Historical plate of the California prune, the Prune d'Agen, introduced to California from France by Louis Pellier in 1856. Source: Lelong 1892.

California Industry Trends

California's horticultural crop acreage expanded rapidly beginning in 1880, and prune orchard acreage was one of the building blocks of this horticultural age. An all-time high of 193,511 total acres was reached in 1926, followed by a bearing-acreage peak of 174,050 in 1930. Santa Clara County alone had 66,000 acres of prunes at that time. The central coast counties of Santa Clara, Napa, and Sonoma accounted for 80% of California's bearing acreage until about 1950. The coastal counties' prune orchards typically produced lower yields than orchards in the Central Valley, and they required multiple harvests in the coastal climate due to less-uniform fruit maturity and premature fruit drop. With these competitive disadvantages and the urbanization of the Santa Clara Valley, the coastal counties' acreage began decreasing, and production shifted to the Central Valley (fig. 1.3). In recent decades, bearing acreage has fluctuated, with increases occurring in the late 1960s and again during the 1980s and 1990s. After reaching a peak of 86,000 acres in 2000 and struggling with heavy production and lower grower prices, the prune industry and USDA funded a voluntary tree pull program in 2002 and 2003. The industry has experienced an overall decline in bearing acres over the past 10 years, with 2010 bearing acreage standing at roughly 55,000 acres.

After heavy planting in the 1960s, the Sacramento Valley became the dominant prune-growing area in the state. It now produces 81.4% of California's dried prunes, with 73% being produced in the counties of Butte, Glenn, Sutter, Tehama, and Yuba; the San Joaquin Valley produces 18.6%; and the Santa Clara-Napa-Sonoma district is nearly out of the prune business entirely, producing only 0.06% (fig. 1.4). Although 22 of the 58 counties in California report commercial prune plantings, over 78% of the total acreage is situated in five Sacramento Valley counties: Butte, Glenn, Sutter, Tehama, and Yuba (table 1.1). The Central Valley will continue to predominate in California's prune industry since most of the state's nonbearing acreage is planted in the Sacramento Valley, while a smaller percentage of the industry is located in the southern San Joaquin Valley.

Prune trees in the interior valleys are heavier producers than those in former coastal valley plantings, with average yields in the Central Valley close to 2.5 tons per acre. Another advantage of interior valley prunes is that they remain attached to the tree as they mature, allowing a once-over harvest by a mechanical shaker and catching frame. In contrast, prunes grown in cooler climates along the central coast tend to drop as they mature.

Historical Variety Trends

The Petite Prune d'Agen was the earliest variety introduced to California; additional varieties imported from France in the late 1800s included the Imperial and the Robe de Sergeant. Luther Burbank introduced the Sugar prune (a French seedling) in 1899, and the large-fruited Burton prune originated

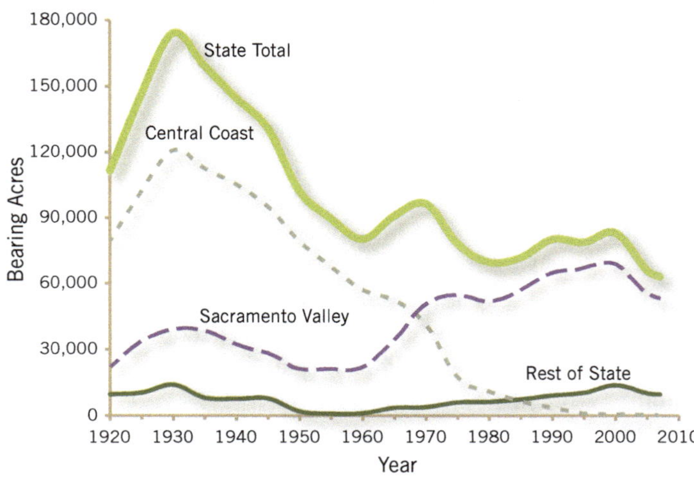

Figure 1.3 California trends in prune-bearing acreage statewide and by districts, 1920–2007. *Sources:* CASS 1982–1991, 1997; CCLRS 1968–1981; Johnston and Dean 1969; NASS 2006, 2008a, 2008b.

Figure 1.4 Prune acreage distribution and production areas in California, 2007. Each dot represents 500 acres. *Sources:* Prune Marketing Committee 2004; NASS 2008a.

from a seedling planted in 1906 by R. E. Burton in Vaca Valley. In 1940, there were 24,500 acres of non-French varieties grown in California, 16.7% of the total acreage. The Silver, Sugar, German, Bulgarian, Hungarian, and other minor varieties that grew principally in Santa Clara County in the early 1900s have all but disappeared.

Several additional prune varieties, including Gerrans Early French, Moyer, Victor Large French, Friedman French, Ross, Punian, and 707 French have been tried in California since the late 1950s, but none of these became significant in acreage. A total of only 2,453 acres of Burton, Friedman, German, Imperial, Italian/Fallenberg, Miro, Moyer, Punian, Ross, Robe de Sergeant, Sierra Sweet, Sugar, Victor Large, 29-C, and 707 were reported to be growing in 2007. This group comprises only 3.5% of the total acreage planted to prunes but accounts for 17.6% of the nonbearing acreage (table 1.2). Sutter and Muir Beauty are two new varieties recently released from the University of California prune breeding program. Although the California prune industry is trying these varieties, their small acreage does not show up in acreage reports. Of the California prune acreage today, the vast majority, 96.5%, is comprised of French types: French, Improved French, and Gerrans Early French. French prune dominates due to its more consistent and greater production, its easier handling in the orchard, its ability to be mechanically harvested by catching frames with less fruit injury, and its ease of drying and processing, including pitting.

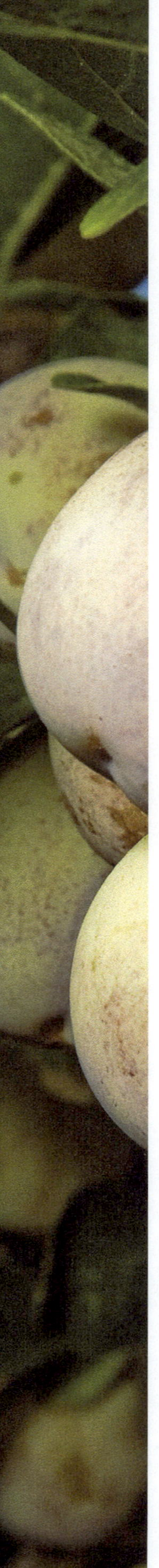

Table 1.1. Dried prune acreage, production, and production per acre in California by county and region

		2007 Acreage			2003-2004 net tons	
District	County	Bearing	Nonbearing	Total	Total dried prune production	Average production per acre
Santa Clara-Napa-Sonoma	Napa	5	0	5	0	0.0
	Santa Clara	85	0	85	97	0.8
	Santa Cruz	15	0	15	0	0.0
	Sonoma	35	0	35	0	0.0
	TOTAL	**140**	**0**	**140**	**97**	**0.8**
Sacramento Valley	Amador	13	0	13	0	0.0
	Butte	8,777	1,165	9,942	25,645	2.7
	Colusa	1,317	39	1,356	4,322	2.1
	Glenn	6,537	311	6,848	14,934	2.1
	Placer	107	0	107	684	2.2
	Shasta	57	0	57	214	4.0
	Solano	1,139	27	1,166	2,260	1.8
	Sutter	17,427	2,780	20,207	45,903	2.3
	Tehama	7,725	622	8,347	17,866	2.1
	Yolo	1,577	175	1,752	7,314	4.3
	Yuba	8,475	693	9,168	23,446	2.5
	TOTAL	**53,151**	**5,812**	**58,963**	**142,588**	**2.6**
San Joaquin Valley	Fresno	3,035	91	3,126	10,861	3.4
	Kern	0	17	17	61	1.0
	Kings	0	34	34	0	0.0
	Madera	1,133	0	1,133	4,402	3.1
	Merced	1,716	125	1,841	5,183	2.6
	Stanislaus	25	0	25	126	2.6
	Tulare	3,795	571	4,366	11,882	2.4
	TOTAL	**9,704**	**838**	**10,542**	**32,515**	**2.5**
STATE TOTALS		62,995	6,650	69,645	175,200	2.0

Sources: Production data from Prune Marketing Committee 2004, 2007. Acreage from NASS 2006, 2008a.
Note: Acreage totals may not add due to rounding; nonbearing acreage includes plantings for 2002–2007.

Worldwide Geographic Distribution

California accounted for an estimated 43% of world prune production in 2011. California's primary historical competitor, France, now produces about 16% of the world supply and has moved into third place behind Chile. Chile and Argentina have increased prune production dramatically, producing an estimated 24 and 15% of the world supply, respectively, in 2011, which led to increased competition in the export market. Other prune-producing countries include Australia, South Africa, and Italy, which collectively produced approximately 2% of world supply in 2011. Yugoslavia's production, once second in the world, has virtually disappeared due to plum pox disease, political trade disruptions, and civil war in the early 1990s.

Economics

Land that is suitable for prune production may also be suitable for a variety of crops. Decisions concerning the appropriateness of development, expansion, sale, or lease of prune orchards must be based on all available data. The entry of Chile and Argentina into the international prune market is a significant factor that must be considered. Bearing acreage of prunes, yields, planting trends, costs of developing and producing a crop, prices received, and current and past market information must be carefully studied before making a planting or purchase decision.

To be commercially successful, a grower must produce acceptable-quality prunes in sufficient quantity to be competitive with other producers in California and around the world. Total costs can be fully recovered only at yields that are above average for the industry and during years with excellent prices. Growers with lower-yielding orchards may cover cultural costs when prices are good but will face real hardship when inventories increase and prices fall.

Cost analyses detailing expected costs per acre to establish and produce prunes are available online at the UC Davis Department of Agricultural and Resource Economics website. Developed with Cooperative Extension Farm Advisors, these cost

Table 1.2. Acreage summary of French varieties compared with all other varieties, 2007

Variety	Bearing acres	%	Nonbearing acres§	%	Total acres	%
French types*	61,708	98.0	5,482	82.4	67,190	96.5
Other varieties†	1,285	2.0	1,168	17.6	2,453	3.5
Total	62,993	90.5	6,650	9.5	69,643	100.0

Source: CASS 2008.
Notes:
*Includes acreage of French, Improved French, and Gerrans Early French varieties.
†Includes acreage of Burton, Friedman, German, Imperial, Italian/Fallenberg, Miro, Moyer, Punian, Ross, Sergeant/Robe De Sergeant, Sierra Sweet, Sugar, Victor Large, 29-C, and 707 varieties.
§Nonbearing acres include plantings from 2002 to 2007.

studies can serve as guides and worksheets to help arrive at costs for a proposed orchard establishment project or to analyze costs in an existing orchard.

Prunes are grown under such variable climatic conditions and soil types that actual costs may vary widely from projections. Major deviations from projected per acre costs may occur based on water costs, machinery, disease and insect infestations, geographic location, and land value. Total cost estimates must include cost of land, planting, irrigation, buildings, labor, and equipment. In addition, taxes, insurance, depreciation of machinery and buildings, and interest on investment must be considered.

Harvesting and drying are the largest cash costs in prune production, with harvest conducted exclusively by mechanical shake and catch techniques. Pruning, the next major cost item, is often done with mechanical toppers to reduce the cost of hand pruning. Fertilization, irrigation, and pest control are the next most significant costs involved in producing prunes.

The size of an orchard has a significant impact on per-acre costs due to economies of scale. The cost per acre to establish or produce prunes decreases as size increases; costs of machines and overhead can be applied to large areas, and the grower can buy products in larger volumes, receiving cost advantages.

Farm investments do not normally produce a rate of return as high as those of many nonfarm investments. However, other considerations such as risk and capital gains on real estate can make agricultural investments attractive. To examine an investment, a present value analysis is helpful. The concept of this analysis is to determine the present value of the total annual expense and income flows that would occur each year until an orchard is developed and while it is farmed over its useful life. This analysis can help determine the estimated rate of return possible on the farming operation. The effects of tax strategies vary widely with each individual and must be considered as well in making a wise investment decision.

References

Amblard, C., and J. M. Delmas. 2006. World prune production: History of evolution and outlook. Proceedings of the 9th International World Prune Association Congress, Cagliari, Italy. Sainte Livra de Sur Lot, France: International Prune Association.

California Dried Plum Board and the Prune Marketing Committee. 2008. California prune industry marketing program 2002–2006 omnibus report. Sacramento: California Dried Plum Board.

CASS (California Agricultural Statistics Service). 1997. 1996 California prune acreage survey. Sacramento: CASS.

———. 1982–1991. California fruit and nut acreage annual reports. Sacramento: CASS.

———. 2008. 2007 California prune acreage survey. Sacramento: National Agricultural Statistics Service, California Field Office.

CCLRS (California Crop and Livestock Reporting Service). 1968–1981. Fruit and nut acreage reports. Sacramento: CCLRS.

Couchman, R. 1967. The Sunsweet story: A history of the establishment of the dried tree fruit industry in California and of the 50 years of service of Sunsweet Growers, Inc. San Jose: Sunsweet Growers.

Johnston, W. E., and G. W. Dean. 1969. California crop trends: Yields, acreages, and production areas. Berkeley: California Agricultural Experiment Station Circular 551.

Lelong, B. M. 1892. Annual report of the State Board of Horticulture of the State of California for 1891. Sacramento: State Printing.

NASS (USDA National Agricultural Statistics Service). 2006. 2005 California prune acreage report. Sacramento: NASS.

———. 2008a. 2007 California prune acreage report. Sacramento: NASS.

———. 2008b. California historic commodity data,

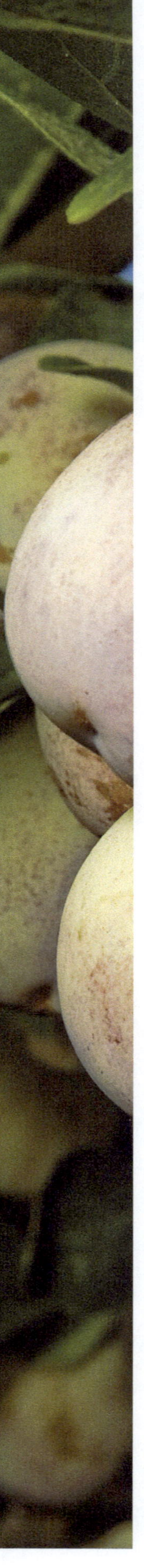

California dried plums (prunes), 1920–2007. Sacramento: NASS.

Niederholzer, F. J., W. H. Krueger, R. P. Buchner, K. M. Klonsky, and R. L. De Moura. 2007. 2007 Sample costs to establish a prune orchard and produce prunes, Sacramento Valley. UC Davis Department of Agricultural and Resource Economics Cost Studies website, http://coststudies.ucdavis.edu/.

Prune Marketing Committee. 2004. Final statistical report 2003–04. Sacramento: California Dried Plum Board.

———. 2007. Final statistical report 2007. Sacramento: California Dried Plum Board.

Wickson, E. J. 1926. The California fruits and how to grow them. 10th ed. San Francisco: Pacific Rural Press.

2 Marketing

• Richard L. Peterson

The World Prune Situation

World prune production has historically been between 250,000 and 275,000 short tons per year, with California enjoying a market share of close to 70%. France has traditionally been California's largest competitor, usually accounting for 15 to 20% of world prune production. French prunes benefitted from European Union (EU) subsidies established in 1978, as well as the ongoing 9.6% EU import duty. The EU paid subsidies to prune processors, who agreed to pay guaranteed minimum prices to French prune growers. These guaranteed grower prices were about twice as high as the prices paid to California growers, which enabled French growers to compete with California's typically larger, more efficient prune-growing operations.

Dramatic changes are under way in the world prune market with production shifting from the Northern Hemisphere to the Southern Hemisphere. A voluntary California tree pull program was funded by the USDA and industry contributions in 2002 and 2003 to reduce the oversupply caused by the dramatic increase in new plantings in the mid-1990s. The reduction in bearing acreage from 86,000 acres in 2001 to 70,000 acres in 2004 led to a 106% increase in the average grower return per ton, from $726 in 2001 to $1,500 in 2004. Not anticipated at the time of the tree pull program were weather-related crop disasters in 2004, 2005, and 2007 that caused California's world market share to drop below 50%.

Meanwhile, Chile and Argentina have been aggressively planting new prune orchards, and Chile's production surpassed that of France in 2006. This poses a serious threat to California's exports, which average 50% of global shipments, because Chile and Argentina have minimal domestic prune consumption and rely on exports to absorb their increased production. Chile and Argentina do not fund marketing campaigns to expand demand in their export markets, but rather to undercut California prune prices to increase their market shares. This has lowered California's profitability in these markets.

Chilean prune exports were aided greatly in January 2003 when the EU awarded duty-free status to all Chilean prune products. This further widened the price advantage Chile already enjoyed versus California prunes due to lower labor costs.

With the discontinuance of USDA foreign prune production reporting due to federal budget cuts, the International Prune Association (IPA) (fig. 2.1) is the only reliable source of information on foreign prune production, inventory, sales, and acreage. The IPA was formed in 1990 by prune growers in the seven largest prune-producing countries: France, Chile, Argentina, Italy, Australia, South Africa, and the United States. The objectives of the IPA are to

- establish a permanent connection among the world's prune producers
- coordinate efforts to safeguard the producers' interests
- increase the worldwide consumption of prunes

Figure 2.1 IPA logo. *Source:* California Dried Plum Board.

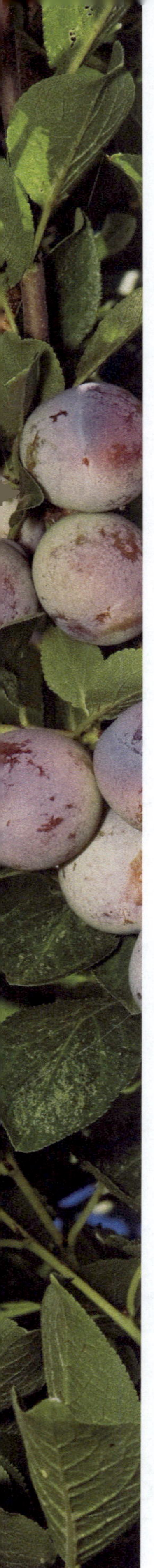

- exchange information on production and sales
- organize industry meetings among the members

The primary benefits of the IPA have been better understanding among the worldwide grower community and improved information on production, supply, and shipment for member countries.

As shown in table 2.1, Chile's prune production is projected to increase by 118% to 120,000 metric tons, and Argentina by 25% to 50,000 metric tons, by 2014.

A paper presented at the June 2009 IPA Conference in Agen, France, revealed that world prune production and sales have increased by an average of 1% annually over the past 40 years. The IPA projected that world prune production would increase by over 54,000 tons (+18%) by 2016, primarily due to South American production increases. It was agreed that developing new export markets that could absorb this increased production would be essential to avoiding global oversupply conditions and reductions in prune prices and profitability.

This projected threat had prompted the IPA member countries to contribute a proportionate amount of funding based on world market share to launch a pilot program in South India to stimulate generic demand for prunes in 2007. The campaign's focus is on communicating the nutritional attributes of prunes as compelling reasons to buy them. Public relations and in-store promotions are the primary vehicles for this generic campaign.

California Marketing Situation

The California prune industry consists of 20 packers and over 900 growers. All are represented by a California marketing order that is administered by the California Prune Board (CPB). The CPB recently completed a comprehensive industry strategic plan that addressed the very challenging competitive global prune marketing situation. An outside facilitator met with growers, packers, CPB staff, and consultants to develop a comprehensive and coordinated global strategic plan that would

- support the long-term health of the industry
- represent areas of common interest
- provide benefits to all industry segments
- be used to make informed decisions about funding and programming

From these sessions, the following mission statements were developed:

Industry: To sell more California prunes to more people worldwide.

CPB: To protect and support the health, growth, and integrity of the California prune industry.

It was agreed that the focus of the CPB's activities should be:

- Generic promotions: Communicate differentiated benefits and encourage large-scale acceptance of California prunes.
- Research: Market research for measurement; production research to improve grower efficiency; nutrition research to identify positioning opportunities and value messages.
- Trade policy: Protect the interests of the California prune industry globally.
- Issues management: Protect the health of the industry.

In addition to the domestic (U.S.) market, which still accounts for about 50% of industry shipments, the following export markets were identified as priorities for promotion:

- Primary markets: Japan, United Kingdom, and Italy.
- Developing markets: China, India, and Greece.

The most critical issues facing the California prune/dried plum industry over the next 5 years were identified as

- shrinking demand

Table 2.1 Prune production forecasts, 2011

	2011 (1,000 tons)	2016 (1,000 tons)	Δ	Δ%
Argentina	44.1	55.1	11.0	25%
Australia	1.8	6.5	4.7	258%
California	131.0	131.0	0	0%
Chile	71.7	110.2	38.5	54%
France	49.6	49.6	0	0%
Italy	2.1	2.3	0.2	0%
South Africa	2.0	2.3	0.3	0%
World Total	302.3	357.0	54.7	+ 30

Source: Amblard 2010.

- competition from exports
- increased production costs
- difficulty of marketing prunes

The objectives of the strategic plan were to

- stop the consumption decline
- increase demand
- improve returns
- maximize financial resources
- address changing markets
- unite the industry under a common vision for greater cooperation and synergy

The strategic planning process revealed the following opportunities and strategies.

Opportunity 1: Export marketing

- Strategy 1: Block foreign competition from taking additional market share by reestablishing credibility and trust with the trade through under-the-radar "guerrilla" marketing tactics that support relationships with retailers but don't indirectly benefit foreign competitors. In the longer term, stop the decline and begin to reclaim lost market share by targeting trade and influencer segments, since the more market share gained by low-price competitors the lower the profitability in the export market.

Opportunity 2: Economic factors

- Strategy 2: Production research; increase grower efficiency and differentiate the product for a competitive advantage.
- Strategy 3: New applications and uses; increase demand through new value-added products sold into industrial markets.
- Strategy 4: Communications; influence demand through communication of the differentiated value of California prunes.

One of the ways the industry is attempting to increase promotional spending is by implementing a credit-back for market promotion program. This program is designed to stimulate increased promotional spending by California prune processors and has been successfully used in the California almond and raisin industries.

The program's objective is to effectively promote the sale, use, and consumption of California prunes. This voluntary program allows processors to conduct and submit eligible marketing activities in return for credit up to their total amount of processor assessments that fund generic CPB promotion activities. Processors must spend $2.00 on eligible marketing activities in order to receive $1.00 in assessment credit as an incentive for higher processor marketing spending.

Figure 2.2 Dried plum logo. *Source:* California Dried Plum Board.

Generic Marketing Programs

Domestic Marketing

In June 2000, the California prune industry was granted approval by the U.S. Food and Drug Administration (FDA) to begin using the name "dried plums" on retail packages as long as "prunes" was also used. This dual labeling approach was required by FDA for 2 years to avoid confusion among consumers.

The California Prune Board (subsequently changed to the California Dried Plum Board, CDPB) had requested this name change because research showed that many U.S. consumers didn't know that prunes were dried plums. The name "dried plums" was also perceived more positively by research participants than "prunes." This name change was made only for the domestic market, since prunes do not have the same image problem overseas (fig. 2.2).

Nutrition research and public relations have been the cornerstones of the CDPB's U.S. marketing programs in recent years. Prunes have been positioned as the number-one fruit for digestive health due to their antioxidants, fiber, sorbitol, and potassium. The CDPB's initial goals were to create awareness about the importance of digestive health among health professionals and dietitians and to increase awareness that dried plums are an essential component to digestive health.

More recently, the emphasis has shifted to consumers with a target audience of women 25 to 54 years old. Frequent travelers, fitness enthusiasts, and new and expectant mothers have been targeted through due to their digestive health challenges and receptivity to the digestive health messages of the

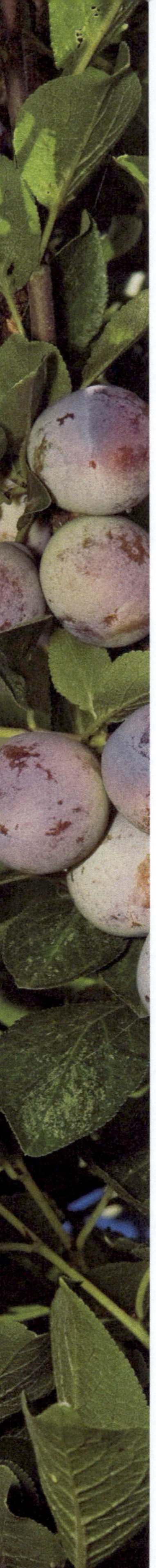

CDPB. Media spokespersons communicated the California dried plum digestive health story through radio, television, magazines, newspapers, and online at the California Dried Plum website Tummywise.com.

After establishing California dried plums as the leading fruit for digestive health due to their sorbitol, fiber, antioxidants, and potassium, the CDPB broadened its communications in 2008 to a Superfruit campaign that promoted dried plums' multiple health benefits (table 2.2).

The CDPB sponsors nutrition and medical research to discover, characterize, and quantify nutrient and nonnutrient components and health-promoting functions of prunes. The research program works directly with scientists at major research institutions to develop an objective research database on which to build credible health communications and generic marketing programs.

The nutrition research program maintains prunes' role in a balanced and health-promoting diet and expands knowledge about the role of dietary fiber, sorbitol, and phenolic compounds in digestive health. To achieve this, the program prospects and funds nutrition research; leverages opportunities in ongoing research funded by other institutions and organizations; maintains effective communications with the nutrition research community; monitors regulatory changes in federal food programs for opportunities to include prunes; supports conferences, summits, seminars, roundtables, alliances, and memberships in organizations as appropriate; and provides timely research-based information to nutrition educators, health and fitness practitioners, medical practitioners, industry members, and consumers.

A nutrition advisory panel helps maintain the scientific integrity and credibility of the nutrition research program, develops the research agenda, and reviews project proposals. Studies have been conducted on prunes' role in lowering the risk of colon cancer, osteoporosis, foodborne pathogens in meat, and oral and dental health problems.

The nutritional attributes of prunes also position them well in the industrial market. With the food industry's focus on issues related to health and nutrition, such as reducing fat, saturated fat, trans fat, carbohydrates, calories, and sugar, prunes' abilities to address these concerns are affording them renewed interest as a natural food ingredient.

Even larger opportunities are offered by food processors' problems with shelf life and shelf stability, since many of the properties of prunes that contribute to their nutritional benefits also directly impact food product shelf life and shelf stability. The antioxidants in prunes suppress the normal growth of bacteria that impacts the shelf life of refrigerated meat products. Prunes' fiber and sorbitol help maintain moisture in meat, bakery, and many snack products. Prunes' organic acids retard mold development in baked goods while also improving the flavor of reduced-fat bakery products. The objectives of the industrial program are to

- increase awareness of dried plums as a natural, innovative way to raise the value of meat cuts with higher profit margins
- demonstrate the effectiveness of dried plums as ingredients
- achieve adoption in existing and new products

International Marketing

The export market was identified through the strategic planning process as the primary opportunity for the California prune industry. This market normally accounts for about half of industry shipments, with California prunes being exported to over 60 countries. The California prune industry has been actively promoting its products in Western Europe and Japan since the mid-1980s with the financial assistance of the USDA's Foreign Agricultural Service. The CDPB has received millions of dollars in match-

Table 2.2. Prune nutrient information

	Blood sugar control	Heart health	Bone health	Digestive health	Cancer protection	Antioxidant power
dietary fiber	•	•		•	•	
sorbitol	•			•	•	
potassium		•				
copper	•	•	•			
vitamin K		•	•			
boron		•	•			
phenolics		•	•		•	•

Source: Stacewicz-Sapuntzakis 2001.

ing funds through the Targeted Export Assistance program, the Market Promotion Program, and the current Market Access Program (MAP). The CDPB must commit industry funding in order to qualify for MAP funds.

The California Dried Plum Board's international market development strategy has been to use generic marketing activities to supplement California packer marketing activities in priority markets identified by the industry, including the core export markets of Japan, the United Kingdom, and Italy and developmental markets such as China, India, and Greece.

The CDPB's activities generally address both the trade and consumer segments of its international markets. Communications emphasize the nutritional attributes of California prunes and position the product as a healthful snack, with a secondary positioning in some markets as a convenient, versatile cooking ingredient. Trade activities keep importers, wholesalers, distributors, rebaggers, and retailers apprised of CDPB activities and California prune industry news such as research findings in crop size and health benefits. In all export markets, trade communications focus on why buyers should "Taste the Quality" and choose California over its low-priced foreign competitors (fig. 2.3).

Figure 2.3 International prune logo. *Source:* California Dried Plum Board.

Prune distribution is excellent worldwide, and export demand has expanded in recent years. Continued expansion of demand is expected with the more aggressive marketing that will follow as worldwide prune production increases, leading to the projected world surplus of 100,000 metric tons. Clearly, the California prune industry's fate is more closely tied to the global marketplace than ever before.

The future for prunes worldwide will continue to be challenging due to economic threats, environmental restrictions, the increasing availability of substitute fruits and snacks, and changing consumer food preferences. The California prune industry's ability to meet these challenges will depend on its success in adapting to changing conditions, improving the quality of its products, developing more value-added prune products, and communicating compelling reasons why consumers should eat prunes.

For the latest information about dried plums, or prunes, see the websites of the California Dried Plum Board, http://www.CaliforniaDriedPlums.org, and the International Prune Association, http://www.ipaprunes.org.

References

Amblard, C. 2010. World prune production: Statistical update. Proceedings of the 12th International World Prune Association Congress, Cape Town, South Africa: International Prune Association.

Stacewicz-Sapuntzakis, M. 2001. Chemical composition and potential health effect of prunes: A functional food? Critical Reviews in Food Science and Nutrition 41(4): 251–286.

Part 2 Prune Biology

3 Setting the Crop: Flowering, Pollination, and Fruit Development

• Vito S. Polito, Matthew DeCeault, and Maxwell V. Norton

Prunes are included in the species *Prunus domestica*, the European plum. There are several prune- and plum-producing cultivars in the species. The French cultivar is the most important in terms of prune production in California, although recent introductions have led to some cultivar diversification. The genus *Prunus* is included in the Rose family (Rosaceae). In addition to prunes and other species of cultivated plums, *Prunus* includes the other stone fruits (peach, nectarine, cherry, and apricot), almond, and numerous other species cultivated as rootstocks and ornamentals.

Prune flowers are perfect, or hermaphroditic, that is, both male and female reproductive structures (stamens and pistil) are present in the same flower (fig. 3.1). Each flower has five sepals, five petals, numerous stamens, and a single pistil. The bases of the sepals, petals, and stamens are united to form a floral tube or cup partially enclosing the central pistil and lined with a nectar-secreting area. The flowers are formed in buds containing flowers but no leaves.

The buds form during the growing season prior to bloom. They form laterally in the axils of leaves on the new growth of spurs or, in some cases, in leaf axils at the older nodes of the current season's shoot growth. Structurally and developmentally, the spur shoots are similar to the long shoots, except a spur shoot typically produces fewer nodes (the leaf- and bud-producing regions of a stem) during a season's growth, and internode elongation of the spur shoots is greatly suppressed relative to that of the long shoot (figs. 3.2 and 3.3). Leaf (vegetative) buds are formed terminally on both spurs and the long shoots. Most lateral buds on the long shoots are vegetative as well; these can develop as spur shoots during the next season. Often, more than one bud forms at a given node on the long shoots; in such shoots

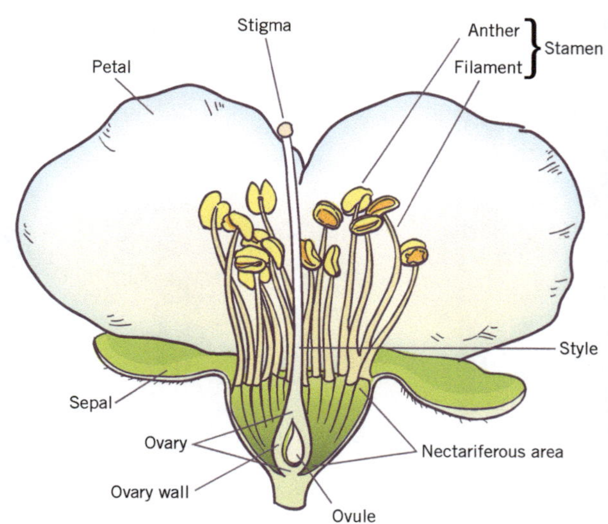

Figure 3.1 The prune flower.

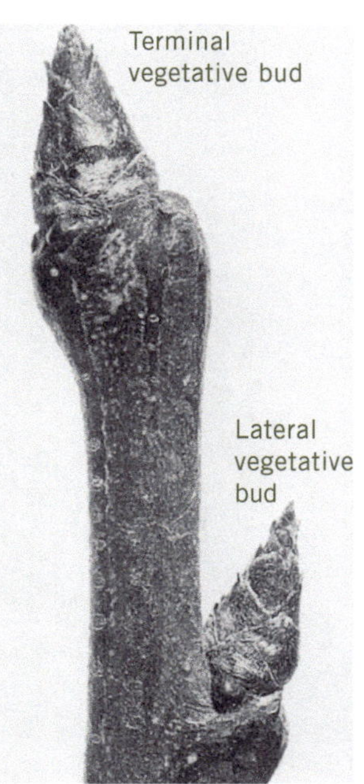

Figure 3.2 Long shoot showing the terminal vegetative bud and one node with a lateral vegetative bud. *Photo:* V. S. Polito.

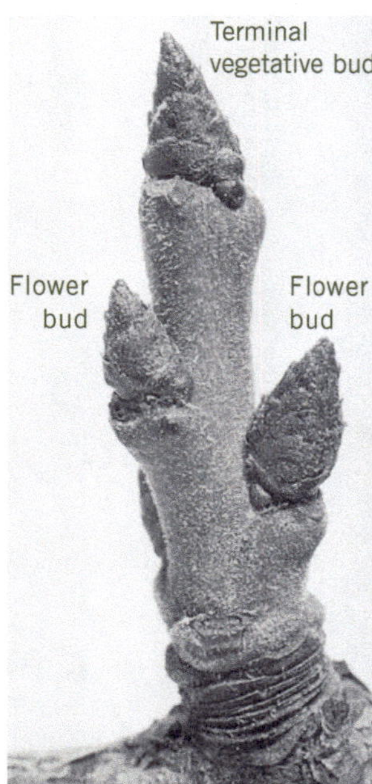

Figure 3.3 One-year-old spur shoot with a terminal vegetative bud and three nodes, each with a lateral flower bud. *Photo:* V. S. Polito.

one bud is floral and the other vegetative (fig. 3.4). Mixed buds containing both flowers and leaves do not occur in prune.

Bud Development

When lateral buds are first initiated, they consist of a dome-shaped group of cells important to the subsequent development of the plant. These cells retain the ability to divide, and from the clusters of dividing cells either leaves and new buds or flowers will ultimately form. These growth regions are called apical meristems. They begin their activities by initiating bud scales as minute primordia on the flanks of the dome.The scales subsequently grow to enclose the bud (figs. 3.5 and 3.6). During this stage of development, vegetative and floral buds have a similar appearance. Only after the initiation and development of the bud scales does the identity of the bud become apparent at a microscopic level. Vegetative buds will continue to produce lateral primordia. At first they appear similar to bud scale primordia but instead of developing into scales, they continue development, forming the first leaves to emerge at bud break the following spring. New buds will be initiated in their axils (fig. 3.7). The apical meristems of the floral buds undergo a series of changes leading to the initiation and development of flowers.

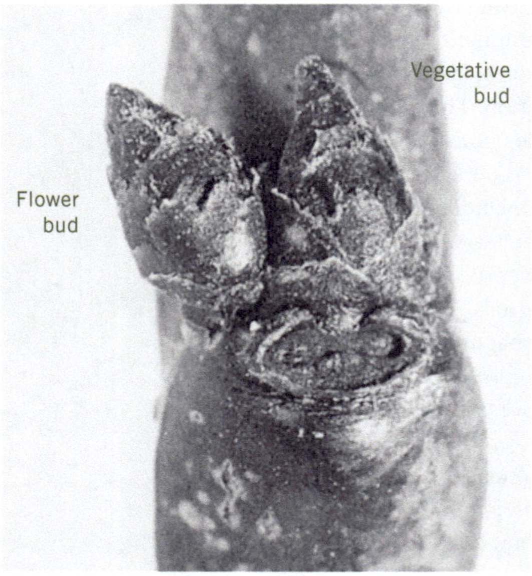

Figure 3.4 One node of a long shoot with a vegetative bud and a flower bud. *Photo:* V. S. Polito.

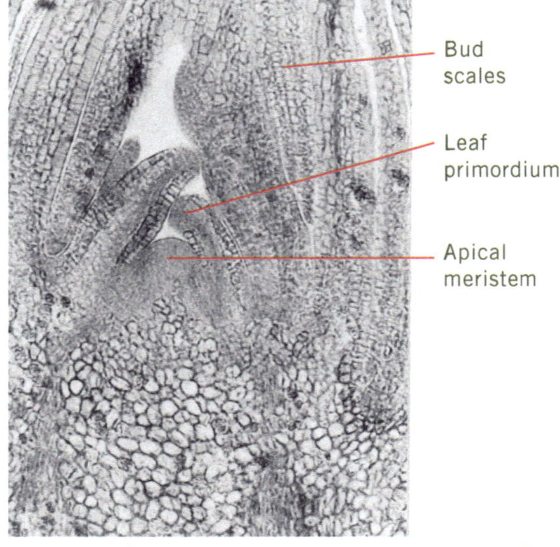

Figure 3.5 Section through a vegetative shoot apex. The pointers indicate (top to bottom) bud scales, leaf primordium, and shoot apical meristem. *Photo:* V. S. Polito.

Figure 3.6 Scanning electron microscope image of a shoot apex enclosed by a primordial bud scale. *Photo:* V. S. Polito.

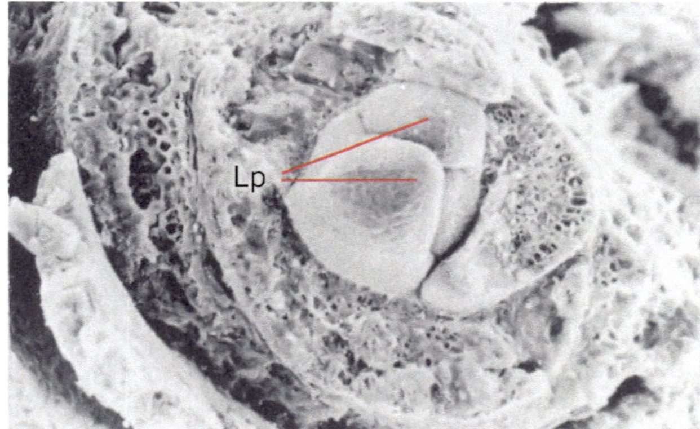

Figure 3.7 Scanning electron microscope image of a shoot apex with the bud scales completely removed to show the young primordial leaves. *Photo:* V. S. Polito.

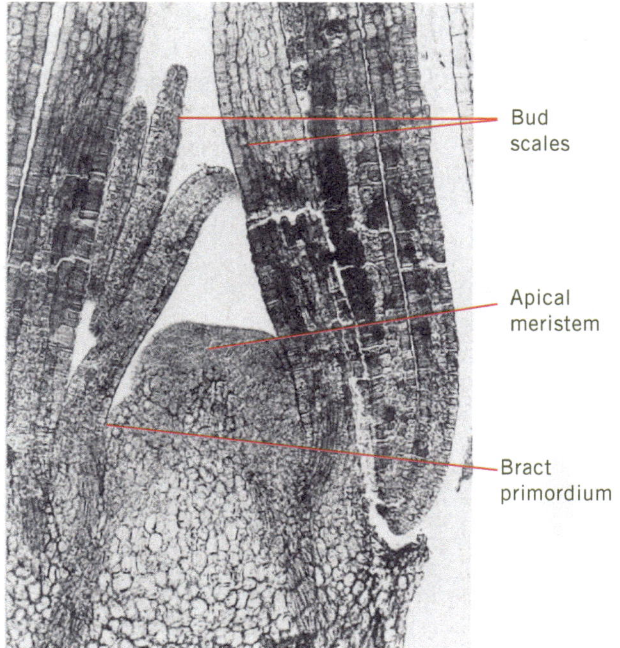

Figure 3.8 Section through an early-stage reproductive shoot apex. The apex is enclosed by bud scales (top two pointers). The shoot apical meristem (pointer 3) has produced a lateral bract (lower pointer), which is the site of the initiation of a flower primordium. Note that the shoot apical meristem is broader and flatter than the vegetative shoot apical meristem seen in figure 3.5. *Photo:* V. S. Polito.

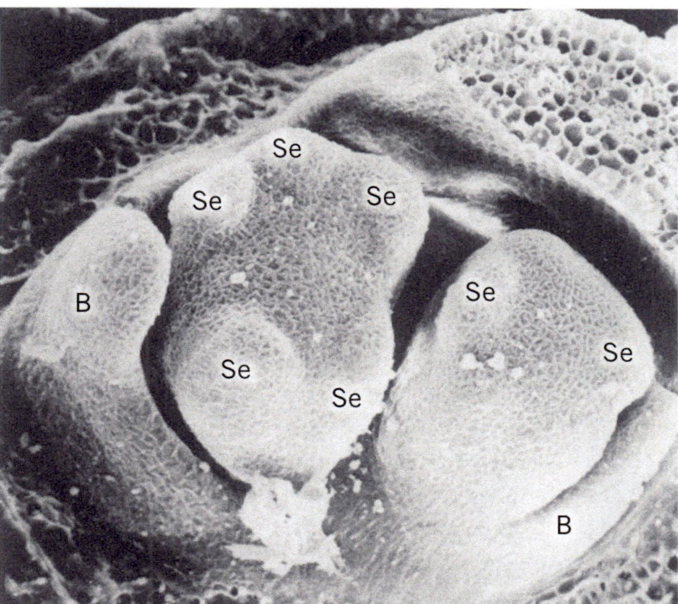

Figure 3.9 Scanning electron microscope image of a flower bud with two developing flower primordia. The flowers arise in the axils of the bracts (B). The flowers in this image have initiated the sepal primordial (Se). The flower on the left, which is the first formed flower, has all five sepals initiated. The one on the right has only two at the time this image was made. *Photo:* V. S. Polito.

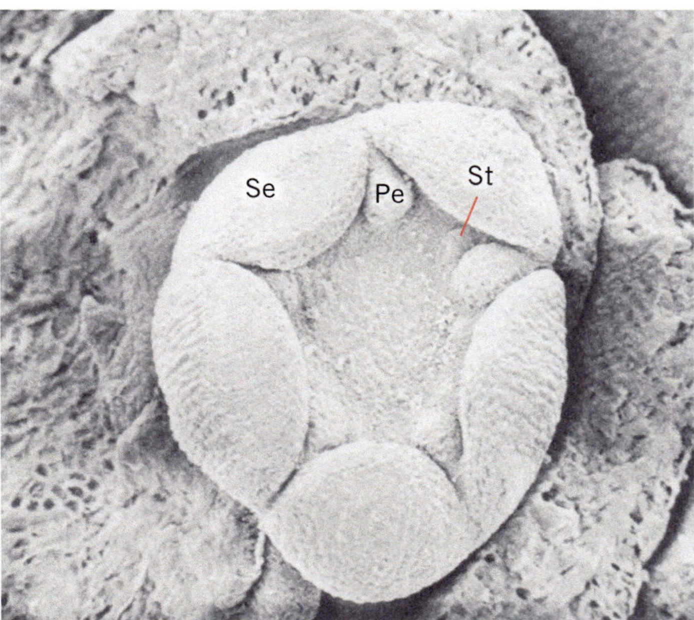

Figure 3.10 Scanning electron microscope image of a prune flower that has initiated sepals (Se), petals (Pe), and the first stamens (St). *Photo:* V. S. Polito.

Flower Initiation and Development

Early in summer, the floral buds undergo a transition to the reproductive state. One or two small, leaflike structures called bracts are initiated in each floral bud in a manner similar to bud scales or leaves. Flower-producing meristems arise in the axils of the bracts (fig. 3.8). The apex of the floral meristem becomes enlarged and flattened (figs. 3.8 and 3.9), and the flower parts are initiated beginning with the outermost whorl of sepals and proceeding inward to the petals, stamens, and pistil (fig. 3.10). Ultimately, the entire meristem is used up in the production of the flower. This is in contrast to vegetative buds, whose meristems retain the capacity to produce leaves indefinitely.

There can be considerable variation in the timing of flower development, particularly in the early stages, which vary not only from season to season but among the flower buds on a single tree as well. Thus, the timing of the various stages given in the following discussion is approximate. About mid-July, the five sepals are initiated as protrusions on the periphery of the flat floral meristem (see fig. 3.9). As these sepal primordia expand, five petals arise in a similar fashion and stamen primordia form interior to the petals (see fig. 3.10). As the bases of the sepals, petals, and stamens grow together, they form the floral cup, which partially encloses the pistil in the fully developed flower. The pistil is the final structure to be initiated. Arising late in September from the center of the floral apex (fig. 3.11), this tissue grows upward and develops into the mature pistil.

Flower Structure

Figure 3.1 illustrates a flower at bloom. The stamen primordia have developed into mature stamens consisting of the filament, a slender stalk that remains attached to the floral tube at its base, and an anther at the tip of the filament. During the weeks preceding bloom, numerous pollen grains develop in the anthers. When the flower bud opens, the pollen sacs split open along their margins to release the pollen grains.

The pistil consists of three parts: a terminal stigma, an elongate style, and a basal ovary (see fig. 3.1). Pollen is transferred to the surface of the stigma, where germination occurs. The pollen tubes grow through the style to the ovary. The fruit develops from the ovary, consisting of an ovary wall enclosing a single locule. The ovary wall has three layers: an inner layer (endocarp) forming the pit, a middle layer (mesocarp) forming the fruit flesh, and an outer layer (exocarp) forming the skin of the fruit. In the locule formed by the ovary wall are two ovules, one of which usually disintegrates; the other is fertilized and develops into the seed. Each ovule is enclosed by two integuments that develop together into the seed coat. Contained in the integuments is the nucellus, which functions as a source of nutrients during the early stages of seed development, and the embryo sac, which is discussed below.

Pollination and Fertilization

Insect pollinators, usually honey bees, carry pollen to the receptive surface of the stigma, where the grains germinate to form pollen tubes. French is a self-compatible variety, which means a flower can be pollinated and fertilized with its own pollen or pollen from other flowers on the same tree. Other prune cultivars may be self-incompatible. For self-incompatible cultivars, pollen from another compatible cultivar must be transferred to the stigma for fertilization and fruit set to occur.

The pollen tubes grow into the stigma and down the style to the ovary. Pollen tube growth through the style takes several days. As the pollen tubes grow through the style, further development occurs in the ovule to complete the formation of the embryo sac, a complex structure containing eight nuclei arranged in seven cells, one of which is the egg. When a pollen tube reaches the ovule, it enters through an opening called the micropyle, a small pore in the integuments that enclose the ovule. Each pollen tube contains two sperm cells. After a pollen tube penetrates the ovule, its contents are released into the embryo sac and fertilization occurs. One of the sperm cells fertilizes the egg cell to form a zygote, and the second sperm fuses with a pair of embryo sac nuclei known as polar nuclei. This double fertilization causes the formation of two distinct structures: the zygote, which develops into an embryonic plant in the seed, and the endosperm, which is formed by cell proliferation following the fusion of the second sperm with the two polar nuclei. The endosperm nourishes the developing embryo as the seed matures; its development precedes that of the embryo, and it is consumed as the embryo attains maturity.

Figure 3.11 Scanning electron microscope image of two flowers at the time of pistil initiation (arrow). *Photo:* V. S. Polito.

Effective Pollination Period

Flowers are capable of supporting pollen germination, tube growth, and fertilization for a limited period of time. Pollen must arrive at the stigma in the effective pollination period (EPP) in order for fruit set to occur. Three factors contribute to EPP. The first is stigma receptivity. Pollen must arrive at the stigma during its limited receptive period in order to germinate and form pollen tubes. Pollen grains coming into contact with the receptive stigma surface germinate rapidly, forming the pollen tubes that grow into the style. Pollen reaching the stigma after this period of receptivity passes will not germinate. The second factor is pollen tube growth from the stigma through the style on to the ovule. This takes several days. When the pollen tubes arrive at the ovules, one of the two ovules present must be viable and receptive to the pollen tube. The period of ovule receptivity is also limited. Thus, EPP is determined by the timing and duration of each of these three factors: stigma receptivity, pollen tube growth

Table 3.1. Optimal temperatures (°F) for pollen germination and pollen tube growth determined from the regression curves shown in figs. 3.12 and 3.13

	Optimal germination temperature, °F	
	Improved French	Muir Beauty
pollen germination	72.7	72.0
pollen tube growth	75.2	75.7

to the ovules, and ovule viability. Pollen may arrive at a receptive stigma too late to have sufficient time for tubes to reach the ovary before the ovules have lost their ability to become fertilized.

The duration of each of the three determinants of EPP—stigma receptivity, pollen tube growth, and ovule viability—varies from year to year, generally depending on temperature conditions. In general, each of the three events has a temperature optimum. At temperatures below this optimum, events slow, while at higher temperatures they speed up. At excessively high temperatures, the growth of pollen tubes and the development of the ovules may terminate completely. To a great extent, the temperature responses of stigmas, pollen tubes, and ovules are coordinated during a temperature range bracketing the optimum. Problems can occur when temperatures fall above or below threshold levels for each of the critical events. Excessively high or low temperatures may lead to developmental failures, severely limiting fruit set.

In recent years, field observations have suggested that high temperatures during prune bloom have adversely affected set. Recent research by DeCeault and Polito has addressed the issue of temperature effects on pollen germination and pollen tube growth in prune flowers. Table 3.1 shows optimal temperatures for germination and tube growth for Improved French and Muir Beauty based on pollen responses to temperature gradients in laboratory cultures. Figures 3.12 and 3.13 illustrate temperature responses for pollen germination and pollen tube growth of Improved French and Muir Beauty prune. Note that both germination and tube growth fall off rapidly at temperatures above their optimums. Although we do not have experimental data on the duration of ovule viability, data from other species strongly suggest that ovule viability is lost more rapidly at high temperatures. These data tend to support reports, based on field observations during the exceptionally warm flowering seasons of 2004, 2005, and 2007, that high temperatures can markedly reduce fruit set.

In 2007, DeCeault and Polito conducted field experiments at ambient and elevated temperatures. The results of these experiments support the laboratory observations that high temperatures inhibit pollen tube growth sufficiently to adversely affect fruit set. They also conducted a small-scale experiment in which an attempt was made to moderate high ambient temperatures by applying microsprinkler irrigation at 10 gallons per hour from noon to 5 p.m. during the bloom period. Figure 3.14 shows the number of hours above three threshold temperatures for each of these treatments. Note that the microsprinkler irrigation was effective at reducing temperature, suggesting that this may be a potential management tool for moderating the adverse effect of high temperature on fruit set.

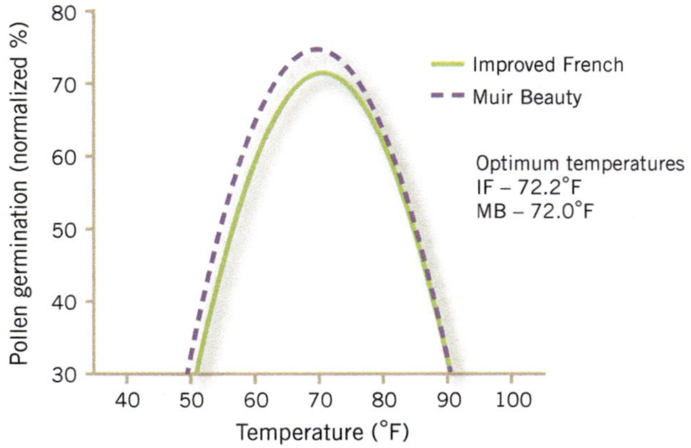

Figure 3.12 Germination of Improved French and Muir Beauty prune pollen in response to temperature. The curve represents a parabolic regression for pollen germination in laboratory cultures incubated on a temperature gradient apparatus.

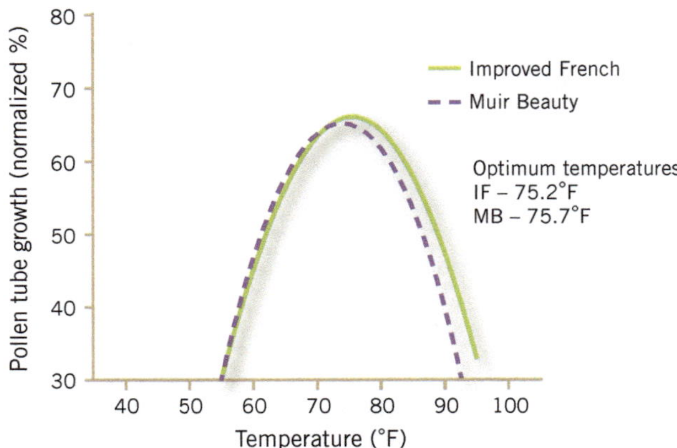

Figure 3.13 Growth of Improved French and Muir Beauty prune pollen tubes in response to temperature. The curve represents a parabolic regression for pollen tube growth in laboratory cultures incubated on a temperature gradient apparatus.

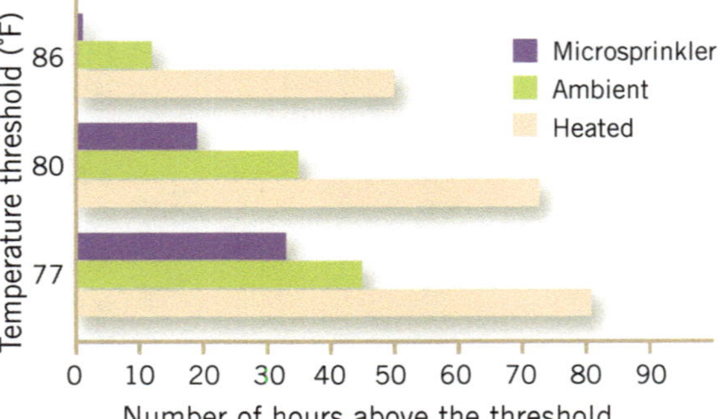

Figure 3.14
Hours recorded above temperature thresholds for 2007 field treatments.

Fruit and Seed Development

The entire fruit, including the stone or pit, develops from the ovary wall (pericarp) (fig. 3.1, 3.15). The kernel contained in the pit is the seed and is derived from the fertilized ovule. Seed growth is a complex process involving the sequential development of its various component parts. Although the seed itself has no commercial significance, its growth greatly influences fruit development. During the first month after fertilization, the ovule and nucellus expand while the microscopically visible endosperm grows slowly in the basal portion of the developing seed. By mid-May, endosperm is detectable with the naked eye (see fig. 3.15). The embryo, which is first seen in the basal portion of the developing endosperm, does not begin to enlarge until fruit growth slows during the middle of May. From mid-May through early June, the endosperm continues to grow in the nucellus, which is consumed as rapid embryo growth occurs. As the embryo enlarges and differentiates, it is nourished by the surrounding endosperm. Ultimately, the embryo will grow to nearly fill the entire seed.

Fruit growth is characterized by two phenomena: cell division, by which the cells of the ovary wall increase in number, and cell enlargement, or the increase in cell size. Taken together these two events account for the large increase in size attained as the ovary of the flower develops into the mature fruit.

The Pit

The innermost cell layers of the pericarp, a region referred to as the endocarp, develop into the pit. Intense cell division activity occurs throughout this region for the first weeks after flowering, at which time the cell walls at the tip of the endocarp begin to harden. As pit hardening proceeds from the tip toward the stem end of the fruit, cell divisions become much less frequent and finally cease altogether. By 8 weeks after flowering, the cells at the tip of the endocarp have attained their final size (fig. 3.16) and hardening continues throughout the entire pit area. When fully formed, the pit consists of three regions: an innermost layer consisting of long narrow cells, a middle region composed of longer and wider cells arranged at right angles to the inner layer, and a third region made up of cells that are approximately spherical. At maturity, the walls of these cells are thick and hard, giving the pit its stony character.

The Flesh

The middle portion of the pericarp, called the mesocarp, is the source of the fleshy portion of the fruit. Cell division occurs in this region for up to 4 weeks after fertilization. When cell division in the mesocarp ceases, all the cells of the flesh are present. As a result, subsequent increase in fruit size results entirely from the enlargement of these cells. The average cell diameter increases nearly twentyfold from the time cell divisions cease until the fruit attains its final size. The cells in the fleshy portion of the fruit are much larger than those in the pits, have thinner walls, and contain large, water-filled vacuoles (compare figs. 3.16 and 3.17). At maturity, the flesh cells begin to separate, and many of the thin walls break down, leaving cavities in the flesh. The fleshy mesocarp also contains the vascular bundles that supply the developing fruit with the water, sugars, and nutrients necessary to support its growth.

The Skin

The exocarp, or outermost portion of the pericarp, is covered with a thin layer of wax deposited on the surface of a single layer of epidermal cells. This waxy cuticle, along with the epidermal cells and the cell layers immediately below them, forms the skin of the fruit (fig. 3.18).

Cell divisions occur in the skin layers for about 6 weeks after flowering, after which the cells elongate and develop thickened walls, especially on their outer surfaces. The waxy cuticle overlying the epidermal cells becomes progressively thicker until about 12 weeks after flowering, at which time it attains its maximum thickness, about 4 micrometers (0.0002 inch). The layer of wax on the fruit's surface helps prevent water loss through the skin.

Setting the Crop

The current year's crop began developing in the previous summer when the flower and leaf buds began developing. Those buds continued developing through the winter and were ready to grow into a flower or leaf in the spring. Environmental conditions and water stress in the prior summer and fall influence the number of flower buds produced. For example, shaded conditions favor leaf bud production over flower production. That is why it is important to use pruning to maintain adequate light in the lower half of the tree.

Prunes, like peaches, almonds, apples, and apricots, must have a viable seed developing for the fruit to develop. Fertilization and a viable ovary are needed for a seed to develop. On rare occasions, a fruit may grow for a few weeks without a viable ovary or an ovary that aborts, but it will soon stop developing and drop. The fruit is simply the end of the long continuum that began with the fruit bud beginning development the summer before to blooming, seed development, and fruit development.

Spring flowering is influenced by the number of winter chilling hours at or below 45°F, and it is also influenced by spring temperatures. Both can influence time of bloom and whether the blossoms bloom together or bloom is spread out.

Pollination

Most commercial prune varieties do not need pollenizer varieties (like almonds do) because they are considered self-fruitful, which means that a French prune flower can be pollinated and fertilized with pollen from its own anthers or those of another flower of the same cultivar. This eliminates the need to plant pollenizer rows. Some prune varieties will benefit from pollenizers; those varieties are described in chapter 9, "Varieties."

Because the pollen of prunes, almonds, peaches, and other stone fruit species is heavy and sticky, it is not carried very far by wind and must be carried by insects. The flowers of these species are showy in order to attract flying insects. Insects forage for pollen as a rich protein source, and they collect nectar as an important calorie and carbohydrate source. In the process of collecting pollen and nectar, tiny pollen grains stick to the insect's body and are inadvertently rubbed onto the sticky surface of the stigma of another flower. While existing wild insects can pollenize flowers, in many situations they do not exist in sufficient numbers to set a commercial crop. In such cases, domestic honeybee *(Apis mellifera)* hives can be brought into the orchard at the beginning of bloom to increase the number of pollinators to transfer pollen. Walnuts and pecans are wind-pollinated and can possibly benefit from artificially

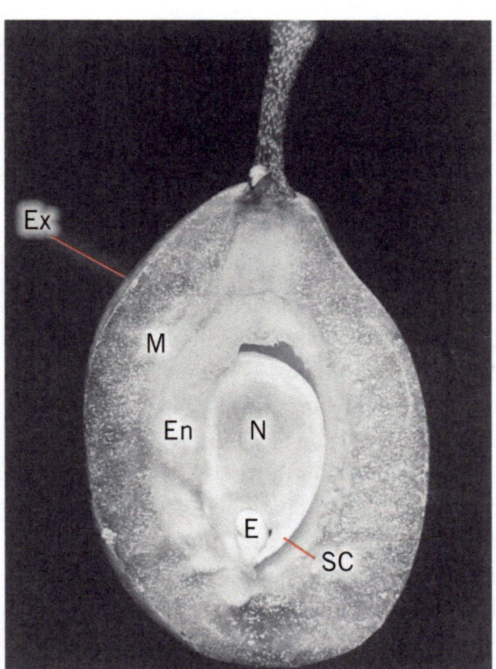

Figure 3.15 A young fruit at reference date in mid-May. The pericarp has differentiated into its three parts: an outer skin, or exocarp (Ex); the middle, fleshy tissue, or mesocarp (M); and the inner stony pit, or endocarp (En). Within the pericarp, the developing seed is present. The seed is enclosed by a seed coat (SC) and contains the nucellus (N) and the endosperm (E). The nucellus is a thin, watery tissue. The endosperm is enclosed as a saclike structure. It is also a thin, translucent tissue, somewhat denser than the nucellus. At this stage, the embryo is present but not visible to the naked eye. It will arise within the endosperm as a dense, solid tissue. *Photo:* V. S. Polito.

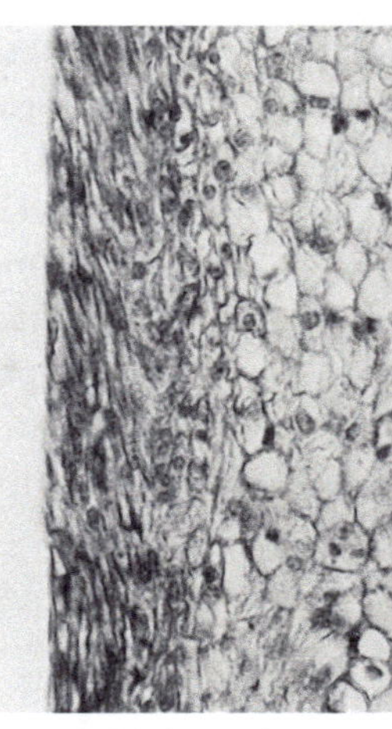

Figure 3.16 Cells of the endocarp region of the developing prune. These cells will form the pit. *Photo:* V. S. Polito.

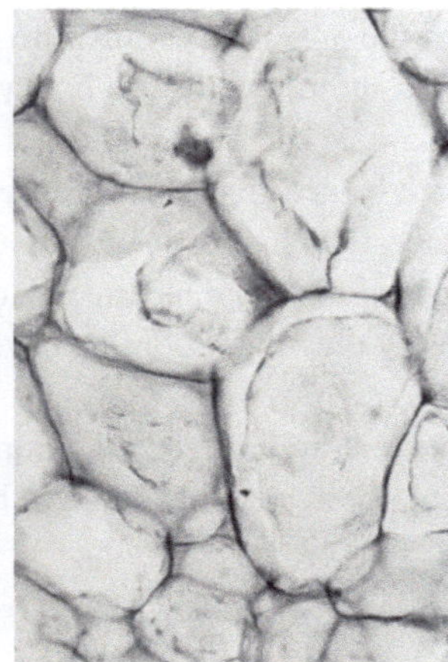

Figure 3.17 Mesocarp cells from the same prune as in figure 3.16. Note the greater size of these cells, which form the fleshy portion of the prune. *Photo:* V. S. Polito.

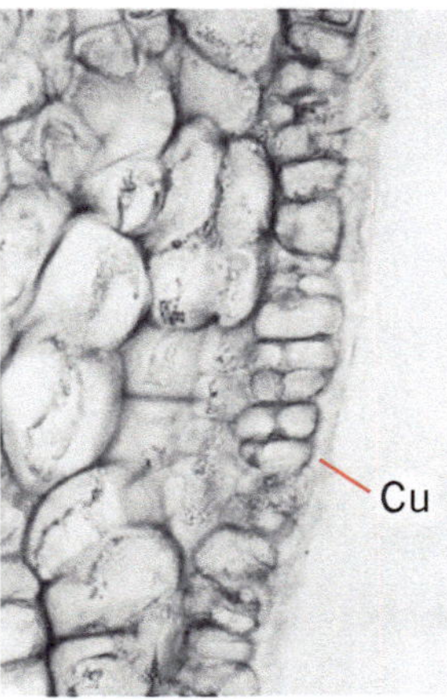

Figure 3.18 Exocarp cells and cuticle (Cu), which develop into the skin of the prune. *Photo:* V. S. Polito.

blowing pollen into the trees, but this is not the case with prunes, almonds, and plums.

When evaluating the effectiveness of a pollination program it is important to consider all of the other factors affecting fruit set. Weak, shaded fruitwood produces less fruit and fruit of smaller size. Young, highly vigorous trees set fewer fruit in favor of vegetative growth. The most important factor is pruning. Wood selection can have a significant influence on flowering and fruit set.

Whether bees are needed for a given block depends on its history and surrounding conditions. Increasing the number of bees in the orchard will increase set, but there may not always be an economic benefit from doing so. If the block historically sets good crops, adding bees is probably not needed and may only increase the thinning costs. If the manager is not satisfied with the set, one-half to one hive per acre may be rented to increase pollination. Because of the large numbers of bees that are introduced to almond orchards that bloom just before prunes, large numbers of honeybees often appear in nearby prune orchards as well.

Hives are normally set in the orchard at or just before bloom begins. Setting the hives several weeks in advance will encourage many of the bees to establish foraging routes outside the target orchard. Foraging in neighboring or distant fields makes the bees much more susceptible to being killed by insecticides that are applied to those crops. Hives can be removed at the beginning of petal fall. Cover crops should be mowed close and weeds in the tree strip removed to eliminate competing flowers. Convincing neighbors to mow blooming cover crops will make their orchard less attractive to the bees set in your orchard.

Before applying any pesticide when honey bees are in or near the orchard, consult with your pest control adviser about its safety to honey bees and natural pollinators. The effectiveness of beehives for pollination can be greatly reduced when bees are exposed to insecticides and even some fungicides. Neighboring farms can reduce the risk of injuring foraging bees by mowing blooming cover crops or weeds before spraying.

Attractants that are sprayed on blooming trees to attract honey bees are considered to be not very effective. They may be somewhat effective if there is a highly competitive bloom nearby, but mowing these competing blooms is much more effective.

Hives should be placed in small groups in locations that receive morning sun that can warm the hives and enable the foragers to begin flying. Remove high weeds that can shade the hive entrance from the placement site. Hives should be kept away from public roadways where they may be hit or be attractive to vandals. Blocks larger than about 20 acres should have the hives distributed through the center of the block. Blocks smaller than 20 acres should have the hives distributed around the perimeter.

To be effective for commercial pollination, beehives should have an actively laying queen and a large number of foragers. The strength of beehives is commonly expressed as the number of frames that are approximately 75% covered with bees on a warm day. For commercial purposes, a hive should have six or more frames of bees, which will contain a total of about 30,000 bees. The strength of the hive can be verified only by removing and examining the frames. Ask the beekeeper to remove the frames and show you or request, for a small fee, that the county agriculture commissioner inspect the hives. Make this request before the hives are delivered so the inspection can be performed in a timely manner.

Beehive rental agreements should be completed many months in advance of bloom. The contract should contain all the important features of the agreement, such as the number of hives, minimum strength, inspection provisions, delivery and pickup, payment schedule, and liability insurance.

Researchers and some growers have been experimenting with other bee species to use as pollinators. While some trials have been encouraging, the hot, dry climate of the Central Valley makes long-term management of these other species difficult. *Osmia lignaria*, also known as the blue orchard bee or the orchard mason bee, is found all over the United States and has been studied extensively. Leafcutter bees *(Megachile rotundata)* have been used in seed alfalfa production for many years. Bumble bees (*Bombus* spp.) are available commercially and have been used in glasshouse systems.

4 Carbohydrate Assimilation and Water Use

• Theodore M. DeJong, Ken A. Shackel, and Bruce D. Lampinen

A prune tree can be viewed as a massive network of solar energy collectors. The individual solar cells are located in chloroplasts, microscopic structures in green cells of the leaves. Each leaf contains thousands of these solar cells, and the tree, in turn, has thousands of leaves. The biological solar energy cells in leaves work only if they are in aqueous solution, so leaves are specially designed to maintain the solar cells in a hydrated state even though they are usually exposed to very dry ambient conditions. In this analogy, the woody framework of the tree can be viewed as providing the structure by which the tree is capable of exposing the maximum numbers of solar cells to the rays of the sun. In addition to providing the structural framework for optimal light exposure, the wood and bark provide a vascular tissue for transporting water and nutrients to the leaves and chemical energy from the solar cells (chloroplasts) in the leaves to other parts of the plant. The efficiency of the prune tree as a solar energy collector network depends on the capture and conversion of light energy into chemical energy (photosynthesis) and the subsequent transport, storage, and use of that chemical energy. An orchard manager's concern is to maximize the efficiency of these processes to produce the prune crop. This requires a basic understanding of the processes and the factors that influence them.

Photosynthesis

Simply summarized, photosynthesis is the process by which energy from the sun is trapped in green pigments (chlorophylls), converted into chemical energy, and used to convert carbon dioxide (CO_2) and water (H_2O) into a simple carbohydrate that eventually becomes sugar ($[CH_2O]_n$). Oxygen (O_2) is given off in the process. The reverse process, whereby all plants and animals recover energy from these simple carbohydrates, is called respiration.

Figure 4.1 shows a simple summary of the overall photosynthetic process. The actual process is a complex set of reactions involving many of the nutrients green plants require. For example, nitrogen (N) is a constituent of photosynthetic enzymes and chlorophyll; phosphorus (P) is important in the energy transfer process; magnesium (Mg) is an essential part of the chlorophyll molecule; and potassium (K), iron (Fe), manganese (Mn), and other nutrients play important roles in specific photosynthetic reactions. The carbohydrate products of photosynthesis are collectively called photosynthates. A principal product is glucose, a six-carbon sugar. It is transformed into other simple sugars such as fructose, sucrose, and sorbitol, a sugar-alcohol. In most plants, sucrose is the predominant carbohydrate that is transported from the leaves to other parts of the plant, but in prune trees sorbitol is the primary transported carbohydrate. All such compounds are classified as carbohydrates because, in addition to carbon (C), they contain hydrogen (H) and oxygen (O) at a ratio of 2:1, respectively, as in H_2O. This class of chemicals

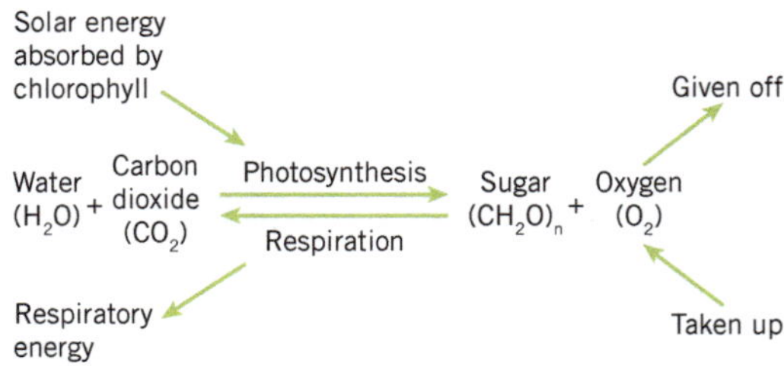

Figure 4.1 In photosynthesis, radiant solar energy converts water (H_2O) and carbon dioxide (CO_2) into an energy-rich sugar ($[CH_2O]_n$) and oxygen (O_2). In the reverse reaction, respiration, living cells use $(CH_2O)_n$ and O_2 to release respiratory, or chemical, energy. CO_2 and H_2O are the by-products.

makes up much of the dry matter of trees. Glucose is the building block for starch and cellulose and is a substrate for synthesis of hemicelluloses, pectins, and gums.

In the reaction shown in figure 4.1, six molecules of carbon dioxide (CO_2) and six of water (H_2O) are required to form one molecule of the six-carbon sugar (glucose) and six molecules of oxygen. The carbon dioxide for this reaction must come from the air surrounding the leaf, and the water comes up from the soil through the plant's vascular system. Carbon dioxide, which makes up only about 0.037% of the earth's atmosphere, diffuses through the stomata (specialized pores) located in the lower epidermis of leaves (fig. 4.2C). The stomata not only allow entry of carbon dioxide into the leaf but also allow water vapor to escape from the leaf. The loss of water vapor is called transpiration. To minimize water loss from the leaf, the stomata have an active mechanism for controlling their opening to permit just enough carbon dioxide into the leaf to allow photosynthesis to continue.

Interactions between Photosynthesis and Water Use

During daylight hours, when photosynthesis occurs, the tree loses (transpires) a substantial quantity of water from its leaves into the air (fig. 4.2C). The water that is lost from the leaf is replenished by transport of water through the tree from the soil (fig. 4.2A). If the soil around the tree is moist, water also evaporates into the air directly from the soil. The total quantity of water lost by the tree and the soil is called orchard evapotranspiration, or crop evapotranspiration (ETc), which is discussed in more detail in chapter 12, "Irrigation Scheduling and Tree Stress." Tree transpiration usually accounts for most of crop water use, and under hot, dry conditions a tree can transpire its entire fresh weight in water each day. The water lost by each leaf is continuously being replaced by water that the leaf pulls from its stem (much as a wick or a siphon pulls water from a reservoir) and that the stem in turn pulls from the branch, and so on through the plant to the roots and ultimately the soil (fig. 4.2D). Most of the transport of water in the plant is through the xylem, which makes up the woody tissues in trees and acts essentially like pipe conducting water.

The sequential pulling of water through the plant, from the soil to the atmosphere, is called the soil-plant-atmosphere-continuum, or SPAC. It is a fundamental concept in plant physiology and a concept vital to the understanding of water use by trees and the physiology of water stress. Since the driving force for transpiration begins with evaporation of water from the leaves, environmental factors that increase evaporation from any wet surface can cause a parallel increase in tree transpiration and ETc. These factors include high temperature, low relative humidity, and high wind speed. In fact, the relationship between transpiration and evaporation is the primary reason why ETc-based methods of irrigation scheduling are useful (see chapter 12). An increase in transpiration, however, also increases the pulling force the plant exerts on the water in it. This pulling force (which is referred to as water potential) is best understood as

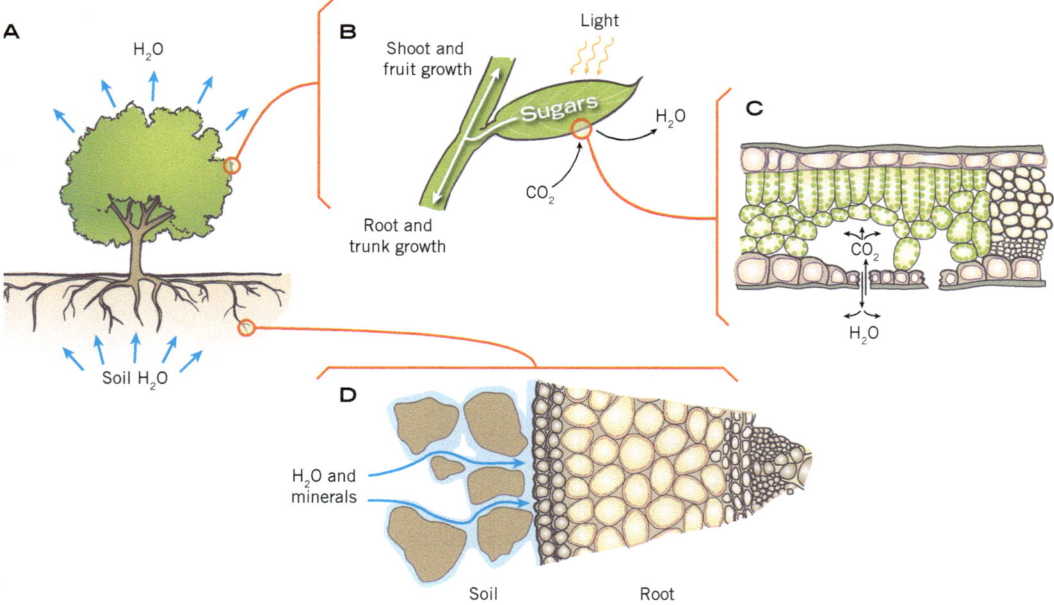

Figure 4.2 The pathway of CO_2 uptake into leaves and water flow through the soil-plant-atmosphere continuum (SPAC). Water evaporates from the same stomatal pores in the leaf (C) that allow the uptake of atmospheric carbon dioxide (CO_2) in the process of photosynthesis (B). Water (H_2O) leaving the canopy (A) must be continuously replaced by water uptake from the soil (D).

tension in the xylem and the walls of all the cells in the tree. Just as more pressure is needed to increase the flow of water through irrigation pipes, more tension is needed to pull water through the plant as transpiration increases. Prune trees are adapted to levels of tension that normally occur as part of daily and seasonal variation in ETc, although some symptoms of too much tension (such as wilting) can occur under very hot and dry conditions when ETc is high, even though the soil is wet. Together with ETc, the main factor that influences the tension, and hence the level of water stress in the tree, is the tension of the water in the soil. As the soil becomes drier, the tree must exert an increasing pull to extract the water held by the soil particles.

Physiologists still do not know all the consequences of the tension, or the level of water potential, in the tree. One important consequence is that when the tension becomes too high, the pressure in the cells of the plant (called turgor pressure) is reduced. Cell turgor is the primary force that keeps the soft parts of the plant erect. The most obvious sign of a loss in cell turgor is wilting, such as wilting of growing shoot tips, or "flagging," of leaves. A less obvious but probably more important result of a loss in turgor is reduced growth rate. Turgor is an important part of the cell expansion process in plant growth and, in fact, for the growth of essentially all plant parts. Hence, if turgor is low, growth will very likely be reduced. This is one of the most significant reasons why overall tree growth, and in some cases yield, is so sensitive to water stress. From a tree survival standpoint, reduction in growth is a beneficial physiological response to water stress, because a small tree requires less water to survive than a large tree. Of course, reduction in growth, especially in the early years of orchard establishment, slows the development of the canopy and decreases canopy photosynthesis and orchard yield. Mature orchards also need a certain amount of yearly growth to maintain active fruitwood.

Another physiological response that controls leaf transpiration, and hence water use by the tree, is the regulation of the size of the stomatal pores. Guard cells are very responsive to many environmental conditions. They can close entirely under severe water stress. Again, from a survival standpoint, stomatal closure is beneficial for the tree because it decreases water use. However, just as with growth reductions, the closing of stomata ultimately reduces tree productivity, in this case by reducing CO_2 uptake and photosynthesis. Figure 4.3 shows measurements of stem water potential (which reflect the amount of water stress a plant experiences), stomatal conductance (which reflect stomatal opening), and photosynthesis (A_l) over the course of a day in the leaves of prune trees under selected irrigation regimes. Under the conditions of this study (Lampinen et al. 2004), stomata were clearly closing and reducing photosynthesis, especially during the middle of the day, in response to the reduced irrigation treatments. Because it reduces transpiration, stomatal closure also reduces evaporative cooling of leaves. Typically, this increases leaf temperatures.

Since the roots are the point of water entry into the tree (see fig. 4.2D), any factor in addition to dry soil that reduces the health of the root system can also cause symptoms of water stress. Trees with diseased root systems may show leaf flagging, reduced growth, and closed stomata, even though the soil may be wet. This can also occur if the soil is flooded and roots are deprived of the oxygen they require to live. From the standpoint of tree irrigation, too much or too little water can have similar effects on the physiology of the tree.

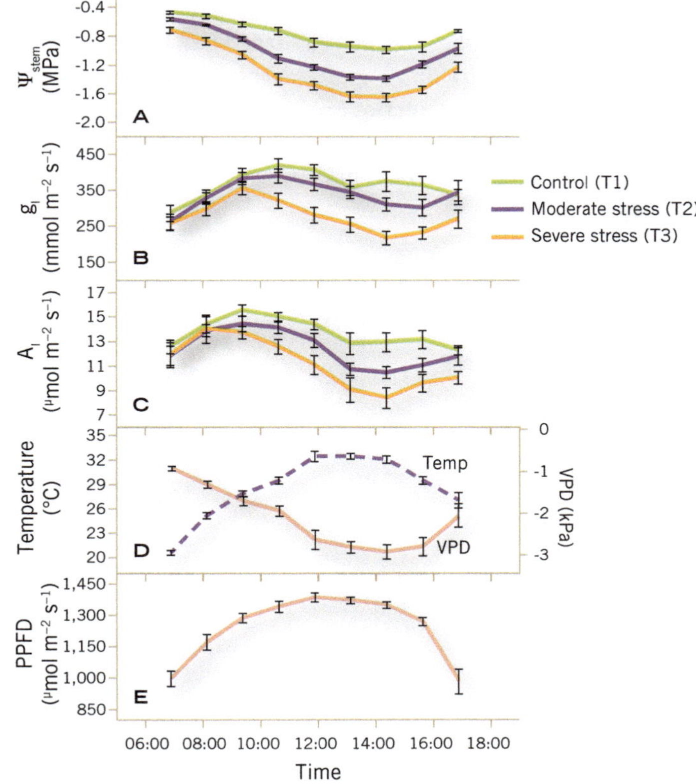

Figure 4.3 Daily patterns of seasonal mean (A) water status (Ψ_{stem}), (B) stomatal conductance (g_l), (C) photosynthesis (A_l), (D) air temperature and vapor pressure deficit (VPD), and (E) direct solar radiation (photosynthetic photon flux density) in prune trees receiving three different irrigation treatments. Data are summaries from daily measurements done on 15 different days from May 28 to October 2. Vertical bars indicate ±2 SE. *Source:* Lampinen et al. 2004.

Other Factors Influencing Photosynthesis

Light intensity

An individual prune leaf exposed to full sunlight can use only approximately one-third of that sunlight for photosynthesis. Therefore, the photosynthetic apparatus of prune leaves is light saturated at approximately one-third full sunlight if a leaf is exposed directly to the sun. Figure 4.4 illustrates this point. Note that the slope of the line in figure 4.4 representing photosynthesis becomes level at photosynthetic flux densities (PPFD) of about 400 to 500 µmols m^{-2} s^{-1}. Only leaves on the outer surface of the tree canopy are ever exposed to direct sunlight for long periods of time. Even those leaves are usually oriented vertically and receive high exposures only for part of the day as the sun moves across the sky. Thus, most leaves on a mature tree function most of the day on the steep portion of the light response curve. Each leaf, located in its zone of the tree canopy, has its own ever-changing light environment. Light becomes more limiting to photosynthesis from the outer edge to the center of the foliar canopy. Thus, if tree canopies are allowed to become too dense, the inner parts become so shaded that leaves can no longer carry on an appreciable amount of photosynthesis. These interior leaves contribute less energy to the spurs to which they are attached than the spurs need; eventually, the spurs die.

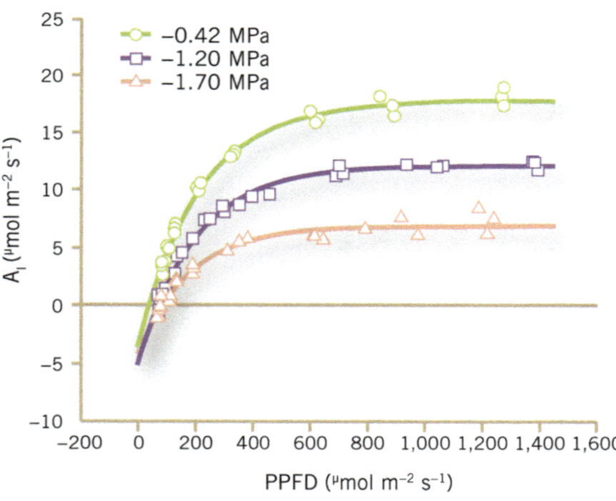

Figure 4.4 Photosynthetic light responses of prune leaves on trees experiencing three different levels of water stress. The upper curve is from a nonstressed tree, and the lower curves are from trees under greater amounts of water stress. All leaves could make efficient use of only about 30 to 40% of full sunlight. Note that responses level off at between 400 and 600 µmol m^{-2} s^{-1}. In a tree canopy, most leaves are exposed to less than 30% of full sunlight at any given time due to leaf angle distribution and internal canopy shading. *Source:* Lampinen et al. 2004.

Temperature

Photosynthesis functions optimally at leaf temperatures between about 70° to 86°F. Because daytime summer temperatures in inland California are rarely less than 70°F, it is unlikely that low temperature during the growing season significantly inhibits photosynthesis in prune leaves. However, temperatures above 86°F occur frequently, and these temperatures can inhibit photosynthesis, especially when they are accompanied with strong winds.

Carbon dioxide concentration

The concentration of carbon dioxide in the atmosphere is fairly constant. It rarely changes enough on a daily or seasonal basis to affect photosynthesis significantly. However, it must enter the leaf to be used in photosynthesis, so any treatment that inhibits its diffusion into the leaf can negatively affect photosynthesis (see fig. 4.3).

Nutrient supply

Each molecule of chlorophyll contains four atoms of nitrogen and one of magnesium. Phosphorus plays a vital role in energy metabolism and transfer. Other elements such as iron (Fe), manganese (Mn), copper (Cu), boron (B), and zinc (Zn) regulate enzyme activity. Deficiencies in any one or all of these nutrients often lead to chlorosis and decreases in photosynthesis. Furthermore, a deficiency of any nutrient, such as zinc, that retards leaf growth also reduces photosynthesis because of decreased canopy size.

Leaf number and exposure

Any tissue containing chlorophyll in the cells, such as young stems and fruit, can carry on photosynthesis. In fruit trees, leaves are the primary organs of photosynthesis. The optimal exposure of the maximum number of leaves to light normally produces the greatest yield of dry matter as long as fruit set is not limiting.

The total leaf area of a tree divided by the land area the tree occupies is called the leaf area index (LAI). The LAI is used as one criterion for land productivity. The optimal LAI occurs when all leaves can contribute to carbon gain. Light intensity and leaf area are critical to photosynthesis, so pruning, training, and spacing of trees are important considerations in starting new orchards. By increasing tree density and minimizing pruning, growers enjoy early returns because of rapid increases in photosynthetically productive canopies. But when trees become crowded or tall enough to cast shadows on adjacent trees, LAI increases beyond its optimum and photosynthetic efficiency decreases.

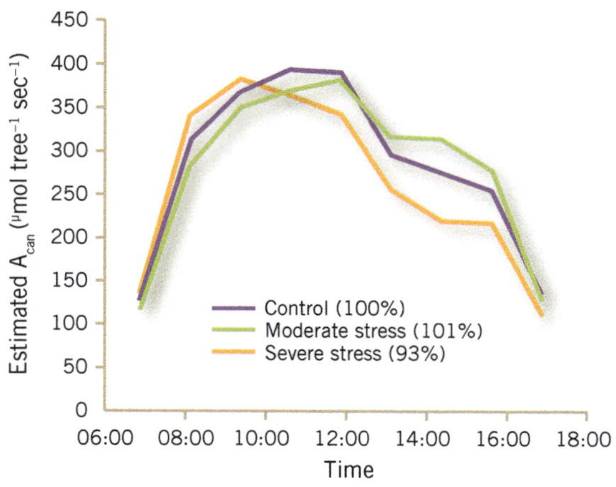

Figure 4.5 Mean canopy photosynthesis (A_{can}) estimated from single-leaf CO_2 exchange data plus leaf angle and light response curve data.

Interaction of factors

Many of these factors interact to cause the pattern of whole tree photosynthesis that is achieved over a day (fig. 4.5). As shown for individual leaves in figure 4.3, in the morning, leaf photosynthesis increases as light intensity increases and generally reaches a maximum level by mid- to late morning. During the afternoon, when temperatures often become relatively high and humidity decreases, the vapor pressure deficit becomes more negative, plant water potential decreases, stomatal apertures narrow, and the photosynthetic rate falls before light becomes limiting (see fig. 4.3). Furthermore, photosynthetic and stomatal responses are accentuated if the tree experiences water stress. During warm afternoon hours, photosynthesis is probably limited by temperature or water stress because the stomata begin to close even though light intensity is still adequate for maximum rates of photosynthesis. However, estimated whole-canopy photosynthesis is not as affected by water stress during the morning hours (see fig. 4.5) as is individual leaf photosynthesis (see fig. 4.3) because water-stressed trees tend to have higher light interception in the morning hours because of the effect of water stress on leaf angles. This indicates the importance of canopy light distribution patterns in determining photosynthetic behavior of tree canopies. Seasonally, leaf photosynthetic rates tend to decrease with leaf age and increasing canopy shading both within and between trees, even when nutrient conditions are adequate.

Translocation and Storage of Photosynthates

Photosynthates produced in the leaves are subject to many possible fates. Some will remain in the leaf for use as an energy source or for building new organic constituents. Some will be temporarily stored in the leaf as starch to be used later. Some will be translocated almost immediately out of the leaf for use or storage in other plant parts.

Physiologists refer to mature green leaves that produce most of the food for the rest of the tree as the sources of photosynthates. Elongating and thickening shoot and root tips and developing fruits that use photosynthates are called sinks. Young, half-expanded leaves, while highly efficient photosynthetically, must retain much of the photosynthates for their own needs and export little. These organs, therefore, could be considered both sources and sinks. The source-sink relationship describes the relative internal competition for food produced by leaves. The direction and steepness of the gradient depend on the degree of competition among the various sinks.

Fruit and nut growers know that in heavy crop years, fruits demand more food and nutrients so that shoot elongation ceases early. They can observe that leaves on heavy-bearing trees are often slightly smaller than those on trees with a lighter crop. But they cannot observe that root growth is curtailed, and the starch, previously deposited in storage cells of branches and roots, is hydrolyzed to soluble sugars and translocated to the developing fruit. This dissolution of starch grains for relocation is called mobilization. Mobilization also occurs in senescing leaves in fall, when carbohydrates, nitrogenous compounds, and some mineral elements are translocated back to the spurs. These substances are reused the following spring for new growth. Reserve food stored in the trunk and roots is mobilized in the spring and translocated to the aerial parts. The opening of flowers, breaking of dormant vegetative buds, and subsequent growth before the new leaves become self-sufficient require much stored carbohydrates. Part of the carbohydrates stored in roots react with nitrogenous compounds absorbed from the soil solution and is assimilated as proteins and other complex organic chemicals.

The translocation of photosynthates out of the leaves occurs in the phloem of the vascular bundles (leaf veins). These join other vascular bundles in the midrib of the leaf blade and form a confluence into the petiole connecting the leaf blade to the spur and limb. In the limb, the phloem tissue is confined to the bark's inner side and consists primarily of specialized

pipelike structures that transport organic substances to developing fruit, shoots, roots, or newly formed cells that add girth to the trunk and roots.

Although photosynthates can be transported over long distances from shoots to roots, the proximity of sinks to photosynthetic sources does influence photosynthate availability to sinks. Recent research (Heerema et al. 2008) indicates that in fruit tree canopies, spurs in shaded positions are more subject to alternate bearing, nonfruitfulness, or death than are spurs in exposed positions. This phenomenon is attributed to a relative starvation of shaded spurs because their leaves do not receive enough light energy to supply the carbohydrate requirements of these spurs. The proximity of fruit to active leaves on exposed spurs places the exposed spurs at an advantage in competing for locally produced photosynthates and leaves an insufficient amount to maintain the productivity of shaded spurs.

In spring, before the phloem is completely activated, some carbohydrates are transported from storage cells in the roots and trunk through the xylem, which is the other major component of the vascular bundles. However, the primary function of the xylem is to conduct water and nutrients absorbed from the soil solution to the leaves and shoots. In stems, trunk, and roots, the xylem tissue is the wood. Each season, as the tree grows in diameter, the new xylem adds another "ring" to the trunk so that the age of the tree can be estimated by counting the annual rings. Mineral elements dissolved in the xylem sap usually remain in the leaf cells, while water evaporates from the cell surfaces and escapes, by transpiration, through the stomata and into the atmosphere.

Use of Photosynthates

Ultimately, all photosynthates are used either as building blocks or sources of chemical energy in the synthesis of compounds involved in tree metabolism, growth, or reproduction. They act as building blocks in the sense that photosynthates are carbohydrates and carbohydrates are the basic compounds in the plant—the compounds from which all other organic compounds are synthesized. The process by which carbohydrates become incorporated into the structural part of the plant is called assimilation. Although assimilates start out as carbohydrates, they are used to make widely different compounds such as

- lignin, a class of chemicals that makes the cell walls of wood, bark, and fruit pits thick and hard
- oils accumulated by the embryo (seed) during its development and subsequently used during germination
- carotenes, which give the leaves a yellow coloration
- anthocyanins, the red and blue pigments found in leaves and fruits
- proteins, which are in the cytoplasm of the cells that carry on the living functions
- waxes and cuticles on the epidermis of fruit and leaves that prevent water loss and serve as a protective envelope

Respiration

Photosynthates also serve as energy sources. Just as a substantial amount of energy is required to synthesize carbohydrates through the process of photosynthesis, the orderly breakdown of those carbohydrates can yield a significant amount of chemical energy for carrying on metabolism in various parts of the tree. This orderly breakdown of photosynthates requires O_2 and is called respiration. Effectively, respiration is the reverse of photosynthesis (see fig. 4.1). In addition to the fact that the substrates for respiration are $(CH_2O)_n$ and O_2, and the products are CO_2 and H_2O and energy, two other major differences between photosynthesis and respiration are that respiration can take place in the dark as well as the light and that respiration takes place in any living tissue cell and does not require chlorophyll to be present.

Respiration is often classified into two types: maintenance respiration and growth respiration. Maintenance respiration goes on continually in living tissues to keep them healthy and functioning. All metabolically active compounds, such as enzymes, undergo a continual process of degradation and rebuilding, so a certain basal level of energy is needed to maintain them in a metabolically active state. The amount of maintenance respiration carried on by a tree depends on its physiological state. During the dormant season, when tree metabolism is at a minimum, maintenance respiration is low. But during the active growing season maintenance metabolism is higher.

Growth respiration is respiration that occurs to supply energy for the construction of new tissues during the growth of the tree. Thus the growth of plant parts not only requires photosynthates for building blocks, it also requires them for respiration energy to be used in converting the building blocks into finished products. Even though respiration is the reverse of photosynthesis, it should not be viewed as a wasteful or undesirable process. It is essential for tissue growth and functioning. An orchard manager's job is to ensure conditions in which respiration can occur efficiently and wasteful respiration is minimized.

Factors Influencing Respiration

Temperature

All living cells respire. The process is, for the most part, catalyzed by several enzymes. These enzymes, being proteinaceous, are temperature sensitive, so the rate at which they carry out a certain reaction is temperature dependent. At near-freezing temperatures, proteins and the water in them change in fluidity so that the reaction rate often slows. At temperatures in excess of 104°F, the enzymes almost become nonfunctional, so enzymatic rates again decrease. Between 32° and about 90°F, respiration increases exponentially with temperature increases (respiration doubles with each 10°C, or about 18°F).

Oxygen (O_2)

An essential substrate for respiration, oxygen is usually not limiting to cells in the aboveground parts of trees. However, in heavy clay soils or under waterlogged situations, lack of oxygen often inhibits respiration and therefore limits root growth and nutrient uptake.

Soluble carbohydrates

In a healthy tree, sugar and sorbitol are produced in abundance and exported from the leaves to all parts of the plant. They are reconverted to glucose at their final destination and are used for respiration or stored as starch. Abnormal situations that affect photosynthesis or translocation of photosynthates—situations such as extended periods of low light intensity, nutrient deficiency, limb girding, and so on—limit carbohydrate supplied to the cells.

Internal factors

Internal changes in the physiological state of tissues significantly influence respiration rates. Dormant branches in the winter respire less than they do in the spring when dormancy is broken. Plant hormones are thought to be active in determining the physiological state of tissue and thus indirectly influence respiration.

Carbon Economy of a Prune Tree

Figure 4.6 is a diagram of the carbon economy of a prune tree. It summarizes how the dynamics of carbohydrate flow and usage are related to yield and cultural practices. Carbon, in the form of carbon dioxide, is taken up by photosynthesis and converted into energy-rich carbohydrates that are then translocated to various plant parts. Translocated carbohydrates are used as energy sources for maintenance or growth respiration, or as building blocks (assimilates) in the creation of new plant parts. Note that there are three types of organic carbon loss in the tree system. The most important to the prune grower is the crop that is harvested. Two other important losses are to pests and diseases and to pruning and litter. Virtually anything the grower does in managing an orchard influences some aspect of the carbon economy of the tree. The goal of efficient management is to maximize photosynthesis and partitioning of photosynthates into economically valuable fruit (good size and quality) while minimizing losses to pests, disease, or excessive growth that is removed by pruning.

The above- and belowground parts of a prune tree form a complex, interdependent production system. Roots depend on shoots for carbohydrates and other organic nutrients, and shoots depend on roots for water and mineral nutrients. Any factor that has a direct impact on either of these systems in the short term can indirectly cause changes in the plant as a whole, and in its productivity, over the long term. Water is a particularly good example of a factor that is involved in both short- and long-term processes, because water plays a key role in almost all plant processes. Since lack of water is a commonly occurring condition in nature, plants have developed many physiological responses to help them survive periods of water stress. Most of these responses cause changes in the carbon economy of the tree through reduced photosynthesis, tree growth, and cropping, but some of these effects can be managed to have minimal impact on overall tree productivity. Whether these responses influence economic production depends on the processes that are occurring at the time of the stress; how important these

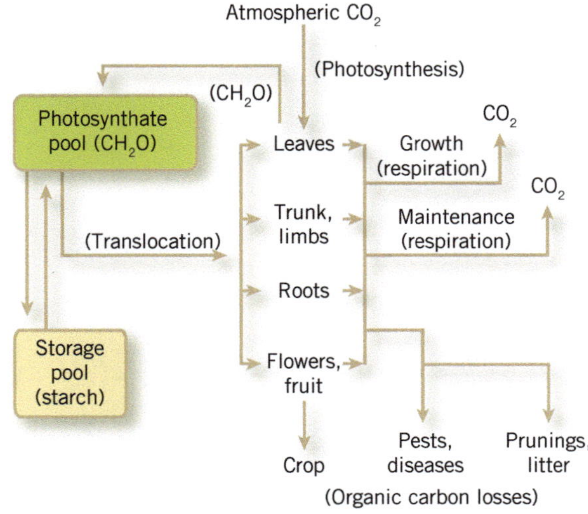

Figure 4.6 The carbon economy of a prune tree.

processes are to tree yield; and whether these processes rely heavily on the current level of photosynthesis or can use stored carbohydrates, like starch, to compensate for the lack of current photosynthesis in the leaf. Not all stresses produce negative outcomes, but managing stress to avoid negative outcomes requires increased intensity of management compared with strategies that are designed to routinely minimize stress.

References

Bradford, K. J., and T. C. Hsiao. 1982. Physiological responses to moderate water stress. In O. L. Lange, P. S. Nobel, C. B. Osmond, and H. Ziegler, eds., Encyclopedia of plant physiology. Berlin/New York: Springer-Verlag. 12B:263–324.

Fereres, E., and D. A. Goldhamer. 1990. Deciduous fruit and nut trees. In B. A. Stewart and D. R. Nielsen, eds., Irrigation of agricultural crops. Madison, WI: American Society of Agronomy, Crop Science Society of America, and Soil Science Society of America. 987–1017.

Heerema, R. J., S. A. Weinbaum, F. Pernice, and T. M. DeJong. 2008. Spur survival and return bloom in almond (*Prunus dulcis* (Mill.) D.A. Webb) varied with spur fruit load, specific leaf weight, and leaf area. Journal of Horticultural Science and Biotechnology 83:274–281.

Jones, H. G., A. N. Lakso, and J. P. Syvertsen. 1985. Physiological control of water status in temperate and subtropical fruit trees. Horticultural Review 7:301–344.

Kramer, P. J. 1983. Water relations of plants. New York: Academic Press.

Lampinen, B. D., K. A. Shackel, S. M. Southwick, W. H. Olson, and T. M. DeJong. 2004. Leaf and canopy level photosynthetic responses of French prune (*Prunus domestica* L. cv. French) to stem water potential based deficit irrigation. Journal of Horticultural Science and Biotechnology 79:638–644.

5 Root Systems and Physiology

• Stephen M. Southwick

Productive fruit trees depend on healthy, functional aboveground parts (trunk, branches, and leaves) as well as belowground parts (roots). Roots have five primary functions that are necessary to ensure success of the entire tree. Performance of those functions can depend on rootstock, scion variety, soil environment, and cultural management practices.

Root Functions

One obvious function of a root system is to provide anchorage. Better anchorage results from deep, spreading root systems, which are more likely when the scion variety or shoot growth is vigorous. Seedling rootstocks are thought to have taproots that may provide better anchorage for prune trees. However, cutting roots when digging trees from the nursery row may alter the branching habit and architecture of taproots. Root systems are comprised of a branching framework. From a relatively few large-diameter roots, laterals of decreasing diameter branch out, terminating with the tips of recently formed fine roots. Root tips (the growing point plus a fraction of an inch to as much as an inch behind it) are the sites where most of the primary functions of roots take place.

The best-known functions of roots, absorbing water and mineral nutrients, are primary activities of extending or recently formed roots. The most efficient absorption takes place just behind the growing point of the root tip. Uptake of ions is not restricted to the area or zone behind the root tip, but is widely distributed over much of the root surface (Eissenstat et al. 2001). Nutrient absorption is a selective and active process requiring the expenditure of respiratory energy. Uptake of water is passive, being dependent on physical forces in the soil and throughout the plant. As roots grow and proliferate throughout the soil volume, the potential for absorption increases. Consequently, well-drained soils lacking shallow hardpans are favored.

Another function provided by roots, which is frequently overlooked, is converting or synthesizing materials. Nitrogen in the form of nitrate is taken up by the roots, converted to ammonium, combined with carbon to form amino acids and amides, transported to other parts of the tree, and finally converted into protein. Each year, considerable nitrogen is removed from the tree as protein in the fruit, leaves, and prunings. Root systems replenish depleted nutrients. Plant hormones that regulate growth and reproductive processes in the tree canopy, such as gibberellins and cytokinins, also are formed in the region of the root tips.

A final function of roots is to provide storage. Roots store carbon, mostly in the form of starch and soluble carbohydrate, as well as nitrogen as protein and amino acids. Carbohydrate reserves support spring growth and the development of flowers, young fruits, leaves, and shoots. Part of the nitrogen absorbed, especially in late summer and autumn, is accumulated and retained in the roots for resumption of growth the following spring. Higher root potassium concentrations have been found in December and January, suggesting that roots are a storage organ for potassium that may be made available in the upcoming growing season.

Roots depend on a supply of photosynthetic products formed in the leaves. In prune, carbon is translocated as a sucrose and sorbitol, a sugar alcohol, to roots, where it is converted to glucose, starch, and many other organic compounds. Glucose is the source of structural material for root growth and is also the source of energy provided by respiration for growth, mineral uptake, synthesis, and many other cellular activities.

Root Growth

Growth Pattern and Periods

Root growth consists of the increase in the diameter of larger, permanent roots that provide anchorage and framework, and the formation, extension, and branching of small lateral roots. These newly formed small lateral roots include the root tips (often referred to as new white feeder roots). As noted above, synthesis and absorption are major activities of growing or recently formed fine roots. The absorbing region of the root may include root hairs, which are lateral extensions of the epidermis or outer cell layer and which greatly increase the root's absorbing surface.

The length of new white feeder roots can vary from a fraction of an inch to over 2 feet, but they are usually less than 1⁄16 inch in diameter. Feeder roots live several weeks, gradually being replaced by newer white feeder roots, provided soil and plant conditions are not limiting. A few white feeder roots may increase in diameter, become suberized (brown and corky on the outside, like outer stem bark), and become part of the permanent root framework. Suberized roots can absorb some water.

Root growth is not continuous throughout the year but occurs in cycles. Root growth is controlled by temperature, available moisture, rootstock, tree age, pruning (Comas et al. 2005), crop load, and the general activity of the scion, making generalizations regarding root growth patterns difficult (Eissenstat et al. 2001) (fig. 5.1). In the California climate, very little if any visible root growth occurs in December and January. As soil temperatures begin to rise in February, roots start growing. Both the number of new roots and their rate of growth increase from 3 to 4 weeks before blossoming until shoot and fruit growth become rapid. This pattern of root growth supports a strategy of high nutrient uptake early in the growing season (Wells and Eissenstat 2003). Nitrate uptake rates decline rapidly with root age, reaching 50% of the starting rate after a single day in grape (Volder et al. 2005). Root nitrogen concentrations also decrease rapidly with root age, with 3-day-old roots having one-half the concentration of 1-day-old roots (Volder et al. 2005). Root growth then decreases as growth of the canopy accelerates.

Reduced root growth appears to be a competitive effect in which photosynthetic products are retained in the canopy and used in shoot and fruit growth. Pruning studies in grape have shown that minimally pruned vines produced a greater quantity of shallow roots 1 month earlier than did heavily pruned vines; early root growth in the shallow soil layers corresponded to early canopy development in minimally pruned vines (Comas et al. 2005). Root lifespan can be affected by pruning, and seasonal rainfall can interact with pruning to alter root lifespan in grape (Anderson et al. 2003). Similar studies have not been conducted for prune. The synthetic function of roots probably plays an important role in regulating the vigor of growth in the canopy. After shoot growth stops and fruit growth slows, a renewed cycle of new root growth occurs. Termination of the second cycle occurs at leaf fall.

Root Activity

In addition to absorption of water and nutrients, root activity includes synthesis, storage, and root growth. Factors that restrict root growth limit root activity and may also limit the performance of the canopy. Readily available soil moisture is necessary for optimal root activity. Roots will not grow in dry soil; mineral uptake can occur only in the presence of available water. (Potassium deficiency, for example, can sometimes be corrected by improving irrigation practices.) Late-summer or postharvest water deficits may inhibit second-cycle root growth. Similarly, insufficient moisture or an incompletely filled soil profile before blossoming suppresses root activity. Research suggests that the most important factor limiting the uptake of nutrients is the total length of the root system, including hairs and mycorrhizal hyphae; however, a comparative study of fruit tree species showed no relationship between

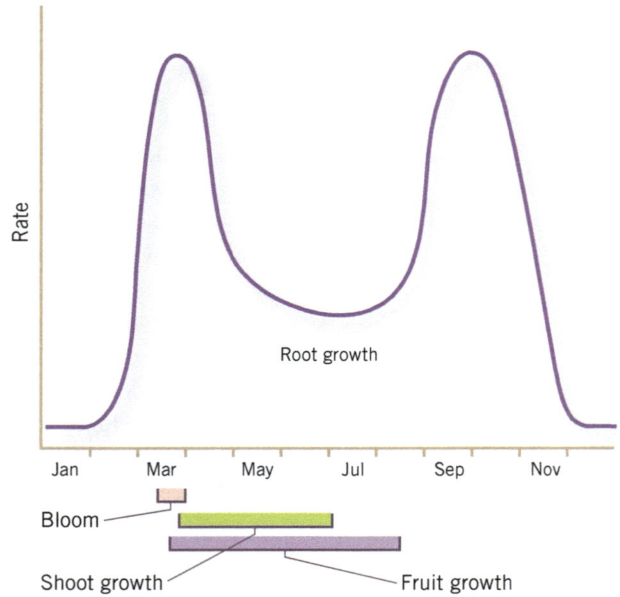

Figure 5.1 Relationship of root growth to fruit growth, shoot growth, and bloom.

root length density and the potential for sustained fruit production under ideal conditions (Eissenstat et al. 2001). Roots from different tree fruit species vary in factors such as lifespan and nutrient uptake (Bouma et al. 2001; Wells and Eissenstat 2001), and most of the research on roots has been done with species other than *Prunus*.

Oxygen

Excessive amounts of water in the soil can damage plants more quickly than insufficient amounts of water. Saturated or waterlogged soils destroy root tips, roots, or even entire root systems by depleting soil oxygen. Although saturated soils are also important in crown and root rots, this discussion is limited to saturation's direct effects on roots.

Air is displaced from soil pore spaces when they become filled with water, as occurs with a fluctuating or perched water table or during flood, seepage, or even irrigation. Because of the extremely low solubility and diffusion rate of oxygen in water, 24 hours of soil saturation are all that is necessary to produce anaerobic conditions in the root zone. Metabolism in the roots begins to shift to less-efficient anaerobic production of energy before oxygen is completely depleted. Terminal portions of roots can die within 1 to 4 days of oxygen starvation, eliminating regions of maximum root activity. Prolonged saturation and high soil temperatures increase the damage. Effects are more severe when roots are actively growing. Injury is more likely in heavy or compacted soils because oxygen reenters more slowly following saturation. A fluctuating water table can be more damaging than a static one because of periodic encroachment on active roots.

Temperature

Root growth does not start and end at the same time every season. The variance depends on soil moisture conditions as well as temperature. Research with apple rootstocks showed that clones MM106 and MM111 reached the highest rate of root growth at soil temperatures of 65.0°F and was lowest at 35.6 to 45.0°F (Carlson 1965; Eissenstat et al. 2001). It is thought that the most intensive root growth for apricot and peach is found at temperatures from 53.6° to 71.6°F and at 53.6° to 55.4°F for plum (Tamási 1986). Root growth in the spring is limited by soil temperature, and the minimum soil temperature required for root growth is 39.2° to 44.6°F. When soil temperatures exceed approximately 95°F, root growth stops. Rootstocks used for prunes probably have similar responses to temperature, but the exact temperature optimums have not been described.

Salinity

Excess salt is a potential problem whenever irrigation is practiced, and it is of increasing concern in California. High levels of sodium, chloride, and boron are of great importance with prunes. Because water is absorbed by the plant preferentially over salts, their concentrations in the soil increase as soil moisture is depleted. Furthermore, increasing concentrations of salt decrease soil moisture availability. Thus, one effect of salinity is on water relations: water stress is more likely under saline conditions.

Another consideration is the action of certain salts in the plant. Effects of specific ions differ with individual salts and also with the species of plant: rootstocks vary in their capacities to absorb and translocate salts to the scion. For example, rootstocks for prune accumulate more chloride than sodium; and less salt accumulates in prune leaves when plum-type rootstocks are used than when other *Prunus* rootstocks are used.

As sodium, chloride, and boron accumulate in the plant, they have an increasingly inhibitory influence on normal cellular processes. Growth and productivity become progressively affected. Ultimately, toxicity symptoms (necrosis of leaf tips and margins, then general necrosis) develop, and the loss of photosynthetic capability further reduces plant performance. Salinity effects become more noticeable as water quality declines and when irrigation frequencies are prolonged.

The levels of salts in leaves are considered injurious when they exceed 0.3% chloride, 0.2% sodium, or 100 parts per million boron on a dry-weight basis. It has been suggested that sodium and chloride levels in the soil should not exceed 3.5 to 4.5 milligrams per 100 grams of soil. In young trees especially, these levels may cause damage.

References

Anderson, L. J., L. H. Comas, A. N. Lakso, and D. M. Eissenstat. 2003. Multiple risk factors in root survivorship: A 4-year study in Concord grape. New Phytologist 158:489–501.

Bouma, T. J., R. D. Yanai, A. D. Elkin, U. Hartmond, D. E. Flores-Alva, and D. M. Eissenstat. 2001. Estimating age-dependent costs and benefits of roots with contrasting life span: Comparing apples and oranges. New Phytologist 150:685–695.

Carlson, R. D. 1965. Responses of Malling Merton clones and delicious seedlings to different root temperatures. Proceedings of the American Society for Horticultural Science 86:41–45.

Comas, L. H., L. J. Anderson, R. M. Dunst, A. N. Lakso, and D. M. Eissenstat. 2005. Canopy and environmental control of root dynamics in a long-term study of Concord grape. New Phytologist 167:829–840.

Eissenstat, D. M., C. E. Wells, and L. Wang. 2001. Root efficiency and mineral nutrition in apple. Acta Horticulturae 564:165–184.

Tamási, J. 1986. Root location of fruit trees and its agrotechnical consequences. Budapest: Akadémiai Kiadó.

Volder, A., D. R. Smart, A. J. Bloom, and D. M. Eissenstat. 2005. Rapid decline in nitrate uptake and respiration with age in fine lateral roots of grape: Implications for root efficiency and competitive effectiveness. New Phytologist 165:493–502.

Wells, C. E., and D. M. Eissenstat. 2001. Marked differences in survivorship among apple roots of different diameters. Ecology 82(3): 882–892.

———. 2003. Beyond the roots of young seedlings: The influence of age and order on fine root physiology. Journal of Plant Growth Regulation 21:324–334.

6 Rootstocks

• Stephen M. Southwick, James F. Doyle, and Richard P. Buchner

Growing prune trees from seed does not produce a tree genetically identical to its parent. Prune seeds are derived from open-pollinated flowers, so seeds are progeny of both parent trees and are not genetically identical to either of the parents or to each other. In fruit and nut production, every tree must produce the same variety. Clonal propagation of fruitwood is one option to achieve this. However, it is much more common to graft the desired variety on a rootstock of choice. This allows a clonal choice for the prune variety and a rootstock choice to manage orchard site factors such as soil type and structure, nematodes, and diseases. Because the life of a prune orchard may be 25 to 40 years, it is important to anticipate rootstock responses to the soil type at the site as well as soilborne diseases and pests that may be present. Certain rootstocks respond differently than others to soil, disease, and pests; selection of the most suitable stock for the proposed site can have a major influence on long-term performance of an orchard. Similarly, with spot or localized replanting, the causes of original tree loss should be taken into account in selecting replacement stocks. The plum rootstocks Myrobalan seedling *(Prunus cerasifera)*, Myrobalan 29C *(Prunus cerasifera* cuttings), and Marianna 2624 cuttings *(Prunus cerasifera* × *Prunus munsoniana)* are the most commonly used in California prune orchards. Other *Prunus* species rootstocks such as peach, almond, and apricot are used for special situations. Marianna 40 plum stock is a relatively new release that may in time replace Marianna 2624.

Plum Rootstocks

Plum rootstocks are adaptable to and are planted on a wide range of soil types. They are especially adapted to soils with high clay and silt content where poor drainage and lack of aeration would impair the growth of or even kill rootstocks such as peach, almond, or apricot. Incompatibility with scion cultivars seldom arises. Check with tree nurseries regarding rootstock-scion compatibility. All plum rootstocks are moderately tolerant of Phytophthora crown and root rot (*Phytophthora* spp.) and Verticillium wilt *(Verticillium dahliae)*. Plum rootstocks are susceptible to root lesion nematodes *(Pratylenchus vulnus)*, peach tree borer *(Synanthedon exitiosa)*, mice, and gophers. Rootstock characteristics are described in table 6.1.

Myrobalan Seedling

Myrobalan plum seedling *(Prunus cerasifera)* is thought to be native to the Caucasus Mountains of southwestern Asia. As the name implies, it is propagated from seed. Each seedling is slightly different genetically, and seedlings show more variation in responses than with rootstocks propagated from cuttings. Examples of this include variability in susceptibility to root knot nematodes *(Meloidogyne arenaria, M. hapla, M. incognita*, and *M. javanica)* and bacterial canker *(Pseudomonas syringae)*, and to a lesser extent crown gall *(Agrobacterium tumefaciens)*. They also are susceptible to oak root fungus *(Armillaria mellea)*. Other advantages of Myrobalan seedlings are that they provide better anchorage than do other plum stocks and are less likely to lean and blow over; they produce few root suckers and only occasionally sucker from the crown; and they are more tolerant of boron and saline soil conditions than are other rootstocks.

Myrobalan 29C

Myrobalan 29C (often referred to as Myro 29C) originated at Marysville, California, as a selec-

Table 6.1. Selected characteristics of commonly used rootstocks for French prune

Rootstock	Soil preferences	Compatibility	Anchorage	Suckering	RESISTANCE OR SUSCEPTIBILITY: Pocket gophers and mice	RESISTANCE OR SUSCEPTIBILITY: Peach tree borer
Myrobalan Seedling	widely adapted to different soil types and moisture conditions	good for most varieties	good	low suckering mostly near crown	susceptible	susceptible
Myrobalan 29C	widely adapted to different soil types and moisture conditions	good for most varieties	roots may be shallow the first 3 or 4 years	some suckering near crown	susceptible	susceptible
Marianna 2624	widely adapted to different soil types and moisture conditions	good for most varieties	roots may be shallow the first 3 or 4 years	some to high suckering near crown and root	susceptible	susceptible
Marianna 40	widely adapted to different soil types and moisture conditions	good for most varieties	more deeply rooted than Marianna 2624 or Myrobalan 29C	low	susceptible	unknown; probably susceptible
Lovell	well-drained, sandy loam soils best; sensitive to wet soil conditions	good for most varieties; not satisfactory for a few, including Robe de Sargeant and Sugar	good	low	moderately resistant	highly susceptible
Nemaguard	well-drained, sandy loam soils best; sensitive to wet soil conditions	probably same as Lovell	good	low	moderately resistant	highly susceptible

	DISEASE RESISTANCE OR SUSCEPTIBILITY*: Bacterial canker	Crown rot	Crown gall	Brownline	Oak root fungus *(Armillaria mellea)*	NEMATODE RESISTANCE: Root lesion *(Pratylenchus vulnus)*	Root knot (*Meloidogyne* spp.)
Myrobalan seedling	variable, generally imparts susceptibility	moderately resistant	seedlings variable, but generally less susceptible than peach	susceptible	susceptible	susceptible	many seedlings susceptible
Myrobalan 29C	imparts susceptibility, but less than Marianna 2624	moderately resistant	more resistant than seedling Myrobalan	susceptible	a little more resistant than Myrobalan, but not as resistant as Marianna 2624	susceptible	resistant
Marianna 2624	imparts high susceptibility	moderately resistant	more resistant than Myrobalan	resistant	moderately resistant	susceptible	resistant
Marianna 40	unknown; early tests suggest it imparts less susceptibility than Marianna 2624	moderately resistant	unknown; probably similar to Marianna 2624	unknown	unknown; probably similar to Marianna 2624	unknown; probably susceptible	unknown; probably susceptible
Lovell	variable, but generally imparts less susceptibility than Nemaguard	susceptible	seedlings variable, but generally susceptible	susceptible	susceptible	susceptible	susceptible
Nemaguard	generally imparts less susceptibility than Marianna 2624	susceptible	unknown; probably similar to Lovell	susceptible	susceptible	susceptible	resistant

Sources: Norton et al. 1963; M. V. McKenry, pers. comm.
Note: *Verticillium wilt: Lovell and Nemaguard highly susceptible, plum less susceptible than peach. Ring nematode: Myrobalan seedling, Myrobalan 29C, Marianna 2624, and Marianna 40 susceptible; Lovell and Nemaguard moderately susceptible.

tion of a vigorous Myrobalan seedling from a population of *Prunus cerasifera* seed imported from France by Marion Gregory. The original Myro 29C seedling was selected in 1915 and released to growers by the Gregory Brothers Nursery Company (eventually located at Brentwood, California) in 1920. It is propagated vegetatively from hardwood cuttings and thus maintains uniform characteristics. It is resistant to root knot nematode and mildly resistant to oak root fungus, but not sufficiently so to be recommended where the fungus exists. It is mildly resistant to crown gall but does impart susceptibility to bacterial canker. The greatest limitation of Myro 29C is poor anchorage. Roots are especially shallow during the first 3 to 4 years.

The Original Marianna

This rootstock cultivar is of unknown parentage, but it is thought to be a naturally occurring hybrid between *Prunus cerasifera* and *Prunus munsoniana*. The seedling was discovered by Charles Fitze at Marianna, Texas, and quite possibly is a seedling of the De Caradeuc variety, a *Prunus cerasifera* variety cultivated for its fruit. The Marianna rootstock was introduced in 1884 by nurseryman Charles Eley at Smith Point, Texas. It was used primarily by nurserymen in the southern United States because of its ease of propagation by cuttings. It was introduced into California in about 1893 and reached some degree of popularity by the 1920s. It eventually proved to be an unsatisfactory stock for some plum varieties.

Marianna 2624

Marianna 2624 (M2624) plum rootstock was released in about 1940 by W. L. Howard of the University of California, Davis. Marianna plum originated in Texas and is thought to be a hybrid between a Myrobalan seedling (*P. cerasifera* Ehrh) and the wild goose plum (*P. munsoniana* Wight & Hedr.). Along with other later cooperators (Leonard Day, Carl Hansen, and others), Howard had begun a rootstock development program that eventually numbered hundreds of seedlings and produced several advanced selections (including 2616, 2623, 2624, and 4001), all of which were open-pollinated seedlings of the original Marianna from Texas. The initial advanced selections were identified in 1926. Marianna 2624 eventually proved to be the best of this group and became widely planted. It is thought to be a triploid plant and normally, in an untopworked condition, produces only a very few prunes. Although successful and popular, Marianna 2624 is generally shallow rooted and often produces an excessive number of rootstock suckers. A vigorous selection of Marianna, it is propagated vegetatively from hardwood cuttings. The main desirable attributes of Marianna 2624 are its resistance to root knot nematode and moderate resistance to oak root fungus. This rootstock is moderately resistant to crown gall but imparts high susceptibility to bacterial canker. Because of the more spreading root system, trees on Marianna 2624 are presumed to suffer from poor anchorage more than trees on Myro 29C. However, these presumptions have not been proven through scientific studies. This stock is not affected by the viral brownline disease, whereas the myrobalan types and peach tree are susceptible.

Marianna 40

The Marianna 40 rootstock originated from a seedling population identified only as "Tennessee" Marianna, presumably from Tennessee. A more exact origin of this seed source was not recorded. The Tennessee source may be the original Texas Marianna or possibly a seedling of it. Tennessee Marianna seed was planted at the University of California Kearney Field Station in 1970. Ten advanced selections were identified by 1977. After considerable field testing, Marianna 40 proved to be the best of the group, and it was released in 2000 by C. Hesse, R. Fenton, and J. Doyle. Marianna 40 has many of the positive attributes of Marianna 2624 but is more deeply rooted and produces substantially fewer rootstock suckers. Although not all *Prunus* species have been tested on Marianna 40, it appears to have the same range of compatibilities as Marianna 2624. Tree productivity also seems to be similar to Marianna 2624. The release of Marianna 40 gives commercial growers the opportunity to choose a rootstock that is similar to Marianna 2624 but may have better anchorage without the expensive nuisance of prolific sucker production.

Other *Prunus* Rootstocks

Prunes have in the past been grown in California on peach, almond, and apricot rootstocks. None of these are currently recommended except in special situations. Only their main characteristics are noted here. These rootstocks are extremely sensitive to inadequate aeration and root and crown rot, so they perform best when planted in well-drained soils.

Peach

Lovell or Nemaguard seedling rootstocks might be considered where bacterial canker is a severe problem. However, Lovell is susceptible to root knot nematode, whereas Nemaguard is resistant. Both rootstocks are susceptible to Phytophthora crown and root rot when planted in poorly drained soils. Prunes on peach roots can sometimes set an excessive amount of fruit, which may contribute to shoot dieback from potassium deficiency and small fruit size. This tendency might be exploited by using peach rootstocks where a history exists of poor fruit set with plum rootstocks. Peach rootstocks are less attractive to pocket gophers than are Myrobalan, apricot, or almond (Day 1953). Not all prune cultivars have been shown to be compatible on peach rootstock. The historical cultivars Robe de Sargeant and Sugar usually do not have satisfactory unions with peach.

Almond

Almond tends to absorb less boron than do other prune rootstocks, but it is not necessarily tolerant of high boron concentrations in soils or water. Field observation has suggested that prunes grown on almond roots are less affected by high levels of lime. Fruit set and size can be similar to prunes on peach rootstock. Compatibility is questionable between almond and scions of the older cultivars Robe de Sargeant, Imperial, and Sugar.

Apricot

Even though apricot is resistant to root knot nematode and more tolerant of root lesion nematode, it generally has no superiority over Myro 29C or Marianna 2624. The only real advantage that might be gained from use of apricot rootstocks is less damage by peachtree borer.

References

Day, L. H. 1953. Rootstocks for stone fruits. Berkeley: California Agricultural Experiment Station Bulletin 736.

Norton, R. A., C. J. Hansen, J. H. O'Reilly, and W. H. Hart. 1963. Rootstocks for plums and prunes in California. Berkeley: California Agricultural Experiment Station Leaflet 158.

Part 3 Prune Orchard Soils, Varieties, Irrigation, and Fertilization

7 Managing Salinity in Soils and Water

• Allan E. Fulton and Blake L. Sanden

Prunes are sensitive to salinity in soils and irrigation water. High yields and fruit quality are best achieved with soils and irrigation water that have desirable levels of salinity and the appropriate composition of salts. Field conditions often associated with salinity include decreased soil water availability, reduced rates of water infiltration into soils, and accumulation of specific elements to toxic levels in plant tissue. These conditions culminate in trees with less vigor and less long-term production potential.

This chapter discusses how to use soil and water analyses as practical and economical management tools to diagnose salinity concerns and to determine appropriate remedial actions to establish and sustain a successful prune orchard. Topics covered include

- collection of representative soil and water samples for diagnostic purposes
- how soil and water samples received from the field are prepared and analyzed for salinity
- understanding terms used in a salinity report and checking the quality of the report
- diagnosing different types of salinity
- making remedial management decisions

Soil and Water Sampling

Sampling Approaches

There are two basic philosophies for sampling soils and water to diagnose and manage salinity in prunes: routine monitoring versus troubleshooting. If the soil or water quality is marginal, sampling is essential before planting an orchard, and it should be conducted routinely at least every 2 or 3 years to guide long-term management. On the other hand, if soils are uniformly low in salinity and the water is of suitable quality for prunes, sampling is necessary only to troubleshoot general production problems and ensure that it is not a contributing factor. Some situations do arise in which a switch in sampling approach may be necessary. For example, if the irrigation water quality changes from higher quality to lower quality, it may be necessary to change from troubleshooting to routine monitoring.

Regardless of which approach is taken, soil and water sampling must represent the condition of concern for the analytical results to be of value. Within reason, sampling must be undertaken using methods that cope with spatial and temporal variability in soils and irrigation water supplies. Results from unrepresentative sampling (for example, soil collected from just one or two sampling sites at inadequate depths) may be misleading and costly. Although obtaining representative composite samples involves some effort and expense, it should not require more than 8 hours of labor every 2 or 3 years for an 80-acre orchard.

Preplant Soil Sampling

Salinity can vary considerably throughout an orchard. The variability often corresponds with changes in physical soil properties that may influence water infiltration and drainage. Figure 7.1 illustrates the nature of spatial soil variability that may be encountered in an orchard site. USDA soil surveys are a good tool for understanding the spatial variability of soils at an orchard site and to identify sampling areas. Alternatively, Geonic electromagnetic conductivity meters (EM38) and Veris electrical conductivity systems are available as diagnostic tools to develop site-specific soils maps to help understand spatial soil variability. Refer to chapter 8, “Soil

Evaluation and Physical Modification" for sources of soil surveys and a description of EM38 and Veris diagnostic methods.

Figure 7.1 illustrates how backhoe pits should be distributed across a 160-acre orchard site to sample the different soils. Each pit provides access to sample soils for salinity analysis. Excavating a backhoe pit 6 feet deep in each area also allows visual examination of the soil profile to assess the drainage and the depth of deep tillage required. For safety reasons, the pit should be as wide as it is deep in the event of a collapse.

Since soils often vary spatially, samples should be composited from more than one backhoe pit within an area of the orchard where soil maps indicate similar soils. This ensures that representative soil samples will be collected for a specific soil type yet minimizes the number of soil samples that require laboratory analysis and costs. Samples should also be collected separately for different soil layers observed within a soil profile. Figure 7.2 shows how soils may vary with depth. Usually, samples are collected in 1- to 2-foot increments. Samples should represent different layers in the soil profile.

Figure 7.1 also provides a hypothetical example of where to dig backhoe pits, how to composite soil samples according to spatially different soil types, and how to separate soil samples based on variability within a soil profile. In this example, the 160-acre site has four spatially different soil zones based on a USDA soil survey: Kimberlina (about 26 acres), Garces (about 80 acres), Lewkalb (about 17 acres), and Milham series (about 37 acres). In total, ten backhoe pits will be dug at the 160-acre site: two pits in each of the Kimberlina, Lewkalb, and Milham soil series, and four in the Garces soil series. In general, it is good practice to dig at least one backhoe site per 20 acres. At a minimum, dig at least one backhoe pit per 40 acres to evaluate soils at sites where soils maps show very little spatial variability. Soil samples will be composited from the backhoe pits within each of the four soil series and separated according to soil depth. In this example, composite soil samples will be collected from four backhoe pits representing the Garces series and from two backhoe pits that represent each of the Kimberlina, Lewkalb, and Milham series. Usually, within a single soil series, a distinct topsoil layer (A horizon) with organic matter and more soil tilth is evident to a depth of about 1 foot; a moderately developed subsoil (B horizon) is often observed at a depth of about 1 to 3 feet; and a much less developed subsoil (C horizon) is apparent at about 3 feet and deeper. Separate samples from each of these three soil horizons should be composited from the backhoe pits in each of the four soil series. On completion, 12 composite soil samples (4 soil series from 3 soil horizons) will have been taken for laboratory analysis.

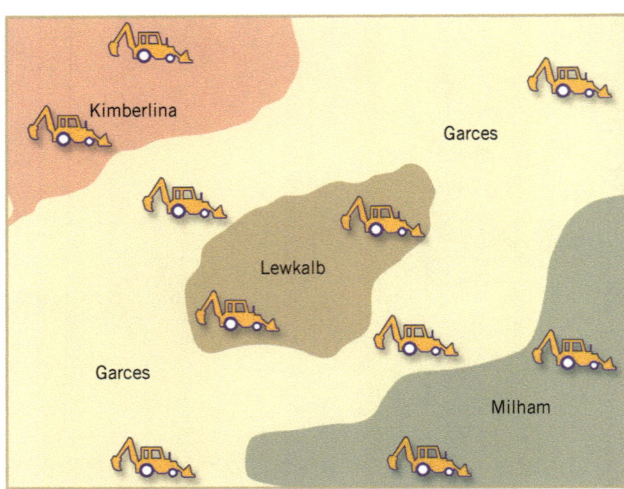

Figure 7.1 Soil type variability and required sampling of areas for a 160-acre orchard development.

A soil auger may be used instead of backhoe pits, but sampling with an auger is not as revealing. If an auger must be used, sample the soil to at least 4 feet deep in four to eight locations (the locations should be 50 to 200 feet apart) within a soil type. An auger with a diameter of 2 to 3 inches works best. Use an open-loop-type Dutch auger for moist clay soils. A closed-barrel auger with opposing cutters is required for sandy or dry soils (see fig. 7.3). A rubber mallet is very useful for loosening and removing soil from the auger. To collect the samples, use four buckets, one for each foot of soil. Take soil at a given depth (e.g., 0 to 1 foot) from different holes and put it into one bucket. Then take soil at another depth (e.g., 1 to 2 feet) from different holes and put it in the second bucket; repeat the process for the third and fourth buckets. Mix the soil thoroughly in each bucket and remove about 1 pound of soil from each as samples to be tested. Take the four 1-pound samples to the laboratory the same day.

If the soil samples cannot be submitted to a laboratory for analysis on the day they are collected, air-dry them by placing each sample in a separate paper bag and setting the bags outside in a dry area exposed to the sun and wind.

Soil Sampling after Planting

After planting, soil sampling and salinity analysis should be routinely conducted every 2 to 3 years in prune orchards that have been planted on marginal soils or where water of marginal quality is used

Figure 7.2 Backhoe pits are the preferred method to evaluate and sample orchard soils for salinity prior to planting (A). They may also be used in established orchards, but their use is more limited. Backhoe pits can be used to distinguish layered soil profiles (B) from uniform soil profiles (C) and to sample for salinity.
Photos: M. L. Poe.

Figure 7.3 Soil augers. The sand auger on the left with an enclosed barrel is designed to sample sandy-textured soils. The Dutch auger (second and third from the left) is best for moist clay soils and is the easiest to clean. The open-sided "mud" auger (second from right) is almost the same as the classic sand auger except that the sides are cut out to facilitate removal of moist soil. The "bit" type auger can be used to sample loam, silt loam, and clay soils. Sandy soils require a high moisture content to sample with a bit auger. The volume of soil sampled with a bit auger is less than the other types of augers shown. Extensions can be added to these augers for sampling as deep as 20 feet in stable soils. *Photo:* B. L. Sanden.

for irrigation. After the orchard has been planted, composite sampling using an auger, as described in the previous section, is more feasible for retrieving soil samples than sampling backhoe pits. Routine evaluation of soil salinity assists with making decisions about irrigation and soil and water amendments.

Year-to-year sampling should be done at the same time each season with respect to rainfall patterns and irrigation scheduling. Sampling from year to year should also be consistent with respect to the method of water application around the tree. This ensures that salinity distribution and accumulation reflect the long-term trends occurring in the orchard and not short-term conditions that are reflected by collecting soil samples at different times in the growing season.

Sampling after harvest provides a salinity assessment in an orchard when root zone salinity is usually highest. Irrigation is commonly delayed or deficit during harvest, allowing salts to accumulate in the root zone (see chapter 12, "Irrigation Scheduling and Tree Stress"). Also, fall sampling gives the most advance notice that additional irrigation may be needed for salinity control during the winter when the ground is cold, trees are dormant, and extra irrigation for leaching is less likely to cause waterlogging and disease.

The method of irrigation and the ability to apply water uniformly must be considered in order to collect representative samples. In flood-irrigated prunes, sample 5 to 10 feet to the side of a tree row for best results. In sprinkler-irrigated prunes, sample across the sprinkler pattern to ensure that about two-thirds of the composite includes soils from the center of the wetting pattern, which receive the most applied water, and about one-third from the edges of the wetting pattern, where salts tend to accumulate. The same method applies to microsprinklers and drip irrigation.

Orchard growth patterns, after planting, provide an additional indicator of where to sample spatially variable soils. Soil samples for salinity analysis after planting should be composited for the different soil series and separated by depth, as described for preplant soil sampling.

Why Sample Soils at Multiple Depths?

Sampling soils and analyzing for soil salinity at multiple soil depths can give insight into irrigation and whether soil and water amendments should be used.

The root zone salinity can be estimated by averaging the electrical conductivity of the saturated soil extract (EC_e) levels for all of the soil depths analyzed. These estimates of average root zone salinity can be compared with critical levels for prune.

If water infiltration is slow or deficit irrigation is occurring, salinity in the top 1 to 2 feet of soil will often be higher than at 3 to 4 feet. Surface soils containing salinity and sodium or chloride levels far in excess of the irrigation water usually indicates that soil crusting restricts water infiltration and that irrigation water quality is poor.

A soil sample from 0 to 3 inches along with a water sample may help diagnose infiltration problems.

Sampling soils at multiple depths helps understand salt distribution and accumulation patterns within the root zone and also helps determine the adequacy of irrigation management for controlling salinity. With flood or sprinkler irrigation, where most of the orchard floor is irrigated, the combination of adequate irrigation and rainfall to meet crop evapotranspiration (ET), good water penetration, and a small amount of leaching should sustain root zone salinity at levels suitable for prune. In this scenario, root zone salinity (EC_e) at 4 feet deep will be about 2 to 3 times the salinity at 0 to 2 feet deep. If the EC_e at 4 feet is four to six times greater than that at 1 foot, there is a good chance that the irrigation management has been too efficient, producing deficit irrigation and little or no leaching of salts.

Similarly, sampling soils at multiple depths can help understand salt distribution and accumulation patterns within the wetted area of drip and microsprinkler irrigation. However, because these low-volume irrigation systems have a smaller wetted soil volume that promotes a more confined root system, salinity distribution and accumulation patterns are likely to change more within a shallower depth or smaller volume of soil. If root zone salinity (EC_e) at 2 feet is only about two to three times the salinity at 1 foot, leaching for salt control may be adequate. If the EC_e at 2 feet is four to six times greater than at 1 foot, there is a good chance that the irrigation management has been too efficient, producing deficit irrigation and little or no leaching of salts.

Sampling Irrigation Water

Sampling irrigation water for salinity assessment is much simpler than sampling soils. To take a sample, rinse a plastic container in the water that is to be sampled. Collect a small sample, 8 to 16 ounces. Completely fill the container with water; this eliminates air, which would otherwise promote calcium carbonate precipitation. Cap or cover the sample after taking it.

Before taking a sample from a well, let the pump run for at least 30 minutes, preferably as long as 2 to 4 hours, to flush the well of static water, stabilize salinity, and establish the pumping water elevation that represents the primary water-bearing strata. If the orchard site is surrounded by deep wells, sample in July or August when groundwater pumping is the greatest. Collect a representative water sample for laboratory analysis to establish a baseline. If the pumping depth to groundwater varies considerably from year to year and has shown a declining trend over the years, invest in an inexpensive portable electrical conductivity (EC) meter and monitor the electrical conductivity of the groundwater (EC_w) in the spring and late summer. For irrigation water supplies with an EC_w above 0.7 dS/m, submit a new sample for laboratory analysis when the EC_w increases about 20%. Otherwise, a sample about every 3 years should suffice to affirm baseline levels.

To establish a baseline for surface water, take samples from canals or ditches with flowing water. Submit the first sample for analysis and use this as a reference point. As described for groundwater monitoring, use a portable electrical conductivity meter to determine how often a laboratory should analyze the water supply.

If possible, submit a water sample for analysis on the same day that it is collected. If the sample must be stored, refrigerate it to minimize changes in salinity. Storage at room temperature will allow calcium (Ca) and bicarbonate (HCO_3) to precipitate and lower the total salinity level of the water.

Soil and Water Analysis

Laboratory Evaluation for Salinity

Figure 7.4 shows the general laboratory steps involved with preparing and analyzing a soil sample for salinity after it has been received from the field. Nearly all laboratories measure soil salinity in a saturated paste extract. The extract represents the soil water at a very high soil moisture content, along with the dissolved salts and nutrients that are readily available to the trees. Decades of research on crop tolerance to salinity have been based on correlations between vegetative growth, reproductive development, yield, and salinity levels in the saturation paste extract. The extract is acquired for testing by following specific procedures to make a saturated soil paste, or mud, and then vacuum suctioning the water from the paste. The water content of the saturated paste (saturation percentage) is measured and reported to provide an understanding of the dilution effect of the saturated soil and to enable the test results to be related to the lower soil moisture levels that typically occur in orchard soils.

The extraction process shown in figures 7.4A through 7.4D is unnecessary for irrigation water samples, but the analytical techniques shown in figure 7.4E are used to analyze irrigation water for

Figure 7.4 Laboratory analysis for salinity in soil and water. *Photos:* M. L. Poe.

7.4A Relatively small samples of about 1 pound of air-dried soil that represent specific soil types and soil horizons are delivered from the field to a laboratory for analysis.

7.4B On delivery to the lab, each air-dried soil sample is ground to homogenize the sample in preparation for extraction procedures and salinity analysis.

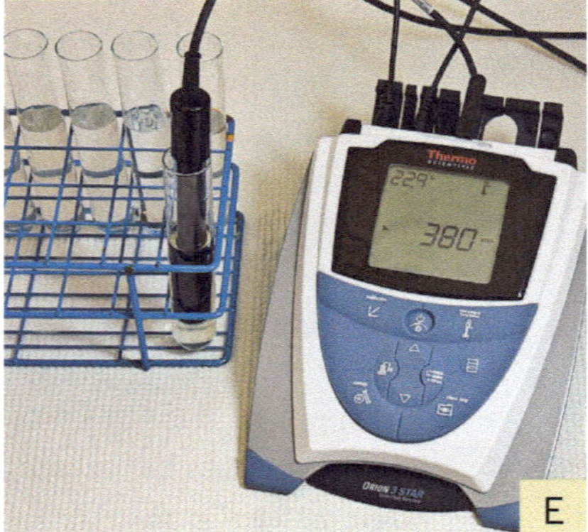

7.4E Various salinity constituents in the extract are evaluated using electrical conductivity meters (above) and other analytical techniques such as a flame photometer (below).

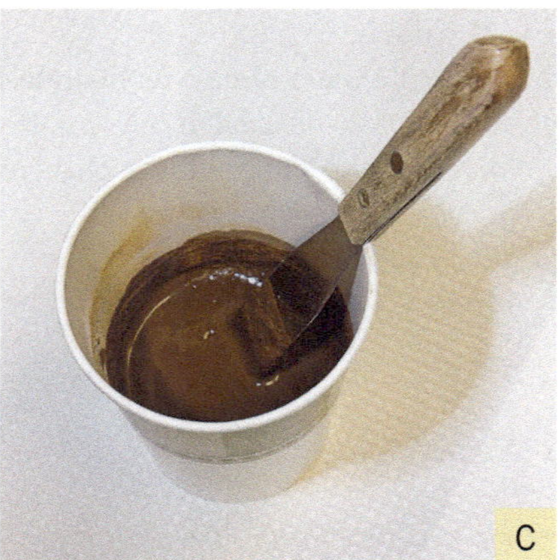

7.4C A smaller subsample of soil is mixed with distilled water to prepare a saturated soil paste, or mud.

7.4D The soil water is extracted from the paste using a suction vacuum and filter flask with a collection test tube.

salinity. An 8- to 16-ounce irrigation water sample collected in a plastic bottle is sufficient to analyze irrigation water quality.

Reading Laboratory Analytical Reports

Soils contain salts in three forms: bulk minerals (e.g., lime, gypsum, dolomite) that release salts very slowly and contribute little to the quantity or intensity of soil salinity; unstable soil-adsorbed salts that constantly seek equilibrium with the salts in the soil water and seek to a lesser extent equilibrium with the bulk minerals; and dissolved salts in the soil water or irrigation water that can readily be absorbed by plant roots. An analysis of an irrigation water sample or a saturated soil extract (the most common method of analyzing salinity in soils of the western United States) measures the dissolved salts in the irrigation water supply or soil water environment. Table 7.1 gives the determinations usually provided in an analytical report relating to a soil extract or irrigation water supply. Analytical results can be lengthy and full of unfamiliar terms; the next section explains what these terms mean.

Terminology and Units

Saturation percentage

The saturation percentage (SP) equals the weight of water required to saturate the pore space divided by the weight of the dry soil. Saturation percentage is useful for characterizing soil texture. Very sandy soils have SP values of less than 20%; sandy loam to loam soils have SP values between 20 and 35%; and silt loam, clay loam, and clay soils have SP values from 35 to over 50%. Salinity measured in a saturated soil can also be correlated to soil salinity at different soil water contents measured in the field. As a rule, the SP soil water content is about two times higher than the soil water content at field capacity. Therefore, the soil salinity in a saturation

Table 7.1. Laboratory determinations and units for evaluating salinity in irrigation water and in soil water extracts

Test	Symbol	Units	Soil	Water
saturation percentage	SP	%	yes	no
acidity-alkalinity	pH	none	yes	yes
electrical conductivity	Ec_e, Ec_w	dS/m	yes	yes
calcium	Ca^{2+}	meq/l	yes	yes
magnesium	Mg^{2+}	meq/l	yes	yes
sodium	Na^+	meq/l	yes	yes
bicarbonate	HCO_3^-	meq/l	yes	yes
carbonate	CO_3^{2-}	meq/l	yes	yes
chloride	Cl^-	meq/l	yes	yes
sulfate	SO_4^{2-}	meq/l	yes	yes
boron	B	mg/l (ppm)	yes	yes
nitrate-nitrogen	NO_3-N	mg/l (ppm)	yes	yes
sodium adsorption ratio	SAR	none	yes	yes
adjusted sodium adsorption ratio	SAR_{adj}	none	no	yes
exchangeable sodium percentage	ESP	%	yes	no
lime requirement	LR	tons/ac	yes	no
gypsum requirement	GR	tons/ac-6 in	yes	no
lime percentage	$CaCO_3$	%	yes	no

extract is about half of the actual concentration in the same soil at field capacity.

pH

The pH of a soil or water supply measures hydrogen ion concentration (activity). Although pH is closely related to bicarbonate concentration and the availability of some macro- and micronutrients, it does not correlate with total salinity in the root zone. It is, however, important when selecting among different soil and water amendments.

Electrical conductivity

The electrical conductivity (denoted EC_e for extracts from soil and EC_w for irrigation water) is a measure of the total salinity in the root zone or in a water supply and is based on how easily an electric current passes through the extract; it does not give any indication of the salt composition. EC is one of the most important values in an analysis because nearly all crop salt tolerance levels are based on EC. The internationally accepted reporting unit for EC is deciseimens per meter (dS/m). This unit is equal to millimhos per centimeter (mmhos/cm), which is still used by some labs. Many labs that do environmental and domestic water testing often report EC_w in micromhos per centimeter (μmhos/cm). Divide these values by 1,000 to obtain dS/m or mmhos/cm.

Total dissolved solids

Another term often found on analytical reports is total dissolved solids (TDS). This is the weight of all soluble salts in milligrams per liter of water (mg/l). About 640 mg/l equals 1 dS/m, depending on salt composition. TDS is not useful in evaluating agricultural salinity because crop tolerance thresholds are correlated with EC_e and EC_w rather than with TDS.

Soluble cations and anions

Salts such as sodium chloride (NaCl) and calcium sulfate ($CaSO_4$) consist of positively charged cations and negatively charged anions bonded together by opposing charges. In irrigation water or soil water extracts, many of the bonds are broken, and the water consists of individual cations and anions. To understand the impact of salinity on crop tolerance and on water infiltration, soil and irrigation water samples must be analyzed for the soluble cations and anions. Calcium (Ca^{2+}), magnesium (Mg^{2+}), and sodium (Na^+) are the major cations in soil extracts and irrigation water. Although potassium (K^+) is important as a nutrient, it is usually a much more minor component of salinity than calcium, sodium, and magnesium salts. Bicarbonate (HCO_3^-), sulfate (SO_4^{2-}), and chloride (Cl^-) are the main anions in most soil extracts and irrigation water. Along with these anions, boron (B) and nitrate-nitrogen (NO_3-N) are commonly reported in analytical results. Boron does not contribute significantly to total salinity and salt stress (osmotic effects) on plants, but it is important in diagnosing specific ion toxicity. Knowing the NO_3-N content of the soil and irrigation water is valuable in making fertilizer decisions, but nitrate does not contribute significantly to the total salinity.

The preferred unit for reporting individual cations and anions is milliequivalents per liter (meq/l). This unit is used specifically in

agricultural salinity evaluation and reporting. Most agriculturists who work with pesticides, fertilizers, and tissue analyses are familiar with parts per million (ppm) and milligrams per liter (mg/l) but are less familiar with meq/l. When all ions are reported in meq/l, this allows comparison of the relative ionic strength of the different cations and anions. Clay particles in soil are negatively charged and adsorb the positively charged cations. It is the concentration of the charges that affects soil salinity and eventually the decisions related to soil and water amendments and leaching.

Reporting cations and anions in meq/l is one of the hallmarks of a quality agricultural lab. Lab reports using meq/l units aid in calculating how much soil or water amendment is needed. Table 7.2 shows how to convert mg/l to meq/l.

Sodium adsorption ratio and exchangeable sodium percentage

The unadjusted sodium adsorption ratio (SAR), adjusted sodium adsorption ratio (SAR_{adj}), and exchangeable sodium percentage (ESP) are calculated from the individual cation and anion determinations. These indices must be used along with EC values to evaluate salinity and sodicity accurately.

The unadjusted SAR indicates the ratio of sodium to calcium and magnesium in a soil water extract or irrigation water sample. An increasing SAR value indicates higher levels of sodium in comparison with calcium and magnesium. Rising levels of sodium reduce soil stability, decrease water infiltration, and increase the chance of sodium accumulating to toxic levels in leaf tissue. Use the SAR rather than sodium cation levels alone to evaluate sodicity. More sodium can be tolerated in a soil extract or water sample when calcium increases proportionally to sodium.

The SAR_{adj} is calculated and reported only for water samples. This index predicts the reaction of bicarbonate with calcium when water is applied to the soil. Irrigation water with low levels of bicarbonate or carbonate (CO_3) anions usually has an SAR_{adj} that is very similar to its unadjusted SAR. Such water will very slowly dissolve lime from the soil and contribute calcium to offset sodium in the soil water. Irrigation water high in bicarbonate or carbonate usually has an SAR_{adj} higher than its unadjusted SAR. Such water precipitates calcium with bicarbonate and forms lime, which reduces free calcium levels in the soil water and increases the proportion of sodium. Prior to 1988, SAR_{adj} was calculated according to an empirical equation using pH constants (pH_c). This has since

Table 7.2. Conversion factors from milligrams per liter (mg/l) to milliequivalents per liter (meq/l)

Cation/anion	Symbol	Conversion factor
bicarbonate	HCO_3^-	61
calcium	Ca^{2+}	20
carbonate	CO_3^{2-}	30
chloride	Cl^-	35
magnesium	Mg^{2+}	12
sodium	Na^+	23
sulfate	SO_4^{2-}	48

Note: mg/l ÷ conversion factor = meq/l.

been proven to overestimate SAR_{adj}. The new procedure is based on the proportion of calcium and bicarbonate in a water sample. If there is any doubt about how the laboratory calculated SAR_{adj} and the laboratory cannot clarify which procedure was used, simply use the unadjusted SAR.

ESP is closely related to the SAR for soil water extracts. These ratios differ in that SAR is an index comparing soluble sodium with soluble calcium and magnesium in the soil water extract, whereas ESP is an index indicating the amount of sodium ions bound to exchange sites in the soil. Today, most laboratories do not measure ESP directly because some of the lab methods have been shown to overestimate calcium by dissolving lime. Also, determining the ESP requires extra analyses to measure the total cation exchange capacity and the exchangeable sodium content. Instead, most laboratories report an estimated ESP based on a correlation between SAR and ESP.

Gypsum requirement

Most laboratory soil analysis reports provide either a gypsum requirement (GR) or a lime requirement (LR). A GR is usually provided for alkaline soils with pH above 7 and SAR above 10 to 15. An LR is provided only for acid soils with a pH less than 7. The most common method of determining the GR is the Schoonover test. The Schoonover test overestimates the GR because it measures how much soluble calcium must be added in the form of gypsum to replace nearly all the sodium on the soil exchange sites. Complete replacement of sodium is unnecessary because even a sensitive crop such as prune can tolerate low levels of sodium in the soil. In addition, the Schoonover test is not appropriate for determining the GR needed to amend surface soils or irrigation water to improve water infiltration. As an alternative, GR can be estimated using other information given in a laboratory report (see the section "Amendment Rates for Soils," below).

Verifying Lab Results

Preferably, a report will give the cation and anion concentrations in meq/l. When these reporting units are used, the accuracy of the analysis can be evaluated by the two methods described below. These checks are illustrated by referring to table 7.3 as a sample analytical report.

Check 1: Cation-anion balance method

Salts such as sodium chloride (NaCl), sodium bicarbonate ($NaHCO_3$), and calcium sulfate ($CaSO_4$) consist of cations and anions bonded by electrical charges. For each cation there is an equivalent charge (meq) of anion bonded to form the salt. This is referred to as the cation-anion balance. Using table 7.3:

$$Na + Ca + Mg \approx HCO_3 + CO_3 + SO_4 + Cl$$

$$10.2 \text{ meq/l} \approx 9.9 \text{ meq/l}$$

When dissolved in a water sample or soil extract, the bonds are broken and the salts exist as individual cations, anions, or neutral ion pairs. The individual cations and ions must be reported in meq/l to perform this check. Omit boron and nitrate (NO_3) in conducting this check procedure, because they are reported in mg/l and are usually an insignificant amount of the total salinity.

Check 2: Comparing total salinity with the sum of cations or anions

If the analysis is valid, the salinity level, EC_e or EC_w, multiplied by a factor of 10, about equals the sum of the cations or anions. Using table 7.3:

$$EC_w \times 10 \approx (Na + Ca + Mg)$$

or

$$\approx (HCO_3 + CO_3 + SO_4 + Cl)$$

$$1.0 \times 10 \approx 9.9 \text{ (sum of anions) or } 10.2 \text{ (sum of cations)}$$

In this example, the sum of the individually measured cations or anions is in close agreement with the value calculated from $EC_w \times 10$.

A report in which the EC multiplied by 10 exactly equals the sum of either the cations or anions, or where the cations exactly equal the anions, indicates that some of the individual cations and anions were probably estimated by subtraction rather than determined by direct measurement. Sulfate (SO_4) and sodium are most likely to be estimated, because measuring them requires additional analytical steps and expense.

Table 7.3. Sample irrigation water quality analysis to demonstrate how to check the quality of a laboratory report

Analysis	Example
pH	8.4
EC_w	1.0 dS/m
Ca	0.5 meq/l
Mg	0.1 meq/l
Na	9.6 meq/l
HCO_3	4.2 meq/l
CO_3	1.0 meq/l
Cl	4.6 meq/l
SO_4	0.1 meq/l
B	0.7 mg/l
NO_3^-	5.2 mg/l
SAR	17.5
SAR_{adj}	16.6

Diagnosing Potential Salinity Conditions

Salinity analyses are used to diagnose three types of salinity conditions in the field: excess root zone salinity surpassing the tolerance of prune; potential accumulation of specific elements to toxic levels for prune; and identifying soils and water supplies likely to have slow water infiltration rates.

Prune Salt Tolerance

A high soil EC_e or EC_w value indicates high salinity. When soil salinity increases in the soil water surrounding the root and exceeds a critical threshold, the gradient between the solute concentration in the root cells and the soil water around the root lessens, reducing water availability to plants. In response to increased soil salinity around the roots, salt-sensitive crops such as prune internally produce more sugars and organic acids to increase the solute concentrations inside the root cells and reestablish the gradient for water to again move freely into the root cells. However, this internal plant process that adjusts for root zone salinity competes for energy that would otherwise be used in photosynthesis, thereby suppressing crop growth and yield. When root zone salts are too high, the trees may display inadequate vigor and overall smaller tree size along with reduced fruit size. Necrotic (brown, dead) tissue along leaf tips and margins may indicate excess salt absorption and accumulation (fig. 7.5). Trees may show symptoms of water stress even though

Figure 7.5 General symptoms of salt effects, reduced tree size and lack of growth (A) in a walnut tree, similar to what would be expected for prune. A close-up of marginal burn (B), which is most often associated with sodium or chloride toxicity. The interveinal chlorosis may be indicative of boron toxicity. *Photos:* B. L. Sanden.

the soil may feel as though it contains sufficient moisture.

The wide variety of crops grown in California are categorized as sensitive, moderately sensitive, moderately tolerant, and tolerant to soil salinity and salinity in irrigation water. Prunes are considered to be sensitive to salinity, very similar to other orchard crops such as almond, citrus, peach, and walnut, and not nearly as tolerant as pistachio or fig. The salt tolerance of prune is considerably less than some common agronomic crops such as barley, wheat, cotton, but similar to other agronomic crops such as alfalfa and corn, which may be important when considering the transition from agronomic crops to an orchard crop such as prune.

Table 7.4. Guidelines to evaluate orchard soils and water supplies for excess salinity for mature prune trees

	Degree of growth or yield reduction			
	Unit	**None**	**Increasing**	**Severe**
average root zone	dS/m	< 1.5	1.5–3.5	> 3.5
irrigation water	dS/m	< 0.7	0.7–2.0	> 2.0

Source: Adapted from Maas 1990, p. 280.
Note: Guidelines assume a 15% leaching fraction.

Table 7.4 provides guidelines for evaluating soils and water supplies for excess salinity for mature prune trees. Land with average root zone salinity less than 1.5 dS/m and irrigated with water supplies containing salinity levels less than 0.7 dS/m are ideal for prune and should not restrict production potential. As either the soil salinity or the salinity of the irrigation water increases above these thresholds, a reduction in orchard vigor and production potential can be anticipated. The tolerance of French prune to salinity has not been evaluated with controlled experiments. However, experimentation with plum has shown that an 18% to 36% reduction in tree growth and production potential can be anticipated when grown on land with average root zone salinity (EC_e) ranging from 2.5 to 3.5 dS/m or by using irrigation water that ranges from 1.2 to 2.0 dS/m (EC_w). More severe restrictions in plum can be expected if average root zone salinity exceeds 3.5 dS/m or if the irrigation water exceeds 2.0 dS/m. This is the best indicator of salt tolerance of French prune that is currently available.

Toxic Ion Accumulation

Toxic ion effects have slow, chronic symptoms that can develop in prunes over several years. Sodium (Na), chloride (Cl), and boron (B) are the primary ions of concern. Trees grown on soils with excessive levels of these elements accumulate the ions in their woody tissue and eventually in their leaves. Foliar symptoms of excess chloride, boron, and sodium in the leaf tissue are marginal leaf burn and then interveinal necrosis, with twisting and curling of the leaves. An accumulation of any of these ions interferes with physiological processes such as photosynthesis and transpiration and ultimately reduces tree growth and production.

It is important to diagnose the accumulation of specific ions before levels become elevated in the woody and leaf tissue. Once sodium, chloride, or boron has accumulated, the trees have no mechanism to rapidly expel these ions. It may require many seasons of tree growth and production, coupled with management steps

Table 7.5. Critical levels for prune production of specific ions in soils (saturated soil extract)

Specific ion	Levels in soil extract Degree of toxicity		
	None	Increasing	Severe
sodium (ESP)	< 5	5–15	> 15
chloride (meq/l)	< 5	5–25	> 25
boron (mg/l)	< 0.5	0.5–3.0	> 3.0

Source: Adapted from Ayers and Westcott 1985.

Table 7.6. Critical levels for prune production of specific ions in irrigation water

Specific ion	Levels in water supply Degree of toxicity		
	None	Increasing	Severe
sodium (SAR)	< 3	3–9	> 9
chloride (meq/l)	< 7	7–17	> 17
boron (mg/l)	< 0.5	0.5–3.0	> 3.0

Source: Adapted from Ayers and Westcott 1985.

Table 7.7. Critical levels of sodium, chloride, and boron in prune leaf tissue (July/August tissue samples)

Specific ion	Levels in leaf tissue Degree of toxicity		
	None	Increasing	Severe
sodium (%)	< 0.25	0.25–0.40	> 0.40
chloride (%)	< 0.3	0.3–0.5	> 0.5
boron (mg/l)	< 30	30–85	> 85

Source: Adapted from Sibbett 1993.

to reduce the level of a specific ion in the root zone, to correct the toxic ion effect. Conditions in which toxic ion effects are likely to occur can be diagnosed through analyses of soil, irrigation water quality, and leaf tissue. Tables 7.5, 7.6, and 7.7 provide guidelines to identify conditions for possible ion toxicities based on soil, irrigation water, and leaf tissue samples, respectively.

If soil, water, and leaf tissue analyses indicate excessive levels of a specific ion, the diagnosis is obvious. If the soil or water analysis indicates excessive levels but the leaf tissue does not, it may be only a matter of time before toxicity develops in the trees. In such a situation, the soil or water analysis may forewarn a problem before it actually affects growth and production. In this way, soil and water monitoring can offer the earliest opportunity to improve management for salinity control.

Rootstock selection when planting a new prune orchard can have some bearing on chloride and boron tolerance. Experimentation has shown that Marianna plum rootstock has higher tolerance to chloride in soil and in irrigation water than does Lovell peach rootstock. Similarly, Myrobalan plum rootstock accumulates less boron in the rootstock and transports less boron to the French variety scion than do Marianna plum or Lovell peach rootstocks.

Nitrate-Nitrogen Toxicity

Nitrate (NO_3^-) toxicity can become a concern when too much nitrogen fertilizer is applied. If overapplication is severe, the first effect may be defoliation; the trees will most likely regrow vigorously. Foliage may grow so large, in fact, that the leaves curl upward. Use soil, water, and plant tissue analyses to avoid overuse of nitrogen fertilizers.

Most labs measure nitrate but report it as the amount of elemental nitrogen (NO_3-N). For irrigation water, a NO_3-N level from 0 to 3 mg/l is considered low; from 3 to 10 mg/l is considered moderately low to moderately high; and above 10 mg/l is considered high. In a soil sample taken from a depth of 1 foot, a NO_3-N level from 0 to 10 mg/l is considered low; 10 to 20 mg/l moderately low to moderately high; and a level exceeding 20 mg/l high.

It is easy to convert a value reported into equivalent pounds of nitrogen available for crop nutrition. To convert the level of NO_3-N in a water analysis to pounds of nitrogen per acre-foot of water, multiply the concentration reported in mg/l by 2.7. (One acre-foot of water weighs 2.7 million pounds.) For example, if the analysis reports that a sample of water contains 2.3 mg/l of NO_3-N, the water supplies 6.2 pounds of nitrogen per acre-foot of water.

Similarly, levels of NO_3-N reported in a soil analysis can be converted to pounds of nitrogen per acre-foot of soil. Find the level of NO_3-N (mg/l) reported for a sample representing a 1-foot depth of soil. Multiply the number by 4. (One acre-foot of soil weights about 4 million pounds.) For example, if the analysis reports that the sample contains 9.7 mg/l of NO_3-N, the amount of nitrogen in the soil is 38.8 pounds per acre-foot.

Poor Water Infiltration

Poor water intake prevents sufficient infiltration to adequately supply water to prunes, and these conditions can lead to insufficient root zone aeration. Orchards hindered by poor water infiltration typically show reduced shoot growth and increased incidence of sunburn, smaller prunes, and crown and root rot.

Soils and irrigation water with very low EC levels or a high SAR are more likely to develop poor infiltration rates. Both conditions contribute

Table 7.8. Guidelines to diagnose water infiltration problems based on soil and irrigation water quality

Sodium adsorption ratio (SAR)	Potential for water infiltration problems to develop	
	Unlikely if EC_e or EC_w is	Likely if EC_e or EC_w is
0–3	> 0.7	< 0.3
3.1–6	> 1.0	< 0.4
6.1–12	> 2.0*	< 0.5
12.1–20	> 3.0*	< 1.0
20.1–40	> 4.0*	< 2.0

Source: Adapted from Oster et al. 1992, p. 67.
Note: *Even though these salinity conditions are less likely to develop slow infiltration rates, the EC values exceed tolerable levels for prunes.

to unstable soil structure. Under these conditions, soil aggregates swell and disperse into individual particles when irrigated. After the applied water recedes, the soil particles settle and the finer-textured clay and silt particles fill the pores between the larger sand particles to form a slowly permeable surface crust (figs. 7.6 and 7.7).

The total salinity of a soil extract or irrigation water (EC_e or EC_w) must be considered together with a measure of sodicity (the SAR of the soil extract or the SAR_{adj} of water) to assess the extent that salinity is contributing to poor infiltration in an orchard. Table 7.8 presents guidelines for evaluating the effects of salinity and sodicity of soil water and irrigation water quality on water infiltration rates. As a rule, the interpretation of this table can be simplified. There is little chance that soil water or irrigation water quality is contributing to slow water infiltration rates when the ratio of SAR to EC is less than 5:1; an increasing likelihood when the ratio of SAR to EC ranges from 5:1 to 10:1; and a high probability when the SAR to EC is greater than 10:1. Table 7.8 and these general rules provide guidelines applicable to San Joaquin Valley soils and irrigation water supplies, with few exceptions.

However, in some soil conditions these guidelines may not identify soil and irrigation water qualities prone to poor infiltration. For example, soils and water supplies that are relatively high in magnesium and relatively low in calcium are sometimes reported in the Sacramento Valley to have slower infiltration. In these conditions, magnesium may behave similarly to sodium, causing soils to disperse and form impermeable crusts on drying. Criteria to diagnose these soil and water conditions are not nearly as well understood as are the effects of sodium on infiltration.

Through anecdotal experience and uncontrolled experimentation, it has been suggested that when magnesium content exceeds a 1:1 ratio to calcium content in soils or irrigation water, soils may begin to develop poor infiltration. Another reported condition not diagnosed by the criteria in table 7.8 is soil with a very high exchangeable

Figure 7.7 Surface sealing of a clay loam in a flood-irrigated orchard is caused by soil and water chemistry that is either very low in salinity (low EC_w) or dominated by sodium (sodic conditions). Water penetration is only to 7 inches deep after a 32-hour irrigation set (inset). The surface remains saturated 8 hours after irrigation ceases. *Photo:* B. L. Sanden.

Figure 7.6 Surface sealing of sandy loam caused by very low salinity in soil and irrigation water or a predominance of sodium in the soil and water chemistry. *Photos:* B. L. Sanden.

potassium level. Potassium may have effects similar to those of sodium on soil stability and porosity when it is predominant in the soil water environment. However, this has not commonly been reported in California.

Reclaiming Saline Soils and Managing Salinity

Prevention of salinity implies that an orchard has been established on nonsaline soils suitable for prune production. The goal then is to manage the soils and irrigation water to avoid the accumulation of salinity or toxic ions and the development of slow infiltration rates. By contrast, reclaiming an orchard implies that an orchard has been diagnosed as having excess salinity, toxic levels of specific ions, poor infiltration rates, or some combination of these, and remediation is needed.

Whether the objective is prevention or remediation, proper irrigation water management is key to successful salinity management. In some situations, the application of soil and water amendments will also be an important management component.

Irrigation or Rainfall and Leaching

Root zone salinity increases when salts or toxic ions are transported into the orchard with irrigation water. The only way of decreasing salinity is by transporting salts out of the root zone with deep percolation, or leaching. This is an important function of irrigation and winter rainfall.

The leaching fraction is the percentage of irrigation water or rainfall needed to transport salts out of the root zone to maintain tolerable salinity levels. The leaching fraction does not contribute to meeting the crop's water needs; rather, it is in excess. It is expressed as a percentage rather than as a specific quantity of water because it is calculated for orchards in which various rates and qualities of irrigation water are applied. As the quantity of the salt in the applied water increases, as is often the case with lower-quality water supplies, more salinity is transported into the orchard; therefore, more leaching is required to percolate salts below the root zone. A sufficient leaching fraction prevents salt accumulation in the root zone.

Leaching can also be expressed as a leaching requirement or the depth of infiltrated water needed to reclaim a saline soil to a salinity level that is tolerated by prunes. It is commonly expressed as inches of water required per foot of soil in the root zone.

Basics of effective leaching

Five factors are especially influential in achieving effective leaching and salinity control.

First, effective salinity management is much easier on soils with deep, well-drained profiles. A well-drained soil provides a zone to which unwanted salts can be transported and in which they can accumulate away from the root system of the trees. Successful leaching for salinity control is much more difficult on poorly drained soils because there is no soil zone in which salts can accumulate that is separate from the root system. The depth to the shallow water table often fluctuates throughout the season: it is nearest the soil surface in the spring and deepest in the fall. As a result, salinity that may be leached to deeper soil depths in the fall is often transported back into the root zone in the spring, when the water table rises. Growers are therefore well advised to avoid planting prunes in poorly drained soils.

Second, the soil water content must exceed field capacity throughout the root zone for leaching to occur. Otherwise, irrigation water in the amount equal to the soil water content depletion will be retained in the root zone and will not actually transport salts below the root zone.

Third, smaller quantities of irrigation water applied more frequently, such as with sprinklers or by winter rainfall, will more effectively transport salinity below the root zone than will an equal quantity of irrigation water applied in one flood application. One-time applications of large quantities of water tend to infiltrate and percolate through the larger porous pathways in the soil, failing to transport salinity from many of the soil's small pores.

Fourth, leaching does not have to be accomplished every irrigation, perhaps not even every season; whether to leach will depend on the specific soil and water conditions. Leaching is necessary only when the average root zone salinity level of a specific ion approaches or exceeds the critical level for prunes. Periodic soil and irrigation water testing will help determine the extent to which leaching is needed.

Fifth, for leaching to occur, the amount of applied water must exceed crop evapotranspiration (ET_c). When ET_c exceeds the amount of applied water or rainfall, the net movement of water and salts is upward into the root zone. Therefore, leaching is most efficient when the trees are dormant and ET_c is minimal.

Preventing Excess Salinity in Soils

Variations in irrigation water quality and soil salinity mean that different leaching fractions may be needed from one orchard to the next. Table 7.9 provides the leaching fractions required for irrigation water qualities ranging from 0.5 to 2.0 dS/m to maintain a desirable root zone salinity ranging from 1.0 to 1.5 dS/m (below the critical level for prunes given in table 7.4).

The leaching fraction needed to maintain root zone salinity within tolerable levels for prunes increases as the salinity level in the irrigation water increases. Also, the leaching fraction increases if the goal is to maintain root zone salinity well below the economic threshold for prunes. The economic threshold for root zone salinity is synonymous with the critical level of root zone salinity for prunes (see table 7.4 and the section "Prune Salt Tolerance," above), which is the point where prune tree growth and productivity is expected to begin declining due to excessive salt accumulation in the root zone. An example application of the information in table 7.9 is given below.

Example: Maintaining salinity using table 7.9

Mature prunes grown with resident vegetation on the orchard floor consume about 38.0 inches of water annually. Assume that the irrigation water has an EC_w of 0.75 dS/m and that the goal is to maintain the average root zone salinity (EC_e) at 1.5 dS/m, the economic threshold for prunes. From table 7.9, we see that a leaching fraction of 12% is required for salinity control. This works out to a total seasonal water application of 43.2 inches: 38.0 divided by (1.0 – 0.12) = 43.2 inches. Of that water, 38.0 inches would be applied over the course of the irrigation season to meet crop evapotranspiration needs and refill the soil profile, and 5.2 inches would be applied to leach salinity below the root zone and maintain the current level of salinity. Remember that smaller quantities of water applied more frequently, such as with sprinklers or as winter rainfall, transport salinity below the root zone more efficiently than does an equal quantity of water applied in a single irrigation application. In the prune-producing areas of the southern San Joaquin Valley, where rainfall is relatively low, supplemental irrigation may be needed in the winter to provide this leaching fraction. However, in the prune-producing regions of the Sacramento Valley, where rainfall often exceeds this leaching fraction, precipitation should suffice.

Table 7.9. Leaching fraction (%) required to maintain a specific level of root zone salinity, given variable levels of salinity in the irrigation water supply

Salinity of irrigation water (EC_w, dS/m)	Desired average root zone salinity for prunes (EC_e, dS/m)	
	1.0 dS/m	1.5 dS/m
0.50	6.5	4.0
0.75	20.0	12.0
1.00	32.0	25.0
1.25	35.0	28.0
1.50	55.0	40.0
1.75	65.0	43.0
2.00	85.0	55.0

Source: Adapted from Hoffman 1996.

Root Zone Salinity Modeling Tool

A computer program called WATSUIT (see Oster and Rhoades 1990) can accurately predict long-term root zone salinity and specific ion concentrations for a given irrigation water quality and leaching fraction. This model has been developed into a Windows-based program by Dr. Laosheng Wu at UC Riverside and can be downloaded at the USDA Agricultural Research Service's U.S. Salinity Laboratory website, http://www.ars.usda.gov/Services/.

Reclamation

If prune production is already restricted by salinity in an existing orchard or if the development of a new orchard would be restricted by existing soil salinity levels, reclamation is needed. Table 7.10 provides leaching requirements for reclaiming soils with various degrees of excess salinity to tolerable levels for prune.

Example: Reclamation using table 7.10

Assume that a grower has a parcel of land that is being considered for orchard establishment. Laboratory analyses indicate that water quality of 0.4 dS/m will be used for irrigation and the average root zone salinity is 3.0 dS/m prior to any land preparation. The grower would like to have tolerable soil salinity levels (average of 1.5 dS/m) extend at least 4 feet deep before planting prunes. According to table 7.10, reducing the initial soil salinity from an average of 3.0 dS/m to at least 1.5 dS/m would require about 1.8 inches of water per foot of soil making up the root zone. That is, a minimum of 7.2 inches of water applied to this soil after it has been refilled to field capacity should reduce the average root zone salinity from about 3.0 to 1.5 dS/m to a depth of 4 feet. In the prune-producing areas in the southern San Joaquin Valley, effective rainfall is not likely to be sufficient

Table 7.10. Depth of leaching water required (in inches per foot of root zone) for leaching to reduce an initially high average rootzone salinity to a tolerable level for prune

Desired root zone salinity (dS/m)	Average root zone salinity before leaching			
	2 dS/m	3 dS/m	4 dS/m	5 dS/m
1.0	1.8	3.0	4.2	5.4
1.5	0.6	1.8	3.0	4.2
2.0	0.0	0.6	1.8	3.0

Source: Adapted from Hoffman 1986.
Note: Guidelines are applicable for all irrigation waters less than 1.0 dS/m.

to achieve the level of leaching needed to reclaim this site, and supplemental irrigation during the winter season would likely be needed. Rainfall is more likely to suffice in the Sacramento Valley.

Leaching requirements can be extrapolated for salinity levels that fall between the levels specified in table 7.10. These guidelines assume that the leaching requirement is applied during seasons of low evapotranspiration (ET) and over several irrigations, or is provided by several rainfall events or smaller water application rates with periods of drying between each irrigation. The guidelines also apply to leaching boron, chloride, and sodium to correct a specific toxicity. If leaching is attempted in one irrigation with a larger application rate, the leaching efficiency will decline; it may then take up to three times the quantity of water to achieve the same degree of reclamation. Leaching to reclaim a soil to tolerable levels of salinity for prune also requires adequate drainage.

Resampling Soils

Tables 7.9 and 7.10 present science-based leaching strategies to achieve successful salinity management. Although these guidelines are based on research and are the best available at this time, effective salinity management can be verified only by resampling the soils in the orchard to confirm whether salinity conditions have been improved to the extent expected.

Soil and Water Amendments

Soils with poor water infiltration and permeability rates are candidates for treatment with amendments. The amendments supply exchangeable calcium that displaces sodium in the soil and, in some instances, magnesium and possibly potassium. Exchangeable calcium is scientifically proven to stabilize soil aggregates as well as porosity, which sustains water infiltration and permeable rates. Sodium (also magnesium and potassium, to a lesser degree), in contrast, is proven to cause swelling and dispersion when a soil is irrigated. As a result, when water recedes after irrigation, the finer, dispersed clay and silt particles settle into the larger pores between the sand particles, creating surface crusts and thin subsoil layers that are barriers to infiltration and percolation.

Amendments can change soil or irrigation water composition from one dominated most often by sodium and bicarbonate to one composed of higher calcium (and usually sulfate) levels. Conditions requiring amendments are indicated by laboratory analyses that show high SAR indices and relatively low EC values (see the section "Poor Water Infiltration," above). Amendments are usually unnecessary when salinity analyses indicate that the soil has high EC values and relatively low SAR indices. A favorable composition of calcium relative to sodium already exists in such soils; thus, leaching may be the most appropriate first step to correct the salinity condition.

Applying amendments is no substitute for irrigation practices that achieve the necessary leaching. Adding amendments improves the salinity composition so that infiltration and permeability rates are higher and leaching is more attainable. However, without sufficient leaching after the addition of amendments, the average root zone salinity will be elevated, and the displacement and removal of sodium (and perhaps magnesium or potassium) from the root zone will not be completed.

Types of Amendments

There are two types of amendments: calcium salts and acid-forming amendments. Sometimes calcium salts are referred to as direct calcium suppliers, and the acid-forming amendments as indirect calcium suppliers.

Calcium is added to the soil directly when calcium salts are used as soil amendments. Calcium salts commonly used include gypsum, lime, dolomite, calcium chloride, and calcium nitrate. Each salt has a specific solubility rate in water, with calcium nitrate and calcium chloride being highly soluble, gypsum moderately soluble, and dolomite and lime very slowly soluble. The highly soluble salts are more convenient to apply directly to irrigation water but are typically more expensive. Gypsum is reasonably simple to add to

irrigation water and is typically less expensive than calcium chloride and calcium nitrate. Lime and dolomite are relatively insoluble in water unless the water is acidic (indicated by a pH of less than 7.0). Gypsum, calcium chloride, and calcium nitrate have negligible effects on soil pH, whereas lime and dolomite can increase soil pH when applied to acidic soils.

Sulfur, sulfuric acid, urea sulfuric acid, Nitro-sul, and lime sulfur are some of the more common acid-forming amendments used in salinity management. Since all the materials contain sulfur or sulfuric acid but no calcium, they supply exchangeable calcium indirectly by dissolving lime that is present in the soil. The sulfur compounds undergo microbiological reactions that oxidize sulfur to sulfuric acid. The acid dissolves soil lime to form a calcium salt (gypsum), which then dissolves in the irrigation water to provide exchangeable calcium. The acid materials do not have to undergo the biological reactions; instead, they immediately react with soil lime on application. Since these materials form an acid in the soil reaction, they can reduce soil pH if applied at high enough rates.

Table 7.11. Amount of amendments required for calcareous soils to replace 1 meq/l of exchangeable sodium in the soil or to increase the calcium content in the irrigation water by 1 meq/l

Chemical name	Trade name: composition	Pounds* per ac-6 in to replace 1 meq/L† exchangable sodium	Pounds* per ac-ft of water to get 1 meq/L calcium
sulfur	100% S	321	43.6
gypsum	$CaSO_4 \cdot 2H_2O$, 100%	1,720	234
calcium polysulfide	Lime sulfur: 23.3% S	1,410	191
calcium chloride	Electro-Cal:13% calcium	3,076	418
potassium thiosulfate	KTS: 25% K_2O, 26% S	1,890‡ 3,770	256‡ 513
ammonium thiosulfate	Thio-sul: 12% N, 26% S	807§ 2,470e	110§ 336**
ammonium polysulfide	Nitro-sul: 20% N, 40% S	510§ 1,000**	69§ 136**
monocarbamide dihydrogen sulfate/sulfuric acid	N-phuric, US-10: 10% N, 18% S	1,090§ 1,780**	148§ 242**
sulfuric acid	100% H_2SO_4	981	133

Source: Adapted from Oster et al. 1992, p. 113.

Notes:

*Salts bound to the soil are replaced on an equal ionic charge basis and not on an equal weight basis. Laboratory data show that an extra 14 to 31%, depending on initial and final ESP or SAR, of the amendment is needed to complete the reaction.

†The meq of exchangeable sodium to replace = initial ESP – desired ESP × total meq/100 grams soil cation exchange capacity.

‡Assumes 1 meq K beneficially replaces 1 meq Na in addition to the acid generated by the S.

§Combined acidification potential from S and oxidation of N source to NO_3 to release free Ca from soil lime. Requires moist, biologically active soil.

**Acidification potential from oxidation of N source to NO_3 only.

Table 7.12. The influence of purity and moisture content on equivalent quantities of soil and water amendments

Amendment	Average purity (%)	Average moisture (%)	Tons of amendment equivalent to 1 ton of 100% pure gypsum
pit gypsum	50 Ca	8	2.17
mined gypsum	75 Ca	4	1.39
delivered bulk solution-grade gypsum	92 Ca	3	1.12
ground wallboard	92 Ca	5	1.14
popcorn sulfur	99 S	6	0.20
granular sulfur	99 S	2	0.19
sulfuric acid	98 S	NA	0.58
Thio-Sul (delivered)	N-12, S-26	NA	0.47
lime sulfur[2,3]	Ca-6, S-23	NA	0.67
urea sulfuric acid (N-phuric, US-10)	N-10, S-18	NA	0.63
beet lime	60 Ca	10	1.85

Amendment Selection

Selecting a soil amendment depends largely on the presence or absence of lime in the soil and the relative costs of the materials. As long as lime is abundant in the soil, it is appropriate to consider either a calcium salt or an acid-forming amendment; the choice then depends largely on cost. In order to compare costs of amendments, the material must be converted to an equivalent basis, as each amendment will have a different chemical formulation. For example, about 1,200 pounds of sulfuric acid on a calcareous soil will supply the same amount of exchangeable calcium as 2,000 pounds of pure gypsum. Table 7.11 summarizes the rates of available amendments needed to supply equal levels of exchangeable calcium to the soil or equal calcium content to the irrigation water.

Table 7.12 illustrates the influence of purity, moisture content, and formulation on the costs of soil and water amendments. In general, solid formulations of amendments are less costly without considering freight and application costs. Liquid formulations, however, are often applied by injection into the irrigation water and may have lower application rates and costs. Other aspects of amendment quality deserve consideration, too. The size and uniformity of grind of some solid forms of soil amendments are important. For example, the conversion of elemental sulfur to acid is a microbial process that occurs most rapidly when finely ground sulfur is incorporated into moist soil. Large, pea- or gravel-sized sulfur particles (common with popcorn sulfur) can take up to 10 years or more to react and form sulfuric acid, especially if spread only on the surface. In general, finer-ground soil amendments react more quickly in the soil.

Another factor influencing the choice of amendment is that some amendments add sulfate to the root zone, and others add chloride or nitrate. The existing nitrate and chloride content will limit how much of these materials can be added. Nitrogen levels should not exceed the annual crop needs, and chloride should not accumulate in the soil above critical levels for prunes. Sulfate accumulation to toxic levels in prunes has not been reported and should not limit the use of amendments that supply sulfate.

Using acid-forming materials is discouraged when significant levels of lime are absent in the soil. These soils will be neutral or acidic in pH, so using calcium salts will be most appropriate. Lime and dolomite are preferred to gypsum and calcium chloride as the pH becomes more acidic, especially as it decreases below 6.0. Acid-forming amendments will be more effective than calcium salts on very alkaline soils (with a pH above 8.4), because they will reduce soil pH if correctly applied at very high rates.

Amendment Rates for Irrigation Water

Amendments are most often added to water to improve the water infiltration rate into the surface soil. Application rates equivalent to 1.0 to 3.0 meq/l of calcium are considered low to moderate; those that supply 3.0 to 6.0 meq/l are considered moderate to high. For example, according to table 7.11 (far right column), using either a low application rate of 468 pounds of pure gypsum or the equivalent rate of 266 pounds of sulfuric acid per acre-foot of water would supply the equivalent of 2.0 meq/l of calcium amendment (assuming that lime is abundant in the surface soil to react with the sulfuric acid). In comparison, an application rate of 936 pounds of pure gypsum or an equivalent rate of 532 pounds of pure sulfuric acid per acre-foot of water would amend the irrigation water at a higher rate of 4.0 meq/l of calcium.

If the irrigation water were analyzed after amendment, EC_w should be increased and SAR should be decreased. An appropriate water amendment rate should modify water so that the combined evaluation of the EC_w and SAR suggest little or no probability of infiltration problems developing (see table 7.8 and the section "Poor Water Infiltration," above).

Amendment Rates for Soils

Amendment rates for soils are considerably higher than for irrigation water supplies. One reason for applying a soil amendment is to reduce the exchangeable sodium throughout the root zone, not just on the soil's surface. For soils having potential for prune production, amendment rates may range from 1.0 to 3.0 tons of 100% pure gypsum or an equivalent amount of another material per acre-foot of soil, which can equate to rates approaching 10 tons per acre when a 4-foot soil profile is amended and the purity and moisture content of available amendments are considered. If higher amendment rates are needed, the soils may be inappropriate and too expensive to develop or maintain prune production.

The gypsum requirement as determined by the Schoonover method is commonly provided on analytical reports to indicate the appropriate soil amendment rate. However, this method often overestimates the gypsum requirement because it measures how much gypsum is needed to replace

Figure 7.8. Solution gypsum machine injects a slurry of liquefied gypsum into the irrigation water supply at a constant, controlled rate to modify irrigation and soil water quality. Solution-grade gypsum is conveyed using an auger from a large silo into the injection machine to minimize the labor required. *Photo:* B. L. Sanden.

nearly all of the exchangeable sodium adsorbed by a soil; such extensive replacement is unnecessary and may be uneconomical. A gypsum rate or the equivalent rate of another amendment that supplies 50 to 75% of the gypsum requirement in a lab report should be sufficient to make marked improvements.

Application Methods

After selecting an amendment and determining its appropriate rate of application, the final decision is choosing the method of application. Options include applying the amendment in the water, applying it to the soil surface and irrigating it into the soil, broadcasting it and tilling it into the soil, and applying it in a band in the soil.

Adding to irrigation water

Adding amendments directly to the water is ideal for managing soils with slow water infiltration caused by surface crusting. Research has shown after just one irrigation that crusts develop on many soils, creating a barrier to infiltration in subsequent irrigations. These surface crusts are often less than ¼ inch in depth (see figs. 7.6 and 7.7). Because their formation is associated with the irrigation water quality, a constant application of amendments in the water is most effective for applying the amendment precisely where the soil crusting forms. In addition, relatively small, constant rates of amendments are needed, which are most easily attained and most precisely applied via water. Solution-grade gypsum injected as a slurry at controlled rates is the most common method of adding calcium into irrigation water (fig. 7.8). Liquid formulations of acid-forming amendments may also be injected, but be certain that soil lime is present in the surface before applying them in this manner.

Broadcasting then irrigating

Broadcasting amendments such as gypsum on the soil surface without tilling them into the soil is an alternative to water treatments. This method attempts to achieve the same effect as injecting an amendment into the irrigation water by applying water onto an amendment that is exposed on the soil surface. The primary advantages to this approach are that less expensive gypsum sources can be used and injection equipment is not needed. However, for surface application to be nearly as effective as water treatment, it must be properly timed. If infiltration is poor in the summer months, the amendment should be applied at the onset of those months and not in the preceding fall or winter months. Applying it too early will cause winter irrigations and rainfall to leach the amendment to depths beyond where the crust forms. Surface applications are most effective when applied at lower rates (equivalent to 500 to 1,000 pounds of gypsum per acre) monthly during June, July, and August. More finely and consistently ground gypsum will be advantageous for this use. Growers may find this approach to be more of a nuisance than adding amendments to the water.

An alternative for flood-irrigated orchards using orchard valves is to place 100 to 250 pounds of coarse gypsum in a pile next to the valve and let the water carry it down the field (fig. 7.9). However,

Figure 7.9 Land-grade gypsum (75% purity) side-banded next to orchard alfalfa valves at 250 pounds per check for dissolving into water stream during irrigation. *Photos:* B. L. Sanden.

the rate and distribution of application is much less controlled and may influence the effectiveness.

Broadcasting and tillage

Land applications and incorporation of amendments with tillage are more appropriate than water applications when the objective is to reclaim a sodic soil (or possibly a high-magnesium soil) that is deep in the root zone rather than on the soil surface. This method requires higher application rates and is most affordable with land applications of coarser, less-refined grades of amendments. Incorporating the amendment by plowing, shanking, or slip plowing speeds up the reclamation by getting the amendment to the deeper soil more quickly so that the exchange reaction can occur sooner.

Banded application

Banding amendments is appropriate when the objective is to correct a micronutrient deficiency such as iron or zinc in alkaline soil by lowering or increasing the soil pH. This is actually a fertility practice, unrelated to salinity management. Banding concentrates these amendments into areas of active rooting. It is not essential to modify soils in the entire root zone; it is necessary only to dissolve all the lime (for alkaline soils) or increase lime content (for acid soils) in a small portion of the root zone to increase the availability of micronutrients. When preparing land prior to planting, effective rates of banded sulfuric acid have been reported to range from 1 to 4 tons of acid per treated acre. In established orchards, a single application of sulfuric acid in a band should not exceed 1,500 pounds per treated acre to avoid injuring the trees. An equivalent rate of other acid-forming amendments can also be applied in a treated band of soil. If sulfur is banded it must be incorporated into the soil to be most effective. Banded applications of lime in acidic soils are commonly from 1 to 3 tons per treated acre and will also be more effective if incorporated. Amendment rates of acids, sulfur, and lime are too large and expensive for broadcast applications to be effective.

Treated soil must receive enough amendment to react sufficiently before the soil pH will change significantly. For example, applications to the entire soil surface would require 20 tons of pure sulfuric acid to neutralize 1% lime content in an acre of soil 1 foot deep. Many calcareous soils in the Central Valley have more than 2% lime content per foot of soil.

Applying acid-forming amendments in irrigation water can be effective as long as water is applied with drip or microsprinkler irrigation, which limits the volume of the soil that is irrigated and treated with the amendment. Options for applying amendments with drip and microsprinkler irrigation to increase soil pH are not as readily available and are less proven.

References

Ayers, R. S. 1977. Quality of water for irrigation. Journal of Irrigation and Drainage Engineering 103:IR2:135–154.

Ayers, R. S., and D. W. Westcott. 1985. Water quality for agriculture. United Nations FAO Irrigation and Drainage Paper 29, rev.1. FAO website, http://www.fao.org/DOCREP/003/T0234E/T0234E00.htm.

Hanson, B., S. R. Grattan, and A. Fulton. 2006. Agricultural salinity and drainage. Oakland: University of California Agriculture and Natural Resources Publication 3375.

Hoffman, G. J. 1986. Guidelines for reclamation of salt-affected soils. Applied Agricultural Research 1(2): 65–72.

———. 1996. Leaching fraction and control of root zone salinity. In K. K. Tanji, ed., Agricultural salinity assessment and management. ASCE Manuals and Reports on Engineering Practice 71. New York: ASCE. 237–247.

Maas, E.V. 1990. Crop salt tolerance. In K. K. Tanji, ed., Agricultural salinity assessment and management. ASCE Manuals and Reports on Engineering Practice 71. New York: ASCE. 287–290.

Oster, J. D., and J. D. Rhoades. 1990. Steady-state root zone salt balance. In K. K. Tanji, ed., Agricultural salinity assessment and management. ASCE Manuals and Reports on Engineering Practice 71. New York: ASCE. 469–481.

Oster, J. D., M. J. Singer, A. Fulton, W. Richardson, and T. Prichard. 1992. Water penetration problems in California soils. Oakland: University of California Agriculture and Natural Resources Leaflet 21553.

Sibbett, G. S. 1993. In-a-nutshell: Critical nutrient levels for nut crops and deciduous fruit. Visalia: University of California Cooperative Extension Tulare County.

8 Soil Evaluation and Physical Modification

• Allan E. Fulton, Blake L. Sanden, and John P. Edstrom

Prunes are similar to other deciduous orchard crops in that high yields and fruit quality are best achieved on lands with deep, uniform soils with a loam texture. Those soils provide the optimal combination of permeability with sufficient water-holding capacity and root zone aeration. However, acquiring these lands is more expensive due to their limited availability, so many new orchards are being developed on soils with limitations.

This chapter discusses learning how to evaluate soils, recognizing specific soil problems, selecting effective tillage methods, and considering irrigation methods and land leveling. Other site evaluation considerations such as climate, water availability, salinity, and replant issues including nematodes are addressed in other chapters of this manual.

One goal of this chapter is to help orchard managers conduct site evaluations and consider physical soil modification before planting, when it is most feasible. Correcting a physical soil limitation after orchard establishment is usually not nearly as effective, if not impossible, without removing the orchard. A second goal is to avoid expensive soil modification when it may not be necessary.

Evaluating Soil Physical Limitations

To evaluate the suitability of soils for prune, it is important to understand

- the kinds of soils on the property
- their physical characteristics
- the effects of the soil characteristics

Evaluation and appropriate modification of orchard soils provide the following benefits:

- reduced physical barriers to drainage and root development
- increased uniformity of water infiltration, permeability, and water-holding characteristics
- improved leaching of excess salts
- more uniform and increased vigor in young orchards and less time to achieve full production

Soil Variability

The most productive orchards usually consist of uniform soils throughout the orchard with very gradual changes with depth in physical soil properties such as texture, structure, density, and porosity. Orchards planted on marginal soils, however, have a higher degree of spatial variability, that is, more than one soil type throughout a parcel of land. They may also contain abrupt changes with soil depth. Orchards grown on marginal soils may be exposed to more crop stress because of more variability in infiltration and percolation characteristics, water-holding capacities, root penetration, and soil fertility levels. The net result is a greater likelihood of nonuniform orchard growth and production, especially under surface irrigation.

Soil Survey Data

Begin the evaluation by using the USDA Natural Resource Conservation Service (NRCS) soil surveys. These surveys illustrate the spatial patterns and types of soils associated with a specific parcel of land (fig. 8.1). Along with the aerial maps, these surveys describe the physical features of

Figure 8.1 Sample detailed soil survey data are available online for most counties in California. This erial map shows an area of Tehama County in the northern Sacramento Valley that consists of several il series ranging from deep alluvial soils near the Sacramento River to more marginal soils associated with the terraces and foothills farther away from the river. The yellow contour lines and soil series abbreviations denote different soil types. *Source:* USDA NRCS Web Soil Survey.

Figure 8.2 A magnified aerial view of a parcel of land in Tehama County. The section of the aerial map within the blue boundary lines illustrates approximately 140 acres of land with three soil series. The orange contour lines suggest three soil series within the parcel for further evaluation: the Molinos fine sandy loam series (My), the Vina loam series (VnA); and the Columbia silt loam (CsA). *Source:* USDA NRCS Web Soil Survey.

each soil type and help to discern any limitations. As shown in figure 8.2, a soils map for a particular parcel of land can help demonstrate the extent of the soil variability and indicate where to conduct additional on-site evaluation of the soils.

Since about 1900, soil surveys have been completed for most of the private lands in California. They have been published for most of the counties and other designated areas (i.e., parks, forests, etc.). Calaveras and Tuolumne Counties, southeastern Los Angeles, and the desert regions of southeastern California are the only areas of the state where soil surveys have not been performed. Development of soil surveys is ongoing to complete unmapped areas and to update older surveys. Soil surveys for western Tulare County, Butte County, and western Fresno County are among the most recently completed or updated surveys. Surveys published since 1970 provide maps with higher resolution and give more in-depth description of soils.

Paper copies of soil surveys can be difficult to obtain, especially for older surveys that have not been reprinted for decades. In addition, the federal government has initiated a policy not to distribute paper copies to the general public for free as was done in the past. Paper copies are expensive to duplicate in their entirety, and there is continuous work under way to update them. One alternative is to access printable versions online. The status and availability of printable versions of soil surveys is updated periodically online at the NRCS Soils Survey website, http://soils.usda.gov/survey/printed_surveys/. Access to a printed copy can be arranged at the NRCS soil survey request website, http://www.ca.nrcs.usda.gov/mlra02/ssrequest.html.

As an alternative to online printable versions or requesting a printed copy, all of the most recent soil surveys are available by using web-based interactive programs such as

- the USDA Web Soil Survey site, http://websoilsurvey.nrcs.usda.gov/app/HomePage.htm
- the California Soil Resource Lab website, http://casoilresource.lawr.ucdavis.edu/drupal/node/902

These interactive programs contain many of the soil surveys available in California as well as survey maps and descriptive data for specific parcels of land.

To create customized maps of soil properties, download soil survey data (detailed data are listed under the acronym SSURGO) from the NRCS Soil Data Mart website, http://soildatamart.nrcs.usda.gov/. Because of the complexity of the data sets, the data should be used with the assistance of the Soils Data Viewer site, http://soils.usda.gov/sdv/.

Mapping Soil Electrical Conductivity

In addition to soil survey data, commercial services are available, at a modest cost, to develop "real time" field maps illustrating spatial patterns of soil variability within a parcel of land. This service may be useful if older soil surveys do not provide the desired level of detailed information or if the land has been altered significantly since the soil survey was developed.

The equipment used to develop soil electrical conductivity (EC) maps consists of Geonics EM38 electromagnetic sensors (fig. 8.3), Veris electrical conductivity sensors (fig. 8.4), global positioning systems (GPS), and global information system (GIS) software.

In nonsaline soils, where root zone salinity as indicated by EC_e is less than 2.0 dS/m (see chapter 7, "Managing Salinity in Soils and Water"), there is reasonable correlation between the soil particle sizes that determine soil texture and EC_a (in-situ soil salinity). This makes it possible to develop electrical conductivity maps that reflect spatial variability of physical soil properties for a parcel of land using properly calibrated EM38 and Veris equipment. Figure 8.5 shows that soils higher in coarse-textured fractions of gravel and sand have lower EC_a levels; it also shows that soils higher in fine-textured silt and clay content have higher EC_a values. EM38 and Veris equipment require careful calibration if they are used to map lands with soils of increasing salinity (EC_e > 2.0 dS/m). Both the EM38 and Veris instruments are sensitive to soil salinity and soil texture. When soil salinity is known to exist and without

Figure 8.3 Two EM38 units used in a commercial application to map soil textural variability. The two EM38 devices are housed in PVC pipe and are towed behind an ATV in a serpentine pattern over the soil surface. The global positioning system (GPS) receiver is housed in the box at the front of the ATV, along with a palm computer to log the continuous EM38 measurements. The GPS antenna is visible at the back of the ATV. *Photos:* A. E. Fulton.

Figure 8.4 A Veris implement is towed behind a pickup in a serpentine pattern over the soil surface to map soil variability. The GPS antenna is visible on the pickup cab. A GPS receiver is located inside the cab along with a palm or laptop computer to log the continuous EC_a data. *Photo:* Brian Bassett.

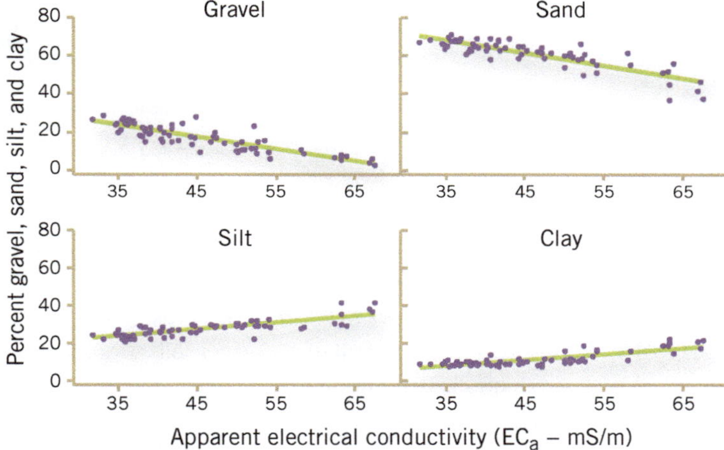

Figure 8.5 Correlation between gravel, sand, silt, and clay content and EC_a in soils from an orchard site in Colusa County. *Source:* Fulton et al. 2010.

proper calibration, it is difficult to discern whether spatial patterns of soil variability is attributable to soil physical properties or to soil salinity and other related soil chemistry properties. It is also advisable to use EM38 or Veris when the soil moisture profile throughout the field is at field capacity to avoid additional variability due to nonuniform soil moisture conditions that are caused by irrigation practices and not necessarily by soil textural differences.

Figure 8.6 shows a soils map generated from EM38 data. The map delineates four soil textural zones in a 22-acre parcel of land: zone 1 (red) was coarsest, with the predominant texture being gravelly sandy loam soils; zone 4 (dark green) was finest, with the predominant texture being loam soils; zones 2 and 3 (yellow and light green) were intermediate in texture, all predominantly sandy loam soils. This type of electrical conductivity map may assist with determining how to proceed with the onsite soil evaluation.

Excavating Soil Profiles to Evaluate Soils

NRCS soil surveys and electrical conductivity maps can show the primary soil types and their spatial patterns in a parcel of land. However, a follow-up evaluation using a series of backhoe pits is necessary to identify physical soil limitations at field scales.

A thorough site evaluation uses a series of excavated soil profiles in different areas of the orchard site. As a rule of thumb, it is advisable to dig at least one backhoe pit per 20 acres at an orchard site and possibly more if evidence indicates a high degree of soil variability. Information from soil surveys, electrical conductivity mapping, and past experience with farming other crops on the land can be used to determine more precisely where to dig the pits. For example, based on the soil variability displayed in the soil survey shown in figure 8.2 for the Tehama County parcel of land, backhoe pits might ideally be excavated in each of the three soil series (Molinos, Vina, and Columbia) denoted on the map to view differences in the soil profile and to collect soil samples from each profile if salinity is of concern. Where a previous cropping history exists, dig backhoe pits in areas that have had desirable as well as poor crop growth for comparison. (For more information on locating backhoe pits to diagnose soil problems, see chapter 7, "Managing Salinity in Soils and Water.")

Backhoe observation pits clearly show the number and nature of soil horizons, the depth of the horizons, how abruptly the horizons change, and the variability of the subsoil throughout the orchard site. This information can help determine the most economical method of soil modification, how to properly set up

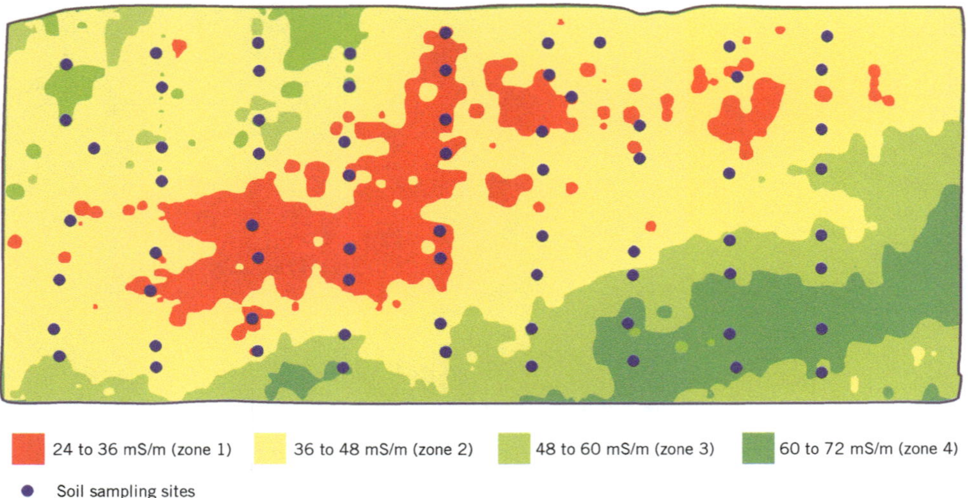

Figure 8.6 EM38 soils map and legend for an orchard in Colusa County. The map shows four distinct soil zones (red, yellow, light green, and dark green). The blue symbols on the map denote the location of soil sampling sites used to determine gravel, sand, silt, and clay content and correlate EM38 measurements with soil texture. *Source:* Fulton et al. 2010.

and use deep-tillage equipment, and when to conduct deep tillage.

One alternative to backhoe pits is using soil coring equipment to pull undisturbed soil cores for evaluation. This option may be more expensive than digging backhoe pits because mechanical coring equipment must be used to obtain undisturbed soil cores, and despite considerable effort, the soil cores are not likely to be as revealing as backhoe pits.

Evaluating Physical Soil Characteristics

Usually, within a single soil series, a distinct topsoil layer (A horizon) is apparent to a depth of about 1 foot. This horizon is most exposed to the climatic elements (e.g., leaching from rain and heat from the sun) and farming activities, and both contribute to the development of this horizon. Typically, there is more organic matter from more plant residue decomposition in the A horizon than in lower horizons. Previously tilled soils may show structural changes in the soil in this horizon. The depth and thickness of the A horizon may have been altered significantly due to cuts and fills from previous land leveling. The B horizon is often observed at a depth of about 1 to 3 feet. This horizon often shows transition in soil texture and structure with depth. Organic matter primarily from plant residue and residual roots may be more evident in the top of the B horizon, but it declines with depth. The upper portions of this horizon may show some change in soil structure from tillage, but the effect is likely to decline with depth. On occasion, where cuts have been made from land leveling, the B horizon may become the soil surface. The C horizon is apparent at about 3 feet and deeper in most soil profiles. This horizon is less developed than the A and B horizons. Usually there is very little evidence of organic matter, and the soil structure appears relatively undisturbed. Soil texture and structure can change with depth in the C horizon.

Soil textural changes in a soil profile are important and, in general, more apparent than changes in soil structural properties. Significant changes in sand, silt, clay, and gravel content in the different soil horizons should be observed and noted. It is important to note the approximate soil textures and the corresponding depths because this affects the extent and method of deep tillage needed to prepare the site for growing prunes.

Changes in soil structure within a soil profile tend to be more subtle and more difficult to discern than textural changes. Soil structure involves the particle arrangement for a specific soil texture and the stability of soil aggregates. Soil structure affects the density and porosity of soils and, ultimately, the water infiltration and permeability rates of soils.

Figure 8.7 summarizes some of the important physical soil characteristics that should be considered when using an excavated soil profile to evaluate a soil for orchard development. Soil structure is commonly categorized as granular, massive, prismatic/blocky, or platy. Generally, granular and massive soils are more common in loamy sand, sandy loam, and loam soils. Granular structure is preferable to massive structure for infiltration and permeability. Prismatic/blocky and platy structures are more common in silt loam, clay loam, and clay soils. Prismatic/blocky structure is preferable to platy structure for infiltration and permeability. Aggregate size and stability should also be considered when evaluating soils. Generally, soils with moderate size and strong aggregation have desirable water-holding capacity with higher and more stable infiltration and permeability. As with textural changes, it is important to note the types and changes in soil structure and the corresponding depths, because this affects the extent and method of deep tillage needed to prepare the site for a prune orchard.

Depending on the soil series, there may be gradual or distinct textural and structural differences in each of these horizons. Figure 8.8 depicts three distinctly different soil profiles that show the horizons and the usefulness of backhoe pits to evaluate and understand the physical features of soils.

Site Evaluation Costs and Safety Considerations

Overall, the cost of a preplant site evaluation is far less than that accrued in establishing an orchard. It should be possible to dig backhoe pits, evaluate the soil profile, and collect soil samples for an 80- to 160-acre parcel of land in about one day; associated costs will be for excavation and laboratory analysis. The principal goal is to avoid establishing an orchard in a manner or at a site that does not perform up to expectations due to pre-existing conditions.

Various safety measures should be considered when conducting a site evaluation using backhoe pits. It is possible for backhoe pits to collapse after excavation. To avoid harm from the collapse of a backhoe pit while examining the soil profile, the width of the backhoe pit should at least equal its depth. Allowing added width provides an area in the center of the pit that should be relatively free from any overburden if the pit collapses. Also, mark open backhoe sites with pylons or stakes and flagging so they are visible to others who may be operating machinery in the field. Fill in the backhoe pits as soon as possible after the evaluation has been completed to minimize the risk of an accident when the pits are unattended.

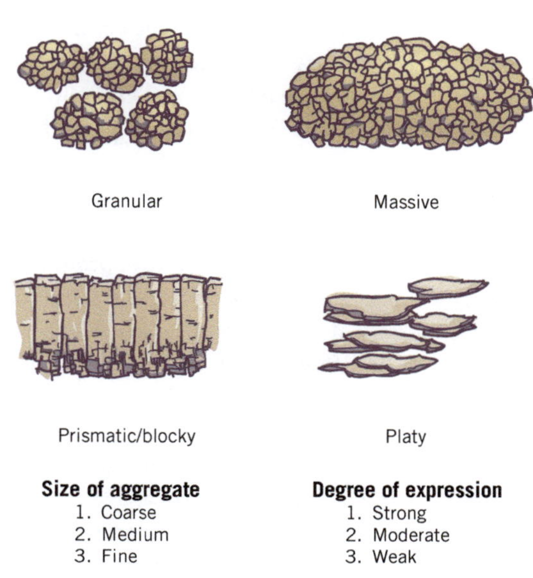

Figure 8.7 Summary of soil physical characteristics for evaluation with excavated soil profiles (backhoe pits). These features influence the method and extent of deep tillage needed to prepare an orchard site for growing prunes.

Figure 8.8 Excavated soil profiles (backhoe pits) for evaluating physical characteristics of soils for growing prunes. In a 6-foot profile of the Tehama soil series (A), a relatively deep soil with only gradual changes in soil texture and structure. In a 6-foot profile of the Perkins soil series (B), the top soil (A horizon) is apparent to a depth of about 1 foot with a relatively uniform B horizon from about 1 to 3 feet. Below about 3 feet, a thick soil strata (C horizon) with gravel is apparent. In a 6-foot profile of the Arbuckle soil series (C), significant changes in soil texture and structure are apparent in this profile; in particular, a thin strata with increasing clay content is apparent at about the 2-foot depth.
Photos: M. L. Poe.

Common Soil Physical Problems

Extensive soil modification is not always necessary to prepare an orchard site for planting. Some soils are naturally deep and relatively uniform in texture and structure and do not require deep tillage (see fig. 8.8A). However, other soils are layered and may need modification. Various types of physical soil limitations may be observed while conducting backhoe site evaluations: stratified soils, claypan soils, hardpan soils, plowpans, and clay-rich soils. Tables 8.1 and 8.2 summarize some common soil series in the San Joaquin Valley and Sacramento Valley known to have physical soil limitations.

Stratified soils have horizons or layers with abrupt changes in soil texture. The layers interfere with the uniform drainage of water, causing zones of poor aeration that may restrict root growth. Modifying stratified soils to improve productivity requires mixing the soil layers. Figure 8.8B illustrates a stratified soil. With this particular Perkins series, the stratification is due to a rather abrupt change from loam to a gravelly loam at a depth of about 3 to 4 feet.

Claypan soils have an abrupt increase in clay content within a very short vertical distance (1 or 2 inches) in a soil horizon. This increase in clay content restricts water movement downward, thus restricting aeration and root growth in the subsoil. Claypans are typically observed at a depth of 12 to 24 inches below the surface. Figure 8.8C shows an example of claypan soil. Modification designed to mix the clay layer with the rest of the soil profile encourages more extensive tree root growth. Ripping has very little effect on a claypan because the claypan is pliable and will not fracture as well as a hardpan. Ripping a claypan is somewhat analogous to cutting soft butter with a knife.

Hardpan soils are similar to claypans, except the soil particles are cemented together by hard mineral matter (silica) that do not soften, even when wet. This hardened layer is usually an absolute barrier to root growth and water percolation. Modification of hardpan soils requires fracturing and breaking rather than mixing.

Soils with plowpans can be found in some orchard sites. Plowpans are caused by tilling a soil at the same depth repeatedly or by repeated heavy traffic, particularly when soils are relatively wet. A soil with a plowpan does not necessarily have different textural layers like those of stratified, claypan, or hardpan soils. Instead, it has subtle changes in soil density and structure in the horizon where the plowpan exists. Plowpans have very low porosity and are usually shallow (6 to 18 inches below the surface) and require only fracturing, rather than mixing, to correct the problem.

Table 8.1. San Joaquin Valley soil series with physical soil limitations but that may be considered for prune orchards

Developed clay subsoil types			
Stratified	Moderate clay content	Claypan	Cemented hardpan
Eastside San Joaquin Valley soils			
Cajon	Borden	Cometa	Academy
Chino	Chiralar	Corning	Dinuba
Foster	Pond	Modesto	Exeter
Grangeville	Ramona	Montpellier	Fresno
Hanford	Ryer	Waukena	Lewis
Kimberlina	Snelling	Milham	San Joaquin
Nord Complex			Whitney
Visalia			
Westside San Joaquin Valley soils			
Camarillo	Avenal	Herdlyn	Dinuba
Columbia	Twisselman	Olcott	
Mocho	Pleasanton	Positas	
Panoche	Rincon	Solano	
Westhaven	Rossi	Waukena	
Vernalis	Garces	Buttonwillow	

Table 8.2. Sacramento Valley soil series with physical soil limitations but that may be considered for prune orchards

Developed clay subsoil types			
Stratified	Moderate clay content	Claypan	Cemented hardpan
Eastside Sacramento Valley soils			
Millrace	Aiken	Berrendos	Berrendos over Tuscan
	Anita	Stockton	Tuscan
	Farwell	Pentz	
		Vina clay	
Westside Sacramento Valley soils			
Arbuckle	Marvin	Arbuckle	Redding
Hillgate	Myers	Capay	
Maywood	Plaza	Kimball	
Perkins	Zamora	Sunnyvale	
Pleasanton			

Clay-rich soils may be found in some orchard sites. These soils typically contain more than 35% clay and appear relatively uniform throughout the profile with subtle changes in soil texture, structure, and hardness. They crack due to clay shrinkage when the soil is dry. The cracks can range from about 1⁄16 inch to over ¾ inch in width and can extend to depths of about 12 to 36 inches, depending on the clay content, clay minerals, and soil moisture. The cracks swell shut when the soil is wet due to expansion of the clay minerals. These soils have initially high water infiltration rates through the cracks when the soil is relatively dry, but the rate of infiltration declines quickly to very low rates upon wetting. Adequate infiltration, permeability,

Figure 8.9 Orchard site being prepared for planting by using backhoe pits to mix and modify an abrupt change in the soil profile deeper than 3 feet. *Photo:* W. E. Wildman.

and root zone aeration can be concerns in clay-rich soils.

Modifying Soil Physical Problems

Several methods are available to accomplish deep tillage prior to planting prunes. Available equipment includes rippers, slip plows, moldboard plows, disk plows, backhoes, and trenchers. Selecting the most cost-effective method of deep tillage or soil mixing requires evaluating the type and severity of the soil problem. Remember, the purposes of modifying a soil prior to planting are to encourage uniform movement of water through the root zone, maintain adequate aeration, and increase the volume of soil available for root growth. Select the method that will most economically achieve these goals.

Stratified Soils

Digging backhoe pits for each tree site is the most effective means of modifying layered soils with stratification down to 4 to 6 feet. Backhoeing is more expensive than ripping and slip plowing because the process is slower. Figure 8.9 shows an orchard site being prepared for planting by digging backhoe pits.

The expense of digging backhoe pits rises with increased tree density. For example, an orchard planted on a spacing of 17 by 22 feet requires 116 backhoe pits per acre, while an orchard planted on a spacing of 15 by 20 feet requires 145 backhoe pits per acre. The required dimensions of the backhoe pits depend on the extent of the soil layering. Generally, a pit 6 feet square and 6 feet deep is the maximum that is required and affordable. However, field research has indicated that the larger the pit, the more successful the results in stratified soils.

Trenching along the intended tree row is an alternative method of mixing abrupt soil layers (fig. 8.10). Trenching is suited to soils with layering within 4 feet of the surface and with soil below 4 feet being more uniform in texture. This is because trenchers that can trench deeper than 4 feet are less available and more costly.

Other methods of mixing stratified soils include slip plowing, moldboard plowing, and disk plowing.

Figure 8.10 This stratified soil was modified using a trencher before planting an orchard. *Photo:* W. E. Wildman.

These methods are more adaptable to larger acreage. More thorough mixing can be achieved with a moldboard plow or a disk plow than with slip plowing, but the depth of mixing is usually limited to 1 to 4 feet. Some modification using large slip plows can go as deep as 6 feet or more. Slip plowing causes less-uniform mixing, but it does break up the continuity of the soil stratification. The final option, ripping, is less beneficial in correcting stratified soils because it is designed to fracture rather than mix soils. Stratified soils eventually reform after ripping.

Claypans

Since claypans typically occur near the soil surface, the moldboard plow and the disk plow are effective methods of thoroughly mixing the pan with the rest of the soil profile; they are also adaptable to larger acreage. Figure 8.11 provides examples of a moldboard plow and disk plow used to modify claypan soils. Slip plowing (fig. 8.12) is an alternative to moldboard or disk plowing for modifying claypans, but it does not mix the entire soil surface, only strips of soil that are 4 feet or more apart. Ripping a claypan soil is not advised unless the ripper shanks are spaced very narrowly to ensure severe fracturing and heaving of the soil. Mixing rather than simple fracturing is needed to destroy claypans. Trenching in the tree row and digging backhoe pits are other alternatives for modifying claypan soils, especially if the claypan is more than 3 feet below the soil surface. If the soil below the claypan is fairly uniform, the added cost of deep mixing achieved with a backhoe may be unnecessary.

Hardpans

Orchards should not be planted on hardpan soils unless the layer is shallow enough that deep ripping (fig. 8.13) can completely break through the hardpan into uncemented, permeable soil below. Backhoe pit evaluations are extremely helpful in determining the required ripping depth and shank spacing. It is beneficial to rip down each intended tree row. Slip plowing is generally not recommended in hardpan soils because it requires more energy than does ripping. Moldboard plowing, trenching, and digging backhoe pits are usually not advised for correcting a hardpan. Thoroughly mixing a hardpan layer with the rest of the profile is unrealistic and expensive if attempted, because of the cemented nature of the layer. Fracturing the hardpan by ripping is all that is necessary to ensure adequate drainage for excess water. If ripping cannot effectively shatter the hardpan, the site should be considered unsuitable for prune production, especially with flood or furrow irrigation systems. Decades ago, dynamite was occasionally used to fracture deep hard pans that could not be effectively reached with tillage equipment. Today, due to safety and liability concerns such techniques are not nearly as feasible or recommended.

Clay-Rich Soils

Orchards may be planted on clay-rich soils as long as these soils are not underlain by a shallow water table that restricts drainage and compromises long-term root zone aeration. These soils have relatively high water-holding capacity, which can be advantageous, but because of their predominantly fine textures slow water infiltration and permeability rates can cause prolonged periods of standing water and concern over adequate root zone aeration. Ripping is an appropriate tillage method to prepare these soils for an orchard, particularly if hard, platy soil horizons are identified in backhoe pit evaluations. Slip plowing is a reasonable alternative but will require more draft, and there may be less need for mixing the soil than fracturing it. Ripping may primarily benefit the early years of orchard establish-

Figure 8.11 A moldboard plow (A) and a disk plow (B) for modifying stratified soils. These implements are typically used to modify shallow to moderately deep soil horizons within 1 to 4 feet of the soil surface. *Photos:* W. E. Wildman.

Figure 8.12 The basic design of a slip plow (A). The back metal plate lifts and turns the soil, resulting in some mixing of soil horizons. Slip plows are among the more commonly used implements to modify stratified soils that are buried deeper in the soil profile. Slip plows are also used in commercial applications (B). *Photos:* W. E. Wildman.

Figure 8.13 A dual shank ripper (A) can penetrate and fracture hardpan soils to a depth of about 5 to 6 feet. Soils are sometimes ripped (B) at a 45° angle to a previous pass with a slip plow. *Photos:* M. L. Poe.

ment in clay-rich soils, as its effect may decline with time. In some situations, if backhoe pits reveal uniform and friable soil texture, structure, and hardness, little benefit may be expected from deep tillage.

Guidelines for Deep Tillage

Soil moisture

The soil water content at the time of deep tillage partly determines the effectiveness of the operation. Although a moist soil requires less draft when tilling, more surface compaction occurs from the heavy equipment. Dry soil conditions break up and mix more readily. Desirable soil water content for clay is 10%, loam 5%, and sandy soil 2.5%, by weight. At these moisture levels, most crops approach permanent wilting. When deep tillage is anticipated to prepare an orchard site, growing a crop such as safflower prior to fall tillage is the best way to dry out the soil to the greatest depth.

Soils with a shallow water table (less than 6 feet from the soil surface) may never dry sufficiently to effectively modify them without installing drainage first. Artificial drainage systems may consist of very deep earthen ditches around the perimeter of an orchard or a grid of perforated plastic pipeline buried in or around the perimeter of the orchard. These drains are connected to a drainage pump that provides a hydraulic gradient and causes the shallow water table to flow through the earthen or plastic drains, ultimately lowering the shallow water table. Artificial drainage also requires a plan for handling the shallow drainwater in an environmentally acceptable manner. In various parts of the Central Valley and other areas of California, shallow drainwater may be of lower quality than the irrigation water due to salinity, nutrients, and trace elements.

Tips for effective ripping

Two factors that specifically influence the effectiveness of ripping or chiseling are the depth of the ripper shanks and the distance between them. Apply the following three reliable rules:

- The shanks should penetrate into the soil 1.5 times the depth of the targeted soil problems.
- The distance between the shanks should be equal to, or preferably less than, the desired ripping depth.
- Rip in one direction with narrowly spaced shanks rather than in two directions with widely spaced shanks.

Insert the shank 50% deeper than the desired ripping depth to ensure sufficient penetration through the restricting zone of soil. Setting the shank spacing to a distance narrower than the desired ripping depth is best, because ripping never breaks the soil straight across between adjacent shanks; there is always a hump of undisturbed soil between two ripper channels. Wider shank spacing yields a greater percentage of undisturbed soil. With extremely wide shank spacing, the zone of undisturbed soil may come all the way to the soil surface. Ripping in one direction with a narrow spacing is more effective than an excessively wide spacing followed by cross-ripping.

The practice of straddle ripping is probably the most common method of using deep tillage to prepare an orchard site for planting. It is among the more economical methods of deep tillage, and it is usually quite effective. To begin straddle ripping, mark the tree rows in the field. Then pull a single-shank ripper, or preferably a slip plow in obviously stratified soils, down the marked tree row to the desired depth. This may require one or two passes, depending on the depth of penetration. If the soil is dry enough, fracturing will occur at an angle of about 45° from the point of the chisel or plow up to the soil surface. A 6-foot-deep penetration will create fracturing 5 to 6 feet from the chisel at the soil surface. (With a slip plow, the subsoil is partially mixed as well.) After the initial pass (or passes) down the row with a single-shank ripper or slip plow, make a pass using a dual-shank ripper. The dual ripper shanks are spaced about 10 to 12 feet apart and set so they straddle the initial pass by about 5 or 6 feet on both sides. Since some fracturing has already occurred from the first pass with the single-shank ripper or slip plow, the straddle rip can usually be done in one pass. This ripping is usually about 1 foot shallower than the initial pass. If the soil moisture was optimal (dry), the final fracture zone at the soil surface should be 18 to 20 feet wide, or 9 to 10 feet wide on both sides of the marked tree row. At a depth of 3 to 4 feet, fracturing should extend over a width of 12 to 14 feet with 6 to 7 feet of fractured soil on both sides of the marked tree row.

Crop Response to Deep Tillage

Deep tillage prior to planting is critical for many orchard sites that are to be irrigated with flood or furrow systems. Thorough soil mixing improves water infiltration and drainage, promotes more extensive root growth, and creates more uniform water storage within the root zone. These factors contribute to a more manageable and productive orchard under a flood or furrow system.

Table 8.3 shows the production and root development responses reported from a tillage study in which a claypan soil was modified before planting almonds (Wildman and Gowans 1978) (similar data are not available for prune, but it is reasonable to expect similar responses). After using a moldboard plow to thoroughly mix the subsoil clay layer prior to planting, production under flood irrigation increased over 40% above the yield of trees planted into nontilled soil, and the root count increased over 120%. Such increases in production will repay the cost of the deep tillage and more. How quickly the investment in deep tillage repays depends on the level of crop response and market price.

Deep tillage may be unnecessary for orchards irrigated with pressurized systems unless some form of claypan or hardpan has been identified close to the surface. Field research conducted in both walnut and almond on stratified and claypan soils on the west side of Colusa County found little response to deep tillage when drip or microsprinkler irrigation was used (Edstrom 2008). Table 8.4 summarizes results from an experiment in microsprinkler irrigated almonds. In this trial, a Class II Arbuckle sandy loam soil, underlain by a clay horizon at 30 to 60 inches, was first slip-plowed 5 feet deep in the direction of the tree rows but not necessarily in line with where the tree rows were eventually planted (Edstrom 2008). Then the land was slip-plowed in a second direction at a 45° angle to the first pass. The orchard planted on this land showed no improvement in almond tree size, yield, or nut size after 8 years compared with land at the site that had not been slip-plowed.

In a long-term study in drip-irrigated walnuts, soil excavations at a test site revealed substantial mixing of stratified soil layers and deeper walnut root development following slip plowing, but no yield or nut quality improvements were found (Edstrom 2008). Low-volume applications of water and nutrients at higher frequency appear to overcome many limitations of marginal soils without expensive deep tillage. While similar studies have not been conducted in prunes, a similar response may be expected, assuming that prunes are not more sensitive to poor drainage.

Shallow soils in environments where 15 inches or more of rainfall are expected with a high likelihood of rainfall in the months prior to and during leafout, and soils located in floodplains, are two conditions in which deep tillage may be beneficial even when microsprinkler or drip irrigation is used. Deep tillage should improve the rate of drainage of soils with physical limitations unless a regionally high groundwater table contributes to poor drainage.

Irrigation Systems and Land Leveling Considerations

Selecting an Irrigation Method

Soils and topography affect the selection and design of an irrigation system, and the type of irrigation system used affects preplant leveling and tillage. Properly designed and managed microsprinkler, surface drip, subsurface drip, and sprinkler systems help minimize physical soil limitations more than do surface irrigation systems.

Table 8.3. Responses to modifications of claypan soil in terms of yield, trunk circumference, and root count (furrow flood irrigation)

Tillage method	Yield* (lb/ac)	Trunk circumference* (in)	Root count† (per 3 ft³)
none	1,009	14.8	78
ripper	1,120	16.6	94
slip plow	1,185	16.7	118
moldboard plow	1,433	17.0	175

Source: Wildman and Gowans 1978.
Notes:
*Measured in fourth year of production.
†Measured in eighth year of production.

Table 8.4. Slip plow evaluation in microsprinkler-irrigated almonds. Nickels Soils Lab, Arbuckle, CA (2000–2008)

Year	Tree age (years)	Nut yield (lb/ac)	
		Slip plowed	Not slip plowed
2000	4	894	830
2001	5	1,070	1,243
2002	6	2,725	2,761
2003	7	2,165	2,323
2004	8	1,869	1,865
2005	9	1,548	1,841
2006	10	2,910	2,862
2007	11	2,770	2,571
2008	12	3,771	3,686
Cumulative Yield		19,722	19,982

Source: Edstrom 2008.

Uniform application of water with a pressurized irrigation system depends more on hydraulic design, aerial distribution of applied water, and system maintenance than on variable soils. If a low-volume micro system has pressure control features, uniform slope is not as critical. Microsprinklers and surface drip systems apply water at low, controlled rates that are usually less than the soil infiltration rate. Uneven ponding of water and runoff are less significant problems. Also, irrigation water is applied in small quantities at high frequencies that closely match the rate of crop water use.

An advantage of a flood or furrow irrigation system over a pressurized irrigation system is reduced system costs. However, the uniformity of applied water, and subsequent prune water use, with flood or furrow systems largely depends on variable infiltration and retention characteristics of orchard soils, which are less manageable than factors affecting pressurized systems.

Typically, irrigation water cannot be applied as uniformly with flood or furrow irrigation systems as with most well-designed and managed pressurized systems. However, implementing the following design and operational guidelines will help maximize the uniformity of surface irrigation systems.

Surface Irrigation System Guidelines

- Irrigate in small orchard blocks that share similar physical soil characteristics to achieve more uniform water infiltration and storage.
- Provide one water discharge valve per tree row.
- Design the surface system so the field length is short enough to rapidly advance water across it, given the available water flow rate.
- Grade to uniform slopes to avoid low and high spots.
- Install a tailwater return system. Steps taken to improve irrigation uniformity in furrow or flood irrigation may cause more tailwater.

Dead-level flood irrigation is sometimes used to irrigate prunes, but it has advantages and disadvantages. The principal advantage is that tailwater runoff from irrigation can virtually be eliminated. However, the main disadvantage is that during seasons of high rainfall it is very difficult to drain excess surface water from the orchard, which can pose risks to tree health, especially during the spring.

Larger irrigation sets are often more attractive to growers to minimize per-acre costs. Unfortunately, flood or furrow systems with large acreage per set usually include more soil variability. This causes lower uniformity of applied water, which usually translates to poorer water permeability, poorer root zone aeration, and less-uniform tree growth and production in the orchard.

Leveling the Land

Orchard soils that are flood or furrow irrigated must be graded to a uniform slope. Prior to leveling activities, heavy crop and weed residues or brush and stumps from a previous orchard must be removed. Incorporating organic material into soil is usually beneficial. However, it can be detrimental when a large mass of material is buried deeply in a compacted soil: after irrigation, this organic residue can remain wet and begin decomposing in the absence of air, producing gases at concentrations that become toxic to roots and can kill trees. Burning (where permitted) as well as mowing and chipping are effective in removing excess crop and weed residues. The remaining vegetative residue can then be incorporated into the soil by disking. Stump removal is best achieved with a backhoe or with a grinding/chipping machine.

After the site has been cleared and prior to leveling, consider the desired final grade, the optimal time to do the leveling, the maximum depth of cuts to be made, and the type of scraper to use.

Information from earlier backhoe pit evaluations may show the types of soil horizons existing below the depth of planned cuts and the characteristics of the soils that are likely to be exhumed into the root zone when leveling cuts are completed.

The final orchard grade for flood or furrow irrigation depends on the soil infiltration characteristics, the length of the field, and the discharge rate of the available water supply. Generally, the final grade, after all land preparation has been completed at an orchard site with medium- to fine-textured soils, should be 0.1 to 0.5 feet per 100 feet (0.01 to 0.05%).

Land leveling should be done under fairly dry soil conditions to minimize soil compaction and avoid breaking down the soil structure. The maximum depth of the cuts depends on the initial slope of the land. If required cuts exceed 0.6 feet, the total cut should be accomplished by making two separate shallower cuts. Two types of scrapers are available for leveling: paddle type and push loading. Paddle-type scrapers create less compaction than do push-loading scrapers, but they tend to pulverize the soil to a powderlike condition that does not settle properly in fill areas.

After completing most of the land leveling, two final questions must be resolved: Is the grade stable? How should the cut and fill areas be managed? Settling time is required to allow the grade to become stable. Where cuts exceed 0.6 feet, the main cuts should be made the first year and the final grade the following year, after an annual crop has been grown. A second year of annual cropping is even more preferable to ensure that the final grade has stabilized.

After land leveling, the possibility of nutritional deficiencies in the cut areas must be considered. Nitrogen, phosphorus, potassium, zinc, and sulfur are commonly lower in exposed subsoil than in the original topsoil because of their lower organic matter content. These deficiencies can be corrected by adding fertilizer and manure and by incorporating plant residues. The exposed cut areas may be compacted into a less desirable soil structure than the original surface horizon, and water infiltration may be lower. In time, cropping, wetting and drying, and the addition of organic matter will rebuild the soil structure and establish desirable soil tilth.

Fill areas may present more difficult problems than cut areas. Cutting, scraping, and moving soil disturbs the soil structure, and a given volume of soil scraped off a high spot in a field will not fill the same volume in a low area; an additional amount of soil is usually needed. Clay loam to loam soils may require the volume of cut soil to be supplemented by about 25% more soil. In coarser, sandy soils, low areas may require 50 to 100% more soil volume. Because of the disturbed soil structure, the porosity of fill areas is often lower than that of the original surface soils. This reduction can cause poor water penetration and root growth, both detriments to establishing a productive orchard.

Filling should be done when the soil is fairly dry (to reduce compaction in the fill areas) but not so dry that the soil becomes powdery. The soil should have small aggregates, or clods, which indicate maintenance of some soil structure. After initial leveling, the fill areas may require additional deep tillage to below the depth of the original soil surface. Ripping is commonly done in this scenario, but other tillage methods such as disk plowing or moldboard plowing might be considered, depending on the physical properties of the soil scraped into the fill areas and the buried topsoil. In some situations, the soil may require more mixing than fracturing. If the fill is deep, tillage after partial leveling is desirable. If the costs of additional tillage and re-leveling seem prohibitive, consider the loss of production, replacing trees that have not progressed or have died, and managing an orchard that never produces to its full potential.

References

Begg, E. L., G. L. Huntington, and W. E. Wildman. Evaluation and modification of soils. In D. E. Ramos, ed., Walnut production manual. Oakland: University of California Division of Agriculture and Natural Resources Publication 3373.

Edstrom, J. P., 2008. Slip plow tillage effects in almonds. University of California Nickels Soil Lab Report.

Edstrom, J. P., B. H. Krueger, W. O. Reil, and J. Yeager. 2008. Walnut production on marginal soils. University of California Nickels Soil Lab Report.

Fulton, A. E., W. E. Wildman, E. L. Begg, and G. L. Huntington. 1996. The evaluation and modification of physical soil problems. In W. C. Micke, ed., Almond production manual. Oakland: University of California Division of Agriculture and Natural Resources Publication 3364. 20–28.

Fulton, A., L. Schwankl, K. Lynn, B. Lampinen, J. Edstrom, and T. Prichard. 2010. Using EM and VERIS technology to assess land suitability for orchard and vineyard development. Irrigation Science doi 10.1007/s00271-010-0253-1.

Wildman, W. E. 1976. Diagnosing soil physical problems. Oakland: University of California Division of Agricultural Sciences Leaflet 2664.

Wildman, W. E., and K. D. Gowans. 1978. Soil physical environment and how it affects plant growth. Oakland: University of California Division of Agricultural Sciences Leaflet 2280.

9 Varieties

• James F. Doyle, Carolyn J. DeBuse, and Theodore M. DeJong

One of the most important sources of commercial plum varieties is the species *Prunus domestica*, known commonly as the European plum. This species is thought to have originated in western Asia, in the area east of the Caucasus Mountains and the Caspian Sea. The *domestica* plum, and especially the prune or dried plum made from it, was for centuries a staple product of the Mongols, Tartars, Turks, and Huns. This dried product was a common commodity of trade and, most likely, a common provision during the migration of these peoples toward Eastern Europe. The first written record of this species in Europe was in the writings of Pliny the Elder in the first century AD. The European plum has a history of cultivation in Europe for 2,000 years, where it has become an important fruit crop. For many years, the prune has been defined as a firm-fleshed plum of the species *Prunus domestica* that has high soluble solids content and can be dried whole without fermentation at the pit. In recent years, there has been a change in preference of nomenclature, now identifying the dried product as a dried plum rather than as a prune.

For the purposes of this manual, "variety" and "cultivar" as they refer to a type of *P. domestica* plant are used somewhat interchangeably, although they do not represent exactly the same thing. A variety is a member of a species of plant that differs from the rest of the species in relatively minor characteristics. "Variety" is an older term than "cultivar" and is widely used in the older publications referenced in this manual to distinguish one species member from another. A cultivar is considered a cultivated variety, that is, a plant that has been selected and given a unique name because of certain desirable characteristics and is maintained under cultivation. It is a more modern and specific term, widely used in scientific publications.

By the time the *domestica* plum became popular in Europe, it was represented by varieties that had a large, highly flavored fruit, an indication of the many years of cultivation before its arrival. In North America, the European plum apparently did not arrive as early as other fruits such as apples, pears, and peaches. References to Damson plums *(Prunus institia)* are present as early as 1629 in Massachusetts, but mention of *Prunus domestica* types does not seem to occur until about 1720. The early immigrants on the Atlantic coast preferred the large dessert-type plum rather than the smaller prune types. The drying of fruit was not easily accomplished in the humid, rainy weather of the eastern colonies (Hedrick 1911).

The first mission in California was established by Spanish Franciscan missionaries at San Diego in 1769, followed by an additional 20 missions in the following 54 years. Records indicate that plums were planted at Missions San Buenaventura, Santa Barbara, San Gabriel, Santa Clara, and San Luis Rey. The British explorer George Vancouver saw plums growing at Mission San Buenaventura as early as 1792 (Butterfield 1938). At Mission Santa Clara, a prune variety named Mission was still growing in 1870, but it has since disappeared and almost nothing is known about it. Numerous plums and prunes of the *Prunus domestica* type were introduced into California from 1851 through the 1880s, but the most important introduction, that of the Agen, or French prune, took place in 1856 (Couchman 1967; Butterfield 1938).

From the time of its introduction into California, the French prune has occupied a dominant position in the California prune industry. The regularity with which this variety bears abundant crops, the high percentage of soluble solids in the fruit, and its ability to hang on the tree well into maturity has, in the past, distinguished the French from most other

prune-type plums. Other non-French varieties developed substantial acreage in the early 1900s, but over time, many have nearly disappeared from production. The USDA National Agricultural Statistics Service (NASS) estimates the 2009 California prune acreage at 64,000 bearing acres. This represents an overall drop of over 45,000 acres from the 109,000 acres reported in 1998. In 2009, French-type prunes accounted for 97% of all prune acreage in the state. NASS currently reports all prune varieties together as a group, at times differentiating between bearing and nonbearing acreage, but not between French and non-French varieties (NASS 2010). The last separation was in 1999, when the only non-French varieties listed separately by NASS were Moyer at 772 acres, Imperial at 168 acres, and Burton at 13 acres, with all other varieties at 306 acres (NASS 2000). As we shall see, since the 1950s there has been an interest in new varieties of the French type and in the evaluation of new French clones. The result of this interest and evaluation is that the specific definition of the "French" prune has been changed to mean a more generic definition of the fruit type rather than a specific variety of prune.

The Agen in France

The origin of the French prune dates back to the thirteenth century. Tradition indicates that Benedictine monks accompanying Crusaders in the vicinity of Turkey or Persia (Iran) brought back what was known in that region as the date plum. This new acquisition was planted on the grounds of the Benedictine abbey near Bordeaux, France. The name *pruneau d'Agen* came into common usage in France because of the extensive plantings of this prune that developed in the region known as Agen near the town of the same name. Several names have come to be associated with this prune. These include Agen, Prune d'Agen, Petite Prune d'Agen, Petite Prune, and the variety name in current usage in France, Prune d'Ente. The prune industry in France is historically associated with the Lot Valley, where the Agen has been grown for nearly 500 years. Production peaked in the mid-1890s, when more than 5 million prune trees were in the ground, with a production of 56,000 metric tons (Carlot et al. 1971).

In modern times, evaluation of various clones of Agen began in France in 1947. Dr. R. Bernhard at the Grande Ferrade Research Station near Bordeaux initiated a survey of the existing old plantings of Agen throughout France seeking naturally occurring mutations. His program had three specific objectives: the improvement of fruit size and productivity, the development of a regular bearing habit, and the improvement of tree vigor. Over many years of culture in France, individual trees of the Agen had developed discernable differences. These included such characteristics as variability in the date of pollen shedding, date of leafing, fruit size, productivity, date of maturity, bearing habit, and tree vigor. Most clones found during this survey are thought to be caused by natural mutations. These mutations likely have occurred on the Agen variety over many years rather than developing from seedlings of Agen, because their differences are small compared with what would be expected from among seedling populations. From these old plantings, Bernhard selected 57 clones of the Agen, each having at least one characteristic different from the general population. This collection of clones was evaluated at a research station at Bourran in the Lot and Garonne Valleys.

By the 1950s, three improved clones of the Agen had been released to French growers. The selections were identified as Prune d'Ente 707, 698, and 626, or more simply as GF (Grande Ferrade) 707, GF 698, and GF 626. All of these clones satisfied the goals of improved size, vigor, and productivity over the general prune population but differed among themselves as to the degree of improvement. The GF 626 clone was less vigorous than GF 707 and had substantial early fruit drop. In addition, the quality of the dried fruit was inferior to that of GF 707. GF 698 was less vigorous than GF 707 but had equal fruit quality. GF 698 was also less productive and harvested later than GF 707. Because of its high level of productivity, fruit quality, and tree vigor, GF 707 rapidly became the standard variety of the prune industry in France. After the removal of all significant viruses from the GF 707, the clone was reidentified as GF 2733 (fig. 9.1). Within a short time, however, it became apparent that self-pollination was poor with GF 2733.

In cool, wet years, pollen shedding occurred after the peak receptive period of the flower pistil, causing poor fruit set and reduced production. To remedy this situation, Bernhard released three additional clones in 1962 that were recommended for interplanting in GF 2733 orchards to enhance fruit set. These were named GF 303, GF 642, and GF 652 and included both earlier and comparable timing of pollen shedding. When prunes were harvested from the ground, maturity differences between these selections were not disruptive to the

Figure 9.1 GF 2733, previously known as GF 707. *Photos:* C. J. DeBuse (top), M. L. Poe (bottom).

harvest. The most common combination of clones for planting in France today is GF 2733 with early-pollen-shedding GF 303 as the pollinator.

Prune d'Ente Clones in California

Original Clones

In 1959, Professor C. Hansen at the University of California, Davis, imported five clones of the Prune d'Ente into California for testing. Included among them were the GF 707, GF 698, and GF 626. Field trials beginning in 1960 and directed by David Chaney, who was then UC Cooperative Extension Farm Advisor in Sutter and Yuba Counties, produced the following information about the clones, as reported by Chaney in 1981:

> **GF 626** flesh firmness softens earlier than French and there is usually appreciable fruit drop before desirable soluble solids are attained. Early harvest is usually required to avoid picking after heavy drop. Drying ratios of the fruit usually exceed those of French.
>
> **GF 698** blooms 1 to 2 days later and harvests a few days later than French. Average fruit size has been slightly larger than French in mature trees. Dried fruit quality has been good. It has been grown only in field trials.
>
> **GF 707** has consistently yielded crops as large as French and produced significantly larger fruit. It produces the largest fruit size of any of the numbered French clones and larger than any of the tested French clones selected from California prune orchards. The skin quality frequently is not as good as that of French. Soluble solids have averaged slightly lower and drying ratio slightly higher than French. It is grown commercially in a small amount. (Chaney 1981, 14)

As a result of the 1959 importation and evaluation, a small acreage of GF 707 was planted in California. Soon after these early plantings began to bear, after dehydration and during processing, GF 707 developed a slight skin peel that caused a dull, grayish appearance, often referred to as a "mousey" color. In addition, in some locations, GF 707 was less productive than the California French prune. Further planting of GF 707 has been discouraged by fruit processors. By 1999, plantings of GF 707 in California had declined to 47 acres (NASS 2000).

Subsequent Clones

In 1985, four additional clones of Prune d'Ente were imported into California by T. J. DeJong and J. F. Doyle of the UC Davis Pomology Department. These clones were established in a field trial at the University of California's Kearney Agricultural Center at Parlier in Fresno County. The Prune d'Ente clones included GF 642, GF 652, GF 303, and GF 2733. At 6 years of age, in a direct comparison with a standard commercial California French clone, all four Prune d'Ente clones were less productive. In addition, with some clones, various problems in adaptation to the hot California interior valley growing conditions became apparent.

GF 642

GF 642 was the best performing and closest in appearance to the California French clone. Tree form, vigor, bearing habit, and date of maturity were all comparable to the standard. The fruit was slightly larger than French, but the soluble solids content was slightly less, causing a slightly lower average count per pound. The most substantial limitation of this clone was its reduced productivity. At 6 years of age, the cumulative productivity of GF 642 was 66% of standard French.

GF 652

GF 652 is a spur-type tree with moderate vigor, few strong upright scaffold limbs, and a relatively dense spur and short lateral limb growth pattern. The tree can be pruned easily and quickly. This clone was one of the two lowest producers of all varieties in the test. At 6 years, the cumulative productivity was 45% of French. The low productivity of this clone has also been reported in France. The fruit of GF 652 does not hang well on the tree. Substantial drop usually occurs when the fruit firmness reaches 4 pounds-force (lbf). This rapid drop, also reported in France, necessitates early harvest of the crop at less than optimal soluble solids content. The size of fresh fruit averages several grams larger than French, but the need for early harvest causes a lower soluble solids content and higher counts per pound than French. The clone is not at all suited to a mechanical shake and catch system of harvest.

GF 303

The GF 303 clone of Prune d'Ente is the most popular pollinator variety interplanted in France with the principal variety GF 2733 (GF 707). In the Kearney evaluation, it was similar in vigor, bearing habit, and date of maturity to the California French but after 6 years produced only 45% of the standard. This cultivar did not appear to be well suited to California interior valley growing conditions. In three consecutive years, toward the end of the growing season, approximately one-third of the fruit darkened internally and began to shrivel prematurely on the tree. Many of these heat-damaged fruit did not drop but remained on the tree and were harvested along with the sound fruit. Soluble solids accumulation was negatively impacted in the shriveled fruit. The presence of these damaged prunes in the bulk harvest produced the lowest soluble solids percentage (degrees Brix equivalent) reading of any prune in the Kearney evaluation. The lack of heat tolerance and low productivity in GF 303 indicate major problems for adaptation in California.

GF 2733

As mentioned previously, GF 2733, the "virus-free" clone of GF 707, is the principal prune variety grown commercially in France. In the Kearney evaluation, the tree was vigorous and very similar to the California French in form and bearing habit. The fruit was large and produced a high-quality dried product. Limited processing indicated skin peel similar to that of GF 707. At 6 years of age, the cumulative productivity was only about 65% of the standard California French. The date of maturity varied from the same as California French to a few days later.

GF 812

At the same time that the four Prune d'Ente clones were being evaluated, another prune from France, GF 812, or Double Robe, was included in the test. In France, it is sometimes identified as a subvariety of Prune d'Ente. The variety originated in France. It was found in an orchard of Agen but is probably a seedling of Agen rather than a mutation. The tree form is distinctly different from that of the Agen: more open, less branched, and with thicker limb caliper. The tree is highly vigorous but can be formed and pruned easily. The leaves are larger and more rugose than those of Agen. Flower characteristics of the GF 812 differ from the Agen in size and anther color but are similar to Agen, appearing to be self-fruitful. At Kearney, the fruit was larger than the standard California French, with a relatively small pit. The tree was shown to be as productive as the standard. The fruit is of good quality; it is lighter colored and matures a few days later than California French. The tree and flowers appear to have a relatively high chilling requirement. Due its large size, the fruit needs a longer drying time than average. It also tends to bleed, slab, and stick somewhat to the drying trays. The light color of the fresh fruit produces a brown dried product. This variety is not widely cultivated in France.

The Agen in California

The Agen prune is reported to have been present on the eastern coast of the United States as early as the 1830s (Prince 1832). Agen was also reportedly imported by the U.S. Patent Office in 1854 (Hedrick 1911). In that same year, however, the Highland Nursery of New Rochelle, New York, and their agent nursery, the Commercial Nurseries of Mission Dolores, San Francisco, listed Agen among the 132 plums offered for sale in their nursery catalog. It would seem that the Highland Nursery must have been propagating Agen for at least several years previous to 1854 (Butterfield 1938). No record has been found that indicates any significant plantings of Agen developed in California from this source, although a small planting of Agen was established in 1850 by Jules Frosiere near Mission San Jose (Chaney 1981).

The credit for the importation of the Agen variety that would develop into one of the most important deciduous fruit tree varieties in California has been given to Louis Pellier (see fig. 1.1 in chapter 1). Born in 1817 in Charente Inferieure, France, Pellier left his native country at the age of 31 on an adventurous journey. Travelling from France to North Africa to Portugal, he eventually arrived in South America. Early in 1849, Pellier considered taking up residence in Chile, but news of the discovery of gold in northern California caused him to reconsider and join the rush for riches. Several different accounts of Pellier's success in the California mines exist. One story indicates that he did well in the mines of Trinity County, but the rigors of the cold winters drove him back to the Bay Area in the early 1850s. Another account attributes failure in his quest for gold as the reason for his return to San Francisco. Regardless of which was true, Pellier eventually turned his attention to the nursery business in San Jose. He had been trained as a horticulturist since his youth, and success as a nurseryman came quickly. In 1853, his brother Pierre Pellier joined him at Pellier's City Gardens in San Jose. The rest of the story has often been retold of how Pierre returned to France to marry and in 1856 returned to California, crossing the Isthmus of Panama with his new bride, brother Jean, and Jean's son and brother-in-law. In addition, at the request of Louis, Pierre brought back from France a substantial collection of seeds, scions, and cuttings of numerous tree fruits and grapes. Among the many varieties was one item of major significance called *la petite pruneau d'Agen* (Butterfield 1938; Couchman 1967).

The Pelliers shared some of these first scions of Agen with other orchardists, most notably George W. Tarleton and J. Q. A. Ballou of San Jose, and another nurseryman, B. Kemp. Fourteen years later, there were 650 acres of prune trees from these first propagations in California. About 300 of these acres were in Santa Clara County, with much of the rest in nearby locations. Undoubtedly, through the late 1800s, there were numerous subsequent importations of Agen, which led to the establishment of several strains of Agen from varying districts in France. By 1875, there was some concern about the authenticity of the Agen clone brought from France by the Pelliers. The dried fruit from California seemed to be smaller than the fruit of the Agen imported from France. It was thought that clones called *la petite pruneau d'Agen* might not have been the true prune of commerce in France. In 1878, William B. West, a Stockton nurseryman, traveled to France with the express purpose of determining whether the Pellier Agen was in fact the real Agen. Eventually the Pellier introduction was deemed correct. At about this same time a general shift began to occur in California away from the names Agen, Prune d'Agen, and Petite Prune d'Agen to the simple designation as French prune or Petite prune (sometimes pronounced "petty prune"). U. P. Hedrick, writing in his definitive *Plums of New York* in 1911, makes note of this change. Hedrick relates the introduction of Agen by Pellier and then states that Agen "soon became and still is the leading plum, though with curious persistency the fruit-growers there (in California) call it the French Prune and the Petite Prune." (Hedrick 1911, 139)

The California Improved French Prune

Almost from the very beginning of the California prune industry, there has been an abiding interest in the development of new and improved varieties of prunes. This search on one hand has taken the form of the importation or development of completely new varieties, with the promise of a larger, more flavorful dried product or improvements in processing or horticultural characteristics. On the other hand, there has also been a continuing desire for improvement of the Agen or French prune itself. This search for improved clones of French or closely related seedling types has been centered on improvements in horticultural or processing characteristics but has also sought to preserve the basic identity of the French variety.

In November 1898, Luther Burbank (fig. 9.2), the famous plant breeder of Santa Rosa, California, sold a prune variety he had named Miller to

Figure 9.2 Luther Burbank (1849–1926). Courtesy Library of Congress.

Leonard Coates, a nurseryman at Morgan Hill, California. Miller had originated as an open-pollinated seedling of the Agen. After 10 years of testing, in 1908, Coates began to sell the variety that he had by then renamed Improved French. The new variety was very similar to the original Agen except that it was reported to be somewhat larger and more uniform in fruit size. Coates began as a nurseryman in Napa from 1875 to 1905, had a short residence at Fresno, and expanded to Morgan Hill about 1910, where he remained until 1920, when he retired to Clear Lake. During his 42-year career, Coates sought to improve the French Prune by the selection of buds on full French trees or specific limbs of French trees that appeared to be superior in size, productivity, or some other horticultural characteristic. By the repropagation of these variants followed by rigorous evaluation, Coates worked to develop the ultimate improvement of the French Prune. In about 1910, E. J. Wickson, the dean of the University of California College of Agriculture and the University of California Experiment Station, addressed a meeting of the California Association of Nurserymen. In a reference to Coates's work, Wickson stated, "Coates then practiced what he preached and was on the trail of improved varieties of the French prune and in such pursuit of the right line, that probably the best variety of French prune that we now have, fifteen years after his initial proclamation, is one of his selection." (Wickson 1921, 52). Ten years later, Coates was propagating several French clones that were the result of natural mutations. In fact, many of the leading nurseries at that time were offering Improved French prunes of one type or another. By then, Burbank's Miller, or Improved French, had undergone another name change to Morganhill (Howard 1945).

In 1904, a limb of a natural mutation was discovered in an Agen orchard at Saratoga. The fruit was reported as considerably larger and somewhat more oval shaped than the original Agen. The limb also was reported as highly productive, producing prunes of high quality. It was repropagated by Leonard Coates in his nursery orchard at Morgan Hill. Eventually, the mutation received the name Coates 1418. At various times the mutation also received the names Cox, Double X, Date Prune, Saratoga, and Saratoga XX (Hansen 1951; Zielinski et al. 1961). It produced large prunes when the trees were young, but the pits were larger than those of standard French. As the trees became mature, however, the fruit size became similar to French. The variety was never widely planted in California, but in the 1960s it was grown to a limited extent in the Willamette Valley of Oregon. No orchards composed of it exist in California today.

It is not clear whether the improved clone of French referred to by Wickson in his address of 1910 was Burbank's Improved French, Coates 1418, or some other clonal selection of the Agen. A. H. Hendrickson, writing in his *Prune Culture* in California in March 1930, stated that "several so-called improved 'strains' of the French prune have been introduced in the past few years. Some of these have much promise." (Hendrickson 1930, 32). Prior to the 1950s, it does appear that there was consensus among some of the leading nurserymen that a single clone of Improved French was the French prune of preference (fig. 9.3).

Since that time, several large California nurseries have listed their French clone as Improved French or French Improved. Experienced nurserymen indicate that one of the primary forces in the selection of a "standard" Improved French in the late 1940s and early 1950s was the shift from harvesting off the ground to the mechanical shake and catch system. This new system necessitated the use of a clone of French that was

Figure 9.3 Improved French. *Photos:* C. J. DeBuse (top), M. L. Poe (bottom).

not only uniform in size but, more importantly, was also uniform in maturity. The Improved French is substantially more uniform than the clones of Agen in both regards. The specific origin of the present-day Improved French is difficult to trace. It is also unclear whether only a single Improved French exists or whether it is a combination of improved clones of French. European researchers and nurserymen generally consider the Improved French grown in California to be a single variety, distinctly different from the Agen or Prune d'Ente grown today in France, and they credit the development of this variety to Luther Burbank (Nicotra 1983). It is certainly reasonable to suggest that the bulk of the prune orchards grown in California in 2009 were made up of Improved French, not the Agen or French Prune of 75 or more years ago. For the curious, it may be left to DNA analysis to sort out whether Improved French is a single variety or a multiple clone and whether it is Burbank's seedling or some mutation of the original Agen.

As grown in California, the Improved French tree is upright and vigorous, with strong branches capable of successfully bearing heavy crops. The tree is self-fertile with average precocity, usually bearing some fruit at third leaf, and, if long-pruned and well grown, it produces substantial fruit in fourth leaf. The tree has a moderate tendency to alternate bearing, especially after heavier than average crops. The blooming period is midseason within the *P. domestica* species. Bloom is abundant and well distributed throughout the mature tree. The medium-sized leaves are medium to light green and moderately shiny. The date of fruit maturity is midseason for the species.

The medium-sized fruit usually matures rather uniformly throughout the tree. The fruit is ovate, with a slight neck. The skin can range from a reddish purple to a full purple and is covered with a light grayish bloom. The dense fruit flesh is yellow to amber. The pit is small, generally oval, thin, and with smooth overall surfaces. The pit is semifree with fibers at times attached along the ventral surfaces of the pit. Soluble solids in prunes are measured in percentage (degrees Brix equivalent) and include sugars, acids, soluble pectins, and other soluble constituents. In Improved French, percentage soluble solids often ranges from 22 to 24% and above with average crop loads. At that soluble solids level, fresh to dry (drying) ratios can commonly average 3 to 1.

In addition to its use for drying, large Improved French fruit are packed and sold as fresh fruit. Much of the product is exported to Pacific Rim countries. Over 238,000 packages of French prunes were shipped fresh in 2001, and 225,000 packages were shipped in 2002 (CTFA 2009). Since that time, fresh production has generally declined; the California Tree Fruit Agreement reports only 23,111 packages of French packed in 2007 (CTFA 2007). No French prunes were reported as fresh packed in 2009 (CTFA 2009). The volume of fresh fruit packed in any specific year often depends on the crop load set throughout the various growing areas in California. Light sets increase the individual fruit size and the economic feasibility of picking for the fresh market.

Mutations of French in California

Since 1950, several mutations of the French prune have been discovered and planted in California. Over time, none have been particularly successful. Many have been planted only in small acreages.

Gerrans Early French

This mutation of French was discovered in the late 1950s in the W. A. Gerrans orchard in Colusa County, California. It was first commercially planted in 1961. The fruit generally resembles French in form but is substantially smaller in size, slightly more narrow, and slightly more necked (fig. 9.4). The Gerrans tree is also similar to French in form, but as it grows to maturity it is most often

Figure 9.4 Gerrans Early French. *Photos:* C. J. DeBuse (top), M. L. Poe (bottom).

smaller. The tree blooms a few days earlier than French and is self-fertile, but its fruit matures up to 2 weeks before French. Cropping is somewhat erratic and is often light. Soluble solids in Gerrans fruit ranges from 2 to 3% higher than in French, and drying ratios are very good, typically ranging from 2.50:1 to 2.75:1. Although initially widely planted, the variety has fallen out of favor due to small fruit size and irregular yields. Very few acres have been planted since 1985, and by 1999 fewer than 400 acres remained in the ground (NASS 2000).

Friedman

The Friedman French clone was introduced and patented by Martin Friedman of Gridley, Butte County, California. The original test planting was established in about 1972. The fruit form is similar to French but with a slightly more pronounced neck. The Friedman tree form is similar to French. The tree blooms a few days later than French, and flowers are particularly abundant and more dense than French. The tree is apparently self-fertile. Fruit quality is good, essentially the same as French. Fruit production has been erratic in some locations. Very few trees have been planted since 1985, and about 300 acres of trees remained in the state by 1999 (NASS 2000).

Victor Large

The Victor Large prune was developed by Victor Locarnini at Morgan Hill. The first commercial plantings were made in California about 1970. The fruit is very similar in form and quality to French, but it may be slightly larger when the tree is young and crops are light. As the tree matures, cropping is identical to French. The tree is vigorous and productive. The tree blooms at the same time as French and is self-fertile. The date of maturity is identical to French. Almost no orchards of this variety have been planted since 1985. In 1999, less than 300 acres remained under cultivation (NASS 2000).

Punian

The Punian prune was discovered in a Yuba City orchard as a mutation of French by Gurbachan Punian in the late 1960s. After repropagation and evaluation, the prune was eventually patented and released in 1990. The vigorous tree is similar to French in form but quite thorny. The fruit matures early, at least 2 weeks before French. The fruit is generally similar to French but has a somewhat darker purple blush and a more green ground color. The flesh of Punian is greener than French and on drying turns rather dark. Soluble solids accumulation in the fruit is good, with low drying ratios. The tree develops abundant bloom, but it has been difficult to set. The tree does not appear to be self-fruitful. Using various pollinators has not seemed to increase the reliability of set. Of the few orchards that were established, most have been removed.

Other Prune Varieties

During the early part of the twentieth century, non-French-type prunes made up a substantial part of California prune acreage. Over the last 50 years, however, the productivity, consistent bearing habit, and ease of handling of the French prune have eventually relegated most other non-French varieties to the pages of history. Varieties such as Bulgarian, Wangenheim, Hungarian, St. Martin, Hungarian Date, St. Catherine, Golden, and others listed in publications of the late 1890s are nowhere to be found in California today (LeLong 1892). The best estimate in 2009 sets the acreage of non-French varieties in California at perhaps only 3% of the total prune acreage of 64,000 acres (NASS 2010). When the comparison is made between 2009 acreage and the 24,500 acres of non-French prunes growing in California in 1940 (16.7% of the total acreage), the downward trend is clear. With the prune crop now completely mechanically harvested, French, and particularly the Improved French with its uniform maturity, has demonstrated its supremacy over the old non-French types. The spread of maturity in many of the older non-French prune varieties was substantial, and the one-pick mechanical harvest reduced the quality of the total harvest. The superior durability of the skin and flesh of French also gave it an advantage over the often soft flesh or delicate skin of the non-French types. Many of the old varieties had large, rough pits that were tightly attached to the fruit flesh. With a high percentage of today's prunes sold as pitted prunes, these large, tight, rough pits presented difficultly in processing. Much of the non-French prune acreage in 2009 is represented by prune varieties grown only for the fresh market. The eight varieties described below, with the exception of Moyer, are either almost commercially extinct in 2009 or will most likely be totally out of commercial production within the next 10 years.

Moyer

The Moyer prune, sometimes referred to as Moyer Perfecto, was discovered as a seedling growing at Roseburg, Oregon, in about 1925. The seed parent is thought to be Coe's Golden Drop, also known as Silver. Originally selected in Oregon as a prune for drying, over the last 15 years Moyer has become one of the primary fresh market shipping prunes in California. Statistics from the California Tree Fruit Agreement Annual Report in 2001 indicate that 223,000 packages of Moyer were packed for fresh shipment from California, and 283,000 packages shipped in 2002. Total production has recently declined, with 126,274 packages packed in 2005, and only 49,000 packages packed in 2009 (CTFA 2009).

The upright-growing tree has strong, thick limbs and smooth, light gray bark. The tree is very vigorous and slow to come into bearing. After cropping commences, the tree can become distinctly alternate bearing. The fruit is substantially larger than French, and production can be quite heavy in the "on" year. The fruit is generally oval and symmetrical, with a uniform blue to dark reddish blue coloration when fully ripe; the fruit is overlaid with a heavy, grayish, waxy bloom (fig. 9.5). The date of full bloom can vary from 2 or 3 days later than French to as much as a week later. The tree is self-fertile. Fruit harvest is about 10 days later than French.

The fruit is difficult to dry. The accumulation of soluble solids is lower than French, and drying ratios are usually in the 4:1 range. The large fruit needs a longer drying time than French, and it tends to slab out and bleed on the trays, making removal from the trays difficult. The internal color of dried Moyer is somewhat orangish, and the dried flavor is different from French. The pit is large and adheres substantially to the flesh. Most Moyers sent to the dehydrators are discards from the fresh fruit packing lines and end up as a by-product. Slightly less than 800 acres of Moyer were reported in the ground in 1999 (NASS 2000). An estimate of Moyer plantings in 2009 is 1,000 acres total, both bearing and nonbearing.

Figure 9.5 Moyer. *Photos:* C. J. DeBuse (top), M. L. Poe (bottom).

Imperial

The Imperial prune is a relatively old prune of French origin. Known in France as Imperial Epineuse, it was discovered in 1870 as a chance seedling growing on the grounds of an abandoned monastery near the French town of Clairac. It was first imported into California in 1883 by the nursery pioneer Felix Gillett and grown at his Barren Hill Nursery at Nevada City. After several years of testing, Gillett first offered it for sale in 1893 under the name Clairac Mammoth. Another California nurseryman, John Rock, of Niles, also imported Imperial in 1886. The two imports were perceived to be two different strains, with the Gillett strain more regular in bearing habit. In more recent years, several strains of varying size and quality have been selected in France (449 and 851), but they have not been tested in California.

The Imperial tree is vigorous with a more spreading growth habit than French. The tree is somewhat brittle and subject to early decay. It is often a shy bearer, but when heavy crops do occur, it becomes distinctly alternate bearing. In heavy crop years, the tree must be thinned to produce an acceptable dried product. The oval fruit is usually somewhat larger than French. The skin can vary from dark red to purple-red. The flesh often has a greenish tint. The pit is somewhat larger than French and semifree. The flesh can darken and develop air pockets when temperatures are elevated near harvest time. The date of full bloom can vary from a few days before French to fully overlapping French full bloom. Imperial is self-incompatible and needs a cross-pollinator; French, Burton, and Sugar are satisfactory. The fruit harvest date is about a week before French, depending on crop load.

Because of its size, Imperial is difficult to dry. Drying times are substantially longer than for French. Soluble solids accumulation is not particularly high, and drying ratios can approach 3.75:1 or higher. The fruit will slab out and bleed, sticking to the drying trays. In 1958, Imperials in California totaled 6,338 acres. By 1999, only 168 acres of Imperials were left in the ground in California (NASS 2000).

Burton

The Burton prune was developed by R. E. Burton of Vaca Valley in 1906. Burton had emigrated from England in 1874 and settled near the present-day town of Vacaville. In 1886, he obtained trees of Imperial from both Felix Gillett and John Rock. In 1906, Burton planted seed that had been produced on a limb of Tragedy plum grafted onto the Imperial tree that he had received from Rock. From this population of 75 to 100 seedlings, the Burton variety was eventually selected.

The tree is vigorous, upright in form, and slightly thorny. The variety is only partially self-fertile and usually needs a pollinator to produce full commercial crops. The timing of bloom of Burton overlaps both French and Sugar. The fruit is very large with a substantial neck (fig. 9.6). The skin varies from a light purple to a darker blue and has a moderate amount of grayish bloom covering the fruit. The flesh ranges from yellow to a darker yellow-gold. The flesh texture can be somewhat coarse, and the pit is very large, semifree, and with a rough surface. Harvest date is about 10 days later than French.

Harvesting and drying are difficult with Burton because of substantial variation in maturity within the tree, the thin skin, and the very large size of the fruit. Drying times are substantially longer than for French. Under standard drying circumstances, the fruit slabs out and bleeds heavily, producing a poor-quality product. Under modified drying conditions, however, where temperatures are reduced and the fruit has attained a high level of maturity, the dried product can be excellent. Very few Burton orchards are left in California; only 13 acres were listed as being in commercial production in 1999 (NASS 2000).

Figure 9.6 Burton. *Photos:* C. J. DeBuse (top), M. L. Poe (bottom).

Sugar

The Sugar prune was released by Luther Burbank in 1899. It is reported to be an open-pollinated seedling of Agen. The tree is upright in growth habit and can grow to be quite large; it is very brittle, and the limbs break readily under heavy crop loads. Sugar is fully self-fertile and can be very productive. Heavy crops must be thinned to produce dried fruit of acceptable size. The tree is extremely alternate bearing. Under normal crop loads, the fruit is somewhat larger than French. The oval fruit is dark red to purple-red (fig. 9.7). The flesh is amber. Mature fruit can develop pockets of air in the flesh prior to harvest. The flesh texture is somewhat coarse, and the pits are large with a rough surface. Many pits have a strong wing along the ventral suture that can crack off or separate from the main body, causing pit fragments. The harvest date of Sugar is a week to 10 days earlier than French.

Although the soluble solids content is reasonably good, Sugar prunes need a longer drying time than French and have a relatively high drying ratio of 4:1. The fruit bleeds excessively under the normal drying temperature (near 165°F). The dried product is often of only fair quality, being rather coarse and stringy. Very few orchards of Sugar are left in California, and fruit from those that remain is usually shipped to the fresh market. The California Tree Fruit Agreement Annual Report stated that a little over 5,000 packages of fresh Sugar prunes were shipped in 2001 and only 484 packages in 2002. In 2009, no shipments were reported (CTFA 2009).

Figure 9.7 Sugar. *Photos:* C. J. DeBuse (top), M. L. Poe (bottom).

Robe de Sargeant

Of French origin, Robe de Sargeant has been grown in California to a moderate extent, with about 1,300 acres reported in the ground in 1958 (Parr and Friesen 1959). By 2009, the variety was very nearly commercially extinct. The oval fruit, when the tree is moderately cropped, can be somewhat larger than French. The fruit has a deep purple skin, sometimes approaching black, overlaid with a thick grayish blue waxy bloom. The flavor of the fresh fruit is sweet and rich, although occasionally with a lingering astringency. The green-yellow flesh is juicier than French and has a large, rough-surfaced stone that adheres to the flesh more than in French. Moderately cropped trees can develop soluble solids of 24% or above, but the fruit begins to drop from the tree badly at fruit firmness of 3.5 pounds-force or below. Drying ratios are higher than those for French, and the fruit on drying trays has a tendency to slab, bleed, and stick because of the excessive juiciness of the ripe fruit. The dried fruit is very black and is of good quality when carefully dried. In the past, Sargeant dried fruit was sometimes blended with French. Sargeant trees are self-sterile and need a pollenizer such as French or Sugar. This self-sterile characteristic can lead to irregular cropping in cool areas or in cool, wet years. The tree is an upright grower and has broad leaves that are large and dark glossy green. The tree is somewhat brittle. The variety is highly incompatible with peach rootstock.

Some confusion exists about the origin of Sargeant in France and its exact name. In 1911, U. P. Hedrick, writing in his *Plums of New York*, indicates that French growers at that time held Sargeant, Agen, and Prunier Datte to be the same variety. John Rock of Niles and W. B. West of Stockton, both early nurserymen in California, imported Sargeant into California prior to the 1890s as a distinctly different variety from Agen. Leonard Coates at Morgan Hill, who had imported Prunier Datte into California, maintained that it was the same variety as Robe de Sargeant. In older literature (1892–1920), the variety can be found listed as Robe de Sergeant, Robe de Sargeant, Robe de Sergent, and simply Sergeant or Sargeant (Butterfield 1938, Hedrick 1911).

Italian

Although no longer grown in California, Italian still remains an important variety in Oregon, where it has been dried, canned, and sold fresh. This variety is known by a number of names: Quetsche D'Italie in France, Prugna D'Italia in Italy, and Fellenburg in other parts of Europe (especially Germany), where it took on the name of the man who first imported it from Italy. The variety originated in Italy around 1800, reputedly from the area around Lombardy. It was reported in the United States in 1831 and was listed in nursery catalogs in California by 1854. Initially, in California, Italian was planted in the Santa Clara Valley, but it has not adapted well to the hot, dry interior valleys of California, where both the tree and the fruit can suffer substantial heat damage. The medium vigor and open growing habit of the tree probably worsen the heat injury due to full exposure of the tree and fruit to the sun.

The fruit is oval, tapering somewhat toward the stem end and is usually larger than French and hangs well on the tree. The skin is dark blue to dark purple-black, often speckled with yellowish dots and covered with very thick, waxy bloom. The fresh fruit, when well grown in adapted areas, can be quite rich and aromatic, with a higher level of acidity than French. The flesh is greenish yellow and relatively firm. The stone is nearly free, although with roughened surfaces, a short tip, and wings along the ventral edge of the stone. The tree is self-fertile. Dried Italian prunes are usually very dark. This variety does not attain the level of soluble solids

content as French, but it does have a higher acidity; the dried product can be quite tart.

Several early-maturing Early Italian types have been grown in Oregon, such as Richards, Milton, Demaris, Reuter, and Greata. All are similar in form to the standard Italian but can mature from 7 to 14 days earlier. The several varieties have not always been maintained as separate varieties but have been sold simply as Early Italian.

German

Numerous strains of this old European prune exist due to the long-standing practice of planting orchards of German from seed. Trees of German produced from seed bear fruit with only moderate variability from tree to tree. German has been one of the most widespread varieties grown in Europe. The source of this variety is obscured by time, but it may be similar to that of the Prune d'Agen, having been carried back to Europe from Asia by Crusaders. German was grown in Hungary in the sixteenth century, with the dried fruit commercialized into Italy, Switzerland, and Germany. The variety then spread widely into Europe by the end of the seventeenth century. The first clear mention of German in America was in 1832 (Hedrick 1911). One of the best of the many German strains, Latz, entered the United States in 1850 from Prussia. In California, German was listed by nurseryman A. P. Smith at Sacramento in his 1856 catalog. Production of the Latz strain occurred at Wright Station in Santa Cruz County beginning about 1900. Production of German was never widespread in California and was limited to specialty packs. Even surviving specimen trees of this old late-season variety are now hard to find in California.

The fresh fruit is generally small to medium in size, long and oval, and slightly tapered at the apical end. The skin is purplish black with some basal speckling and heavy bloom. The stone is relatively free. The fruit flesh is firm, yellow-green, and moderately juicy and sweet with a mild, pleasant flavor. The tree is large, hardy, and vigorous, and is a regular and quite productive bearer.

Silver

The Silver variety, also known as Golden Drop, was grown commercially in small acreage in the early 1900s, primarily in Santa Cruz County (Chaney 1981). The yellow fruit, when dried and sulfured, produced a high-quality light blond product whose large size brought excellent returns. The fruit was also shipped fresh to the San Francisco market.

This late-season variety was developed in about 1809 by Jervaise Coe. Coe was a market gardener living at Bury St. Edmunds, in Suffolk, England. It was raised from a seed of Green Gage that was thought to have been pollinated by the White Magnum Bonum plum (also know as Yellow Egg). The Golden Drop arrived in America in 1823, sent by the noted English pomologist Thomas A. Knight to John Lowell of Massachusetts. At one time on the Pacific Coast, Silver, as it was renamed there, was considered to be a seedling of Golden Drop that originated in Oregon. That theory has been discredited; Silver and Golden Drop are now considered to be the same variety. Golden Drop was listed in the catalog of Warren and Sons Garden and Nursery at Sacramento as early as 1853.

The fruit is large, yellow, and oval, with only a very slight neck. The flesh is firm, juicy and yellow. The stone is semifree, somewhat rough surfaced, and often has a conspicuous wing. Silver matures late in the season and seems to perform best in cooler areas of the Pacific Coast. The flavor of the fresh fruit, when grown in adapted areas, has been regarded as excellent, perhaps reflective of its Green Gage parentage. The dried product, when carefully processed, has also been considered as among the best in sweetness, size, appearance, and flavor. The hardy tree is a slow grower, neither highly precocious nor productive. The variety is not self-fertile. Commercial production of Silver has not occurred in California for many years.

Prune Variety Breeding Programs

The French Programs

Several prune breeding programs have been active in France since the end of World War II and have released new varieties to the prune industry. Their work followed earlier efforts to improve prune varieties by selection of mutants of Agen and importation and evaluation of varieties that had been developed prior to the early part of the twentieth century. The program of R. Bernhard at the Grande Ferrade Research Station near Bordeaux that was initiated in 1947 has been noted previously in relation to the selection of superior clones of the Prune d'Ente. In 1958, a second program at Grande Ferrade was initiated by R. Renaud with the goal of developing new prune varieties by hybridization between existing varieties. This hybridization program had three objectives: first, to develop new varieties with harvest dates both earlier and later than Prune d'Ente, with the intention of spreading the timing of harvest and drying, thereby improving the efficiency of these operations; second, to develop more precocious tree types; and third, to improve the technical quality of the fruit (e.g., lower drying ratios, freestone fruit types). Several thousand hybrid seedlings were developed in 1962 and 1963, selected to 60 by 1970, and reduced to a final 3 by 1982. These final selections were named Primacotes, Lorida, and Tardicotes and released to growers in France (Renaud 1984).

Primacotes is a cross between a seedling of Prune d'Ente and a clone of Prune d'Ente. In France, the fruit is described as maturing 3 to 4 weeks earlier than GF 707. It is medium to large, oblong, and slightly elongated at the stem end. The fruit has a reddish violet skin, yellowish flesh, and a medium-sized semicling stone. The fruit is easy to dry and of good quality, considering its early season of maturity. The tree is of average vigor, slightly weeping in form, a regular bearer, and productive; it is early blooming and self-fertile.

Lorida is a hybrid of Prune d'Ente and the European plum variety Anna Spath. In France, it ripens at the same time as GF 707. The fruit is large, oblong, and rounded at both the apex and base. The skin is red to violet, and the flesh is yellow to orange. The stone is completely free. The fruit dries into a very high quality prune. The tree is of average vigor and upright in form. Full bloom occurs at the same time as full bloom of GF 707. The tree is a productive and regular bearer and is self-fertile.

Tardicotes is a cross of Prune d'Ente and the plum variety Grand Duke. In the Lot and Garonne Valleys of France, Tardicotes ripens 8 to 10 days after GF 707. The fruit is medium to large, oblong, and somewhat elongated. The skin is blue-violet and quite resistant to cracking. The flesh is yellow with slight greenish tones, becoming more orange with advancing maturity. The stone is semifree. The dried fruit is of good quality with a low drying ratio. The tree is vigorous and very upright and early blooming, attaining full bloom at least 4 to 6 days before GF 707. The productive tree is a regular bearer and is self-fertile.

Performance of the French Hybrids in California

In 1985, shortly after their release, Primacotes, Lorida, and Tardicotes were imported into California and established in a comparative trial at the UC Kearney Agricultural Center at Parlier. In addition to these three new varieties above, the trial included four clones of Prune d'Ente, GF 642, GF 652, GF 303, and GF 2733; an older variety from France called Double Robe (GF 812); and the standard California Improved French as a comparison. Ten trees of each variety were established in a randomized and replicated planting. By the end of the 1989 crop year, it was clear that Lorida would not be successful in interior valley plantings in California. The Lorida fruit had cracked badly from the outset of testing, with from 85 to 90% of the fruit split, rotted, and insect infested before attaining full maturity. The variety continued to perform this way under both light and heavy cropping situations, and in 1990 it was eliminated from the test plot as unsuited for California growing conditions. At the end of 6 years of testing (1991), the following report on the performance of the remaining two hybrid varieties was made by T. J. DeJong and J. F. Doyle to the California Prune Board.

Primacotes

The early maturity date and large fruit are positive characteristics of Primacotes, a new prune cultivar. Bearing characteristics, low productivity, and difficulty in harvesting, however, are such negative factors that it is questionable whether this prune can be used successfully by California growers. Production in 1990 and 1991 averaged only about 1.5 dry tons per acre, in comparison with 4.5 dry tons for Improved French. In 1991, as in 1990, the primary fruit-bearing area in the tree was at the end of shoots on 1-year-old wood. Very little fruit was borne on spurs formed on the scaffolds and older secondary limbs. Large areas of blank 2-year-old wood were present. The major portion of the crop was borne in large drooping clusters of 20 to 50 fruit, far out on the terminal of the shoots. Shaker vibration is poorly transmitted to these high-inertia clusters, making fruit difficult to remove. If the long shoots are shortened at dormant pruning time, a much more detailed pruning job is necessary, and the removal of terminal crop-bearing wood reduces yield. This type of bearing habit substantially limits the value of Primacotes in a shake and catch harvesting system. Primacotes produces a large, coarsely wrinkled prune with a more acidic flavor than Improved French. The exterior color of the dried fruit is irregular, with a more brownish cast than Improved French. The flesh is dark brown (DeJong and Doyle 1991).

Tardicotes

Tardicotes is at the opposite end of the maturity spectrum from Primacotes, ripening 2 to 2½ weeks later than Improved French. The cultivar is

poorly adapted to hot interior valley conditions, producing large numbers of double fruit (20%) and exhibiting internal heat damage prior to harvest. The internal damage is characterized by darkening or drying of the flesh near the pit and by a substantial amount of preharvest drop, even at fruit flesh firmness above 3.0 pounds-force. The cultivar produces a dark, dull black prune. The prune flesh is extremely tough and fibrous and somewhat dark. The flavor of Tardicotes is more acidic than Improved French and lacks richness and balance. Lack of productivity and poor adaptation rule out the use of this cultivar in California (DeJong and Doyle 1991).

Development of New Varieties by Mutagenesis

In 1966 at Grande Ferrade, Dr. Renaud embarked on a prune variety development program using artificial mutagenesis. Shoots collected from various clones of Prune d'Ente were irradiated, and buds from these shoots were budded onto rootstocks. Buds that survived and grew from this first propagation were repropagated and grown to fruiting age in the field. From an initial population of 1,200 mutant buds, the Spurdente variety was selected and released by Renaud in 1986. The fruit of Spurdente is reported to be very similar in form and processing quality to that of Prune d'Ente. It flowers about 4 days after GF 707 but ripens from 8 to 10 days before the GF 707. The tree is a spur type with fewer vigorous upright shoots but a proliferation of fruiting spurs. The tree is vigorous and easily pruned. Nurserymen from France have indicated that Spurdente may produce smaller fruit than GF 707, but the variety has not been field tested under California growing conditions.

Prune Variety Breeding at UC Davis

Early Research

In 1975, a study was completed at UC Davis on the inheritance of ripening date, fruit size, and soluble solids in *Prunus domestica* by C. O. Hesse of the Department of Pomology and P. E. Hansche and V. Beres of the Departments of Pomology and Genetics (Hansche et al. 1975).

The study was based on measurements taken by Hesse and D. O. Walker from seedling populations developed from 1948 through 1966. Measurements taken from a second population during the 1973 and 1974 growing seasons were also included in the study. The results indicated that fruit size and ripening dates in *P. domestica* were very highly heritable, with the percentage of soluble solids less so but still in the high heritability range. After an estimate of genetic and environmental components was made, the study concluded that "inter-mating of parents selected on the basis of their own performance could be very effective and efficient in improving the genetic potential of such breeding stocks in bloom date, ripening date, fruit size and probably (though not certainly) soluble solids" and that "these results indicated that the rates of gain to be expected from mass selection on such populations are substantial" (Hansche et al. 1975, 522, 524).

Encouraged by these findings, Hansche and Beres initiated a prune variety development program in 1975. The immediate goal of this program was to develop a prune with all the cultural and processing traits of standard French, but one that ripened about 2 weeks earlier. Using quantitative genetics, the estimate was made that 13,000 seeds of self-pollinated French would need to be grown and fruited to allow for enough recombination to occur so that a seedling (or seedlings) containing all the desired traits could be recovered. The 13,000 seeds were processed, but only about 5,000 seedlings were developed. It rapidly became apparent that French was not a good choice as a parent. Several cultural and genetic problems appeared. Self-pollinated seedlings of French had a long juvenile period and were distinctly nonprecocious. Even after 9 years, many seedlings had not bloomed or fruited. Additionally, over 60% of the seedlings were so weak that they were very short lived or had no commercial potential because of their vegetative form. The trait of high fruit yield was found to have a low heritability, and very few seedlings attained the productivity level of the French parent. Although two seedling selections achieved the goal of ripening 2 weeks ahead of French, both had severe upper stem cracks, weak growth habit, and low levels of productivity. In fact, only one selection out of the 33 selections derived from the entire population had a production level equal to French. This productive selection unfortunately produced a weak tree and had extremely pear-shaped fruit and only fair-quality processed fruit. Only 38% of the population estimated to be required for recovery of the desired seedling recombinant was developed, jeopardizing the chance of success. No varieties were released from this program.

The Current UC Breeding Program

Although unsuccessful in the development of new varieties, the early work of Hesse, Hansche, Walker, and Beres provided substantial insight into potential gains and problems associated with variety improvement of *P. domestica*. As the initial breeding work was terminating, there remained a continuing interest within the prune industry in the development of new prune varieties. In response to this interest and encouragement, primarily by the California Prune Board, a second prune variety development program was begun in 1985. This program was initiated under the leadership of T. M. DeJong of the Department of Pomology at UC Davis, with field management of the program by J. F. Doyle at the UC Kearney Agricultural Center (KAC) at Parlier. Initially the program focused on the importation and evaluation of the several Prune d'Ente clones and the new Primacotes, Lorida, and Tardicotes varieties developed in France. As it became apparent that the new varieties from France were not adapted to California growing conditions and industry practices, more emphasis was placed on development of new varieties with specific adaptation to growing conditions found in the interior valleys of California. The primary goal was to obtain new prune varieties similar to French in appearance and processing ability but maturing substantially earlier or later than Improved French. These new varieties would improve the efficiency of harvesting and drying by spreading out the short high-peak volume of harvest and dehydration currently experienced by the exclusive use of French. A new strategy was developed to overcome the limitations found with the exclusive use of French as a parent in the previous UC program.

The first phase of the new strategy was to grow several hundred seedlings of each of the principal prune varieties adapted to California growing conditions and to test progeny for precocity, seedling vigor, spread of maturity (both early and late), and freedom from heat damage in both the fruit and tree. Selections from Phase 1 of the program would then be used as parents in the Phases 2 and 3 hybridizations that followed. Traditional California varieties do not exhibit a sufficiently wide range of variability in their seedling populations, so in Phase 2 a wider range of parental varieties was used, many not historically grown in California. The purpose of Phase 2 was to incorporate through hybridization a wider range of diversity of desired traits not evident in California varieties, emphasizing such characteristics as spread in dates of maturity, precocity, tree vigor, freedom from heat damage, fruit size, fruit color, freestone pit, and other traits of commercial significance. Because of the diversity of parentage used in Phase 2, it was expected that advanced selections from this phase would, for the most part, not closely resemble French, except in cases where French or related varieties were used in the cross. A third phase was then initiated to use advanced selections from Phases 1 and 2 in a series of crosses. These crosses would hybridize the advanced selections with Improved French to produce seedling populations that had at least a 50% French pedigree. The seedling populations in Phase 3 have a high potential for producing selections that closely resemble Improved French, but with fruit maturities in a broad range and trees that are vigorous, productive, and precocious. Two new prune varieties, Sutter and Tulare Giant, have recently been released to California growers from Phase 2 crosses, and one new variety, Muir Beauty, has come from Phase 3.

Sutter

Sutter is the result of a hybridization in 1987 between the prune variety Sugar as the seed parent and the prune variety Primacotes as the pollen parent. The seedling was first selected from among Phase 2 seedlings growing at Kearney in 1993. Testing of the variety began at Kearney and in grower plots in 1994. The date of maturity of Sutter can range from early to mid-August, about a week to 10 days earlier than Improved French grown in the same location. The fruit of Sutter usually develops at least 2% more soluble solids than Improved French at similar crop loads. The fruit is large, dark purple, and covered with a medium waxy bloom. The fruit shape is similar to that of Improved French but longer (fig. 9.8). The fruit flesh is yellow-amber. The pit of Sutter is a smooth-sided freestone, oval, slightly larger that the pit of Improved French, and is easily mechanically removed. The tree is similar in form and vigor to Improved French and is equal in its precocity to bear fruit. Timing of bloom is nearly identical to Improved French in most locations. The variety is self-fertile. The fruit hangs well on the tree with a minimum of preharvest drop. For premium-quality dried fruit, Sutter should be harvested at higher fruit flesh firmness, in the range of 4.5 to 7 pounds-force. The new variety performs well on both Marianna and Myrobalan rootstock. Peach rootstock or rootstock that is a peach hybrid is not recommended for Sutter because grafts onto peach

Figure 9.8 Sutter. *Photos:* C. J. DeBuse (top), M. L. Poe (bottom).

have a high amount of scion breakage. Leaves of Sutter are similar in form but less glossy than leaves of Improved French.

The fruit of Sutter dries into a very high quality prune. The average sizes are slightly larger than Improved French, and drying ratios are in the 3:1 range. The flesh is golden amber. The exterior appearance of the dried fruit is quite similar to Improved French, but the dried flavor is lighter, with a more complex fruity flavor and a sweeter taste. Sutter was released to California growers in 2000, and commercial plantings came into bearing in 2006. However, it was discovered that overripe fruit is more susceptible to breakdown during processing than Improved French. Growers must be careful during harvesting and drying to avoid fruit breakage. This problem may decrease its acceptance by the California industry.

Some growers are considering Sutter as a dual purpose variety that could produce a high-quality product for the dried as well as the fresh market. Sutter's large size, early maturity, high soluble solids, and dark skin color hold out the promise of success for the variety as a fresh market product.

Tulare Giant

Tulare Giant is the result of a cross between the *P. domestica* plum variety Empress as the seed parent and Primacotes as the pollen parent. The cross was made in 1987, and the seedling was first selected from among Phase 2 seedlings in 1991. Field testing of the selection began in 1993 at Kearney and in grower trials. The date of maturity for fresh market prunes in the central San Joaquin Valley usually begins in the first week of July. Tulare Giant can mature as early as 3 to 5 weeks earlier than Improved French grown at the same location. Tulare Giant fruit is substantially larger than Improved French. The fruit is oval and somewhat tapered at both ends, and the skin is a dark purple under a gray, waxy epidermal bloom (fig. 9.9). At the stem end of the fruit, the skin can be a lighter purple or show a small amount of yellowish ground coloration. The flesh is a very pale yellow with shades of green, depending on the stage of maturity. Trees are vigorous, upright-spreading, and highly precocious, bearing fruit as early as their third year. Unlike many other *P. domestica* varieties, Tulare Giant produces many flowers and sets numerous fruit on 1-year-old wood, as well as on older spurs and branches. The variety is early blooming and has abundant flowers. Tulare

Figure 9.9 Tulare Giant. *Photos:* C. J. DeBuse (top), M. L. Poe (bottom).

Giant is not self-fertile and requires a pollenizer to set adequate crops. It is very important to use another early-blooming European *P. domestica* variety as a pollenizer in order to set regular crops of adequate numbers. The new UC variety Muir Beauty blooms at approximately the same time and is recommended as the pollenizer of Tulare Giant.

Tulare Giant, with its large size and good color, appears to be highly suited for shipping as a fresh market prune. In order to be successful, however, the tree must be well pruned, and the crop usually needs to be well thinned for the fruit to develop adequate levels of soluble solids. If not carefully thinned, heavy crops usually have fruit of adequate size but with unacceptably low levels of soluble solids and a bland flavor (soluble solids of 18.0% and above produce the highest-quality and most flavorful fruit). In 2002, total shipments of Tulare Giant numbered 9,200 packages. Plantings have increased since then, with 89,199 packages packed in 2005 and slightly over 245,000 packages in 2009 (CTFA 2009).

Tulare Giant could be used for drying in specialty packs, but the pit is quite large and is only semifree, making it difficult to remove. At 3.0 pounds-force fruit firmness and with a moderate crop load, soluble solids can approach 22%. Although the dried fruit is glossy black and has good-quality flesh, Tulare Giant is not recommended as a dried fruit because of its large size, long drying time (more than 20 hours), and tendency to bleed, slab, and stick to drying trays.

Figure 9.10 Muir Beauty. *Photos:* C. J. DeBuse (top), M. L. Poe (bottom).

Muir Beauty

A third prune developed from the UC breeding program was released to California prune growers in 2005. The new variety was tested as D6N-72 and later named Muir Beauty (fig. 9.10). The variety was developed from a hybridization made in 1992 of Improved French as seed parent and Tulare Giant as pollen parent. It was selected from among Phase 3 seedlings growing at UC Davis in 1997. Testing of the selection began in 1998 at Kearney and at the UC Wolfskill Experimental Orchard at Winters. Muir Beauty matures approximately 10 to 15 days before Improved French growing in the same location. The large fruit is somewhat intermediate in size between Tulare Giant and Improved French. The fruit is oval with a purple-rose skin covered with a grayish medium waxy bloom. The fruit flesh ranges from a dark gold to a golden orange. The tree is very vigorous, upright in form, and very productive. The fruit is borne on 1-year-old shoots and older spurs, similar to the fruiting habit of Tulare Giant. The variety grows successfully on both Marianna and Myrobalan plum rootstock. The variety has been successfully grafted on to peach rootstock and shows good compatibility. The leaves are moderately large, deep green, and relatively shiny. The tree is more precocious than Improved French, flowering and fruiting at an earlier age. The time of full bloom is 7 to 10 days before Improved French. Muir Beauty bloom overlaps with the bloom of Tulare Giant very well. The tree is self-fruitful. The fruit hangs well on the tree with no more than normal preharvest drop. Limited mechanical tree-shaking tests have resulted in complete fruit removal. Muir Beauty fruit develops soluble solids in the range of 21 to 24%.

The dried fruit of the Muir Beauty is large and dark shiny black with larger but somewhat fewer wrinkles than Improved French. The flesh is golden orange. The medium-sized pit varies somewhat from semifree to completely free. Limited pitting tests have resulted in easy removal of the pit. The skin of Muir Beauty is thinner than that of Improved French and can be problematic. Also, the length of cooking time during rehydration may need to be shorter than the cooking time of Improved French. The dried fruit is thick and meaty, with a very pleasant fruity flavor that is more intense than Improved French. This variety has consistently

scored at the top of flavor evaluations both within the program and in grower blind-tasting sessions. Currently this prune variety does not appear to be suitable for standard high-volume California processing but may fit into a niche market.

New Directions

Beginning in 1995, the prune variety development work at the University of California made a transition from its original location at the UC Kearney Agricultural Center at Parlier to the Department of Plant Sciences pomology orchard research area on the campus of UC Davis. The highest-quality seedlings are selected and moved to orchard blocks at Parlier and Winters for further evaluation. In 2010, over 8,700 unselected hybrid seedlings were in the program along with 166 advanced selections that are in higher levels of evaluation (DeJong et al. 2010). The program at Davis continues under the direction of T. M. DeJong and C. J. DeBuse, with the breeding, selection, and field management of the program carried out by S. J. Castro.

The program priorities have changed throughout the years to include new breeding objectives along with the primary objective of the program, which was to extend the harvest with new varieties having maturity dates before or after Improved French. The new objectives include breeding for pest and disease tolerance, new specialty traits such as dried plums that have different colors and flavors, and improved fruit quality and fruit characteristics that increase efficiency and quality of drying and processing. These new directions will address the new objectives while continuing to broaden and strengthen the germplasm base. The breeding program has created and expanded the UC *Prunus domestica* germplasm into a strong and diverse collection of advanced items and a wealth of seedling families from which to select the next generation of California dried plum cultivars.

References

Butterfield, H. M. 1938. History of deciduous fruits in California. Sacramento: The Inland Press.

Carlot, D., Cartier, and Fabre. 1971. Le prunier d'Ente et … la pruneau d'Agen. Villeneuve-Sur-Lot, France: Bureau National Interprofessionnel du Pruneau.

Chaney, D. H. 1981. Prune varieties past and present. In D. E. Ramos, ed., Prune orchard management. Oakland: University of California Division of Agricultural Sciences Publication 3269. 12–18.

Couchman, R. 1967. The Sunsweet Story. San Jose, CA: Sunsweet Growers, Inc.

CTFA (California Tree Fruit Agreement). 2007. California Tree Fruit Agreement Annual Report. Reedley, CA.

———. 2009. California Tree Fruit Agreement Annual Report. Reedley, CA.

DeJong T. M., and J. F. Doyle. 1991. Prune cultivar evaluation and development. Annual report. Prune Research Reports and Index of Prune Research. Sacramento: California Prune Board.

DeJong T. M., S. J. Castro, and C. J. DeBuse. 2010. Dried plum cultivar development and evaluation. Annual report. Prune Research Reports and Index of Prune Research; Sacramento: California Prune Board.

Hansche, P. E., C. O. Hesse, and V. Beres. 1975. Inheritance of fruit size, soluble solids, and ripening date in *Prunus domestica* cv. Agen. Journal of the American Society for Horticultural Science 100(5): 522–524.

Hansen, C. J. 1951. Prune production in California. Berkeley: California Agricultural Extension Service Circular 180.

Hedrick, U. P. 1911. The plums of New York. Albany: J. B. Lyon Company.

Hendrickson, A. H. 1930. Prune culture in California. Berkeley: California Agricultural Extension Service Circular 41.

Howard, W. L. 1945. Luther Burbank's plant contributions. Berkeley: University of California Agricultural Experiment Station Bulletin 691.

LeLong, B. M. 1892. California prune industry. Sacramento: California State Printing Office.

NASS (USDA National Agricultural Statistics Service). 2000. 1999 California prune acreage report. NASS website, http://www.nass.usda.gov/Statistics_by_State/California/Publications/Fruits_and_Nuts/200005prnac.pdf.

———. 2010. Walnut/raisin/prune report state summary, 2009 crop year. NASS website, http://www.nass.usda.gov/Statistics_by_State/California/Publications/Fruits_and_Nuts/201010wrp.pdf.

Nicotra, A., L. Moser, D. Cobianchi, C. Damiano, and W. Faedi. 1983. Monografia di cultivar di susino. Ciampino, Italy: Tipolitografia Vito Vicini.

Parr, R. D., and H. Friesen. 1959. California fruit and nut crops as of 1959. Sacramento: USDA Marketing Service and California Department of Agriculture.

Prince, W. R. 1832. The pomological manual, or, A treatise on fruits. 2nd ed. New York: T. & J. Swords.

Renaud, R. 1984. L'amelioration et la creation de varietes et porte-greffes de prunier domestique. Pont-De-La-Maye, France: Station de recherches d'arboriculture fruitiere, la Grande Ferrade, Publication 59.

Wickson, E. J. 1921. *California Nurseryman* and the plant industry, 1850–1910. Los Angeles: The California Association of Nurserymen.

Zielinski, Q. B., W. A. Sistrunk, and T. P. Davidson. 1961. Plum varieties for Oregon. Corvallis: Oregon State University Agricultural Experiment Station Bulletin 582.

10 Propagation

• Wilbur O. Reil and James F. Doyle

Prune trees do not grow true to type from seed. Therefore, propagators plant the seed or cutting of a rootstock and then bud or graft the desired cultivar to that rootstock. Most of the budding is done by nurseries in the spring on a rootstock that is only a few months old. The bud is forced to grow, and a complete tree composed of a rootstock of either plum or peach and the aboveground prune top is dug and sold to growers during the winter. Thus the tree sold by nurseries in California is generally 1 year old. Occasionally, a 2-year-old tree is produced with the rootstock being 2 years old: the bud was inserted in the summer and forced to grow only in the spring of the second year. Either type of tree has transplanted well and has produced excellent trees for growers.

Prune trees are grown on rootstocks that provide anchorage, disease and nematode resistance, nutrient and water uptake, and tolerance to adverse soil conditions. The desirable rootstock should help grow a tree quickly to fill the allotted space in the orchard, then stimulate early and continued production. Once a tree has attained the optimal size, it should grow enough new wood for production but not enough to cause shading and require excessive pruning. Rootstocks, cultivars, soils, and management can manipulate the tree for the desired growth and productivity.

Nursery Management

Site Selection

A tree nursery should be established on fertile, deep, well-drained sandy loam or loam soil. The soil should be free of nematodes and significant pathogens, such as the fungi that cause crown rot and root rot and the bacterium that causes crown gall. Neither the soil nor the available irrigation water should contain toxic levels of nutrients such as sodium, chloride, or boron.

Site Preparation

Nursery preparation begins the year before planting, when workers level soil and install drainage. They rip or chisel the soil to break up plowpans, hardpans, and other compacted layers, then cultivate it to the desired texture. In late spring or summer, the site is fumigated. This eliminates most nematodes, soil insects, perennial weeds, and soil fungi, and it kills most weed seeds.

For fumigation to be effective, the soil must be loose and have suitable moisture content. The degree of moisture that constitutes a suitable amount depends on the nursery owner's goal. If the goal is nematode control, fumigation is most effective when soil moisture is moderate. If the goal is weed control, fumigation is most effective when the soil, especially the top 6 inches, is near field capacity. In soil that is too dry, soil particles adsorb the liquid fumigant, reducing its effectiveness. In soil that is too wet, the moisture impairs movement of the fumigant. In addition, variation in the percentage of water within the profile can influence where and how the fumigant moves.

After fumigating, workers remove the tarp to allow the fumigant to dissipate. The next step is forming the planting rows or beds.

Fumigation can suppress the uptake of phosphorus and zinc in nursery trees. To ensure an adequate supply of these nutrients, many nurseries apply phosphorus- and zinc-rich fertilizers below the beds when the beds are formed. Some nurseries apply potassium and a small amount of nitrogen as well.

Irrigation

Seeds, cuttings, and trees should never be so dry that they stop growing. Most nurseries use sprinkler irrigation because it waters evenly. If the slope of the field is uniform enough to prevent water accumulation, a nursery owner may choose furrow or flood irrigation. Whatever the irrigation method, the irrigation plan must meet both expected and unexpected needs.

During the 1 or 2 years the trees are in the nursery, they should receive the best of care. Irrigation should be frequent. During the summer, weekly or even more frequent irrigation may be needed during hot weather. Water management is a major factor in maintaining maximum growth throughout the growing season.

Fertilizer

While trees are growing, fertilizer should be applied as needed, either as sidedressings or through sprinkler applications. Frequent applications of nitrogen fertilizer are necessary to encourage continuous rapid growth of the tree.

Cultural Practices

Weed and pest control is vital to a successful nursery operation. Weeds can rob the small trees of nutrients, light, and moisture. They can be controlled by cultivation, pre- and postplant herbicides, or a combination of these. Finally, insects and mites should be controlled as needed when these pests develop.

Rootstock Propagation

Commercial prune production in California is from varieties budded or grafted to rootstocks. The three methods used to grow rootstocks are propagation by hardwood cuttings, propagation by seed, and micropropagation.

Propagation by Hardwood Cuttings

A tree grown from a cutting has an advantage over one propagated from seed in being genetically identical to its source. Propagating from hardwood cuttings requires four steps.

1. Wood collection: From late October through November, workers collect 1-year-old wood, 0.2 to 1 inch in diameter, from vigorously growing source trees in a heavily pruned stock block.
2. Defoliation and cutting: Workers defoliate the wood and cut it into 12-inch lengths.
3. Dipping: To enhance rooting, the freshly cut basal end of the cuttings is dipped into indolebutyric acid (IBA), napthalene acetic acid (NAA), or a commercially prepared powder. One procedure involves a solution in which 500 to 1,500 parts per million (ppm) IBA is dissolved in 50% alcohol. For a quick dip, workers dip the basal end of each shoot in the solution for 10 to 30 seconds. An alternative treatment consists of soaking the cuttings for 24 hours in an aqueous solution of 100 ppm IBA. Rooting percentages and survival rates increase if the hormone dip is supplemented with fungicide. Cuttings treated with a fungicide are dipped in water first and then in the commercial powder.
4. Planting: Workers plant the shoots about 6 inches deep and 6 inches apart.

Propagation from cuttings is the standard method for growing Marianna 2624 and Myrobalan 29C rootstocks.

Propagation by Seed

During the summer, workers harvest mature fruit from seed trees. After the seed are extracted from the fruit, the seed is washed, dried, stored in a cool, dry location, and stratified. Stratification is the process of exposing seed to cool, moist conditions, causing the seed to break dormancy. Plum seed are stratified at 34° to 40°F for 1 to 2 months. (In comparison, peach seed requires 3 to 4 months for stratification.) Although seed can be stratified in vermiculite or peat moss in a cold room, most nurseries stratify seed directly in the field.

When field stratifying, workers soak seed in water for 24 hours and then plant them 3 inches deep in the nursery row in the fall. Row spacing is from 4 to 5 feet. Some nurseries plant seed 4 to 6 inches apart in a row; others plant seed closer and then thin the seedlings until they are 6 inches apart. In areas where there is a minimal amount of winter chilling or where there is considerable sunshine in the winter, nurseries mound soil to a depth of 6 inches over the seed in the fall. This practice allows the seed to be better insulated from sunlight and heat and receive better stratification. The extra soil is then scraped off to 3 inches over the seed in early spring before the seed sprouts.

Micropropagation

Micropropagation is propagating plant tissues, primarily shoot tips and meristems, by using in-vitro techniques. Because the process often involves the mass production of clonal material, it can supply large numbers of plants in a relatively short time, much faster than conventional propagation methods. Large numbers of new rootstocks can be introduced faster into the industry. In addition, micropropagation allows nurseries to produce and maintain virus- and disease-free clones.

- Workers use aseptic techniques to remove a small amount of plant tissue called an explant from the mother plant, surface-sterilize the explant, and place it in a nutrient medium. A shoot tip, consisting of about ¾ inch of apical shoot or a meristem consisting of the apical dome and two or three primordial leaves, is first established in culture.
- After the explants adjust to in-vitro conditions and increase in size (the meristem up to several hundred times its original size), workers transfer shoots to a multiplication medium where their numbers may increase as much as tenfold. The selection of mineral components and plant hormones determines the growth of the cultured material.
- Workers remove ½- to ¾-inch shoots individually from the proliferating mass, dip them in an IBA solution for root induction, then plant them either in planting medium in greenhouse flats or directly in the field.
- Shoots must be hardened so they can withstand greenhouse or field conditions. In the greenhouse, shoots are kept in a high relative humidity atmosphere with reduced light for up to 4 weeks. In the field, shoots are covered with light-reducing opaque material. To harden the tender plants, the light intensity is gradually increased by removal of some of the opaque material. Decreasing the humidity also helps harden the plants. In conventional micropropagation, cultured plants are the finished product.

Although micropropagation has been used to produce own-rooted trees of several fruit species, it has not been used with prune. European nurseries micropropagate plums and prunes extensively, however, and some California nurseries have recently developed procedures to mass-propagate prunes. As methods continue to develop, micropropagation will probably become more widespread.

Propagation of Complete Trees

Nurseries propagate complete trees through budding or topworking (grafting). These methods of vegetative propagation are the only way to maintain the specific qualities of a particular variety. Budding or grafting also imparts the advantages of the rootstock to the variety to which it is joined.

Specific rootstocks offer qualities related to anchorage, disease or nematode resistance, yield, nutrition uptake, and tolerance of adverse soil conditions. The rootstocks most commonly used with prune trees are Myrobalan plum produced from seed, Nemaguard and Lovell peach seedlings, and Marianna 2624 and Myrobalan 29C plum produced from hardwood cuttings. Recently, other selections of Marianna have also been introduced. Choosing the best plant material in any given case means selecting the rootstock best suited for the orchard site and the most productive scion variety that is compatible with the rootstock and local conditions.

Budding

Three budding techniques are in common use: June budding, spring budding, and fall budding. Because the growing season in California is long, California nurseries that use June budding or spring budding can produce a complete tree by the end of summer. Trees are not budded from mid-June to August because high temperatures can interfere with the healing of the scion-rootstock union and there is insufficient growing weather to produce a desirable tree by fall.

June budding

Most California nurseries use June budding, in which scion buds are inserted into the rootstock starting in mid-May. T-budding (also called shield budding because the bud and the adjoining stem resemble a shield when cut) is the preferred technique used in June budding. For T-budding to be successful, the bark of the rootstock must be slipping (easily removed from the subtending wood).

The process of T-budding begins when workers collect bud sticks at the time of budding and remove leaves from the current season's growth. They cut the bud from the bud stick, being careful to remove the wood within the bud (fig. 10.1). Workers then cut a T-shaped slit into the bark of the rootstock (fig. 10.2). The slit should be a few inches above the soil line. The worker slides a bud into the cut (fig. 10.3) and wraps the union with a budding rubber (fig. 10.4) to hold the bud tightly in place.

About 4 days after insertion of the bud, workers cut off rootstock shoots, leaving 2 to 5 inches above the inserted bud. From one to several leaves are allowed to remain above and below the bud.

Figure 10.1 Cutting the wood from a bud stick.
Photo: W. O. Reil.

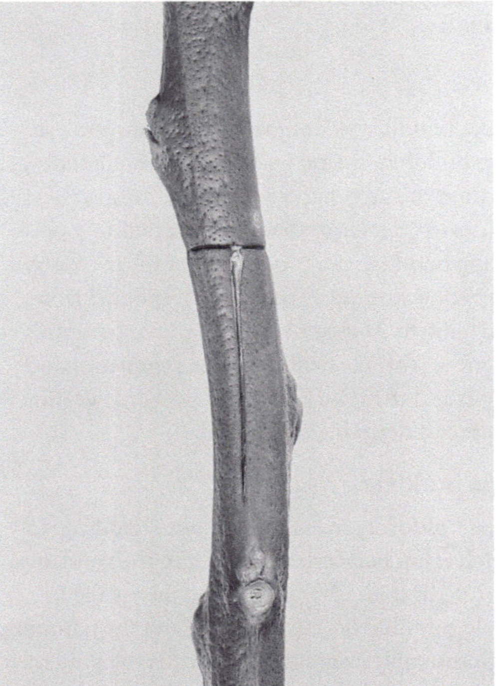

Figure 10.2 A T-shaped slit in the bark.
Photo: W. O. Reil.

These leaves provide the photosynthates needed for growth and offer shade to protect the union from direct sun. From 2 to 2½ weeks after budding, shoots are cut back farther, to just above the bud. This will force the bud, causing it to grow. When the bud grows out to a shoot 10 to 18 inches long, all other shoots and leaves are removed from the rootstock.

Spring budding

The process of spring budding begins in winter, when workers collect dormant wood. The wood is stored at 36° to 38°F until needed in mid-April to mid-May. T-budding is usually used for spring budding. In contrast to buds used in June budding, buds used in spring budding retain some wood under the bud. All shoots are removed

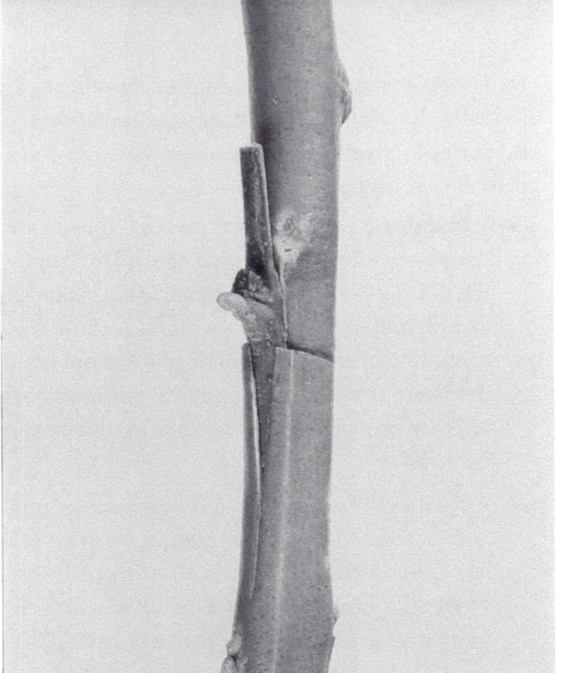

Figure 10.3 A bud partly inserted into the rootstock.
Photo: W. O. Reil.

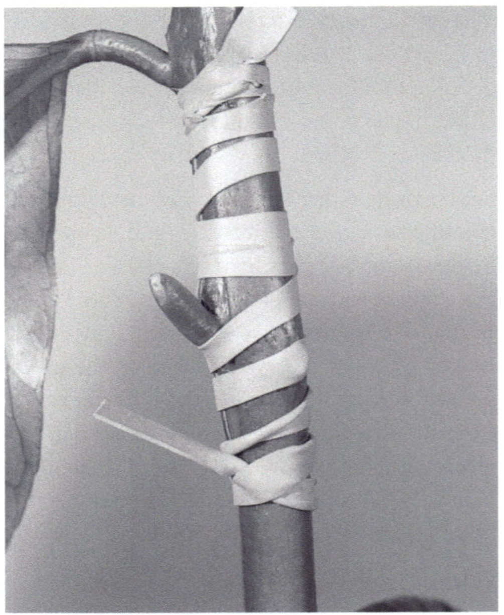

Figure 10.4 A budding rubber holds the bud in place.
Photo: W. O. Reil

from the rootstock as soon as the bud is inserted. Spring budding is usually the method of choice for Marianna 2624 and Myrobalan 29C rootstocks, since, unlike seedlings, these rooted cuttings are sufficiently large by mid-April.

The rootstock shoots are cut back after 4 days, the same as with trees that were budded in June. Care for both the spring- and June-budded trees is the same for the rest of the year. Both June-budded and spring-budded trees are only 1 year old when sold.

Fall budding

A few nurseries insert buds into rootstock in August.

The union heals during the fall, but the remaining shoot (above the bud) keeps the bud from growing. In spring, just prior to growth, a worker cuts off the shoot and the bud begins to develop. When the tree is sold at the end of the season, the rootstock is 2 years old and the scion is 1 year old.

Topworking

Trees are topworked (grafted) to change to a more suitable scion variety. Topworking involves several steps: collecting and possibly storing the scion wood; using one of several methods to graft the scion to the rootstock; and performing appropriate follow-up procedures.

In older orchards, topworking an existing tree offers an advantage over planting a new tree: grafted trees are easier to manage and care for than are newly planted ones. In addition, topworked trees can compete effectively with established trees. The disadvantage of topworking is that the grafting process can introduce viruses from an infected rootstock or scion. The best time to graft trees is when they are less than 6 to 8 years old. Cuts on trees older than 8 years often fail to heal adequately.

Wood collection and storage

Scion wood to be grafted is collected during winter, when buds are dormant. It may be grafted immediately or stored. To store the wood, workers place it in plastic bags and refrigerate it at temperatures slightly above freezing. The wood must not dry out during storage, and the temperatures must be cold enough to keep buds from growing.

Grafting methods

Nursery personnel can choose from a number of grafting methods: whip grafting, wedge grafting (also known as saw kerf grafting), cleft grafting, and bark grafting. Consult Hartmann and Beutel (1981) to learn the specific techniques for each method.

Growers usually use whip grafting or T-budding on young trees. Wedge grafting and cleft grafting are used during the dormant season because they do not require that the bark be slipping. The best time to wedge- or cleft-graft a tree is in early spring, after the buds have swelled but before they have started to grow. Bark grafting is done when the bark slips in spring, usually after bloom.

In wedge, cleft, or bark grafting, the main branches of the tree are cut 6 to 12 inches above the crotch. Each cut is at a right angle to the limb. If these grafting methods are used to topwork a tree that is 3 years or older, a nurse limb is usually kept on the tree. Younger trees do not need a nurse limb.

To bark-graft, the union end of the scion is cut at a sharp angle, with a small cut made on the back side to create a chisel-like tip. On branches that are not vertical, a worker grafts the scion to the top part of the stub, not the side, to ensure support of the weight of the scion and to prevent bark stripping. A vertical cut is made through the bark of the rootstock and the scion is inserted between the bark and wood. If the scion is large, two vertical cuts in the bark of the rootstock just wide enough for the scion to fit between will be needed.

The scion is nailed into place using small uncoated nails or brads. The nails are inserted through the rootstock bark, the scion, and into the rootstock wood. The worker inserts at least two scions in each stub. The graft is then covered with grafting wax to prevent drying and painted with white interior-grade latex paint to protect the graft from sunburn.

Follow-up procedures

After scions are grafted, workers may have to rewax the union, remove suckers, provide support to protect the tree from wind, and prune branches in summer to begin training the limbs. Many growers paint exposed limbs and trunks with white interior latex paint to protect them from sunburn and invasion by flatheaded borers. Larvae of these borers often try to penetrate the bark at the base of the scion, where it may take some time for the graft union to heal over completely. When the grafted limbs are well established, the nurse limbs are removed completely; this usually occurs in late summer to winter the first year after grafting. At the end of the first season's growth, workers remove surplus scions by cutting them back severely and pruning to establish scaffolds.

Removal and Storage of Nursery Trees

Before being dug from the soil, nursery trees must be dormant. To ensure timely dormancy in the fall, nurseries decrease watering in late summer and early fall. Trees should not be removed until most of the leaves have fallen. In November, many nurseries defoliate the trees by spraying them with zinc sulfate. Nursery trees are usually ready for removal in late December; all trees must be removed by late January, when root growth starts.

Removal involves using a curved cutting bar to undercut the root system. The bar is pulled

through the soil at a depth of 14 to 16 inches. An extension on the back of the cutting bar lifts the roots and soil, loosening the soil from the roots. The tree is dropped back into the hole and then pulled from the ground by hand. Each tree is graded according to uniformity and quality, then all trees are sorted by size. Size is determined by the diameter, or caliper, of the stem 2 inches above the bud union. Workers bundle the trees into groups of 5 large trees or 10 small ones. The trees are heeled in using sand or wood shavings until needed. Roots must be kept damp by maintaining the moisture of the sand or shavings.

Prune trees to be planted after late February usually require cold storage while dormant. The temperature of the storage unit should be 32° to 36°F. Trees stored at these temperatures remain dormant well into summer. The roots of stored trees need humidity; workers should wet them down if necessary. Some nurseries store trees in bins with damp wood shavings around the roots. The shavings should be fresh and free of pathogens and resins. Other nurseries pile bare-root trees on pallets and periodically mist the roots to prevent drying.

While trees are in cold storage, nurseries make sure not to expose them to ethylene gas or store them with fruits that produce ethylene, such as apples and pears. Extended exposure to an ethylene concentration as low as 0.5 part per milliom can retard tree growth. For the same reason, trees are not stored in environments that contain fumes produced by gasoline- or propane-powered vehicles.

References

Hartmann, H. T., and J. A. Beutel. 1981. Propagation of temperate zone fruit plants. Oakland: University of California Division of Agriculture and Natural Resources Publication 21103.

Hartmann, H. T., D. E. Kester, and F. T. Davis Jr. 1990. Plant propagation: Principles and practices. 5th ed. Englewood Cliffs, NJ: Prentice Hall.

LaRue, J. H., and R. S. Johnson, eds. 1989. Peaches, plums, and nectarines: Growing and handling for fresh market. Oakland: University of California Division of Agriculture and Natural Resources Publication 3331.

Micke, W. C., ed. 1996. Almond production manual. Oakland: University of California Division of Agriculture and Natural Resources Publication 3364.

Ramos, D. E., ed. 1998. Walnut production manual. Oakland: University of California Division of Agriculture and Natural Resources Publication 3373.

11 Irrigation Systems

• Lawrence J. Schwankl and Terry L. Prichard

An orchard irrigation system should be able to meet tree water demands efficiently. Most of the applied water should go to the beneficial use of meeting tree water demands, while little water should be lost to deep percolation (leaching below the crop root zone) or to unused tailwater runoff.

Irrigation systems commonly used in prune orchards include gravity surface systems and pressurized systems. Surface systems include furrow and border check systems, in which water is applied to the soil surface at the high end, or head, of the field and allowed to move downslope to the tail of the field, infiltrating along the wetted portion of the field. Generally these systems do not apply the irrigation evenly and have significant amounts of irrigation runoff. Surface irrigation systems require a high degree of management to be operated efficiently. Pressurized systems—sprinkler irrigation and microirrigation—allow the operator to manage the quantity and timing of irrigation applications more carefully.

Many irrigation systems are suitable for use in orchards. There is no single best system or method, since land, water, and energy and labor costs—in addition to soil and plant conditions—vary at each site. Irrigation systems are designed and operated to favor optimal crop yield and quality and efficient use of water and energy, but other considerations may dominate the decision of which system is best suited to a given orchard. Total cost—including initial installation, operation, and maintenance—substantially influences the choice. Other factors to consider include soil physical and chemical characteristics, soil type and texture, slope of the land, frost hazard, and the cost and availability of water. Table 11.1 provides general guidelines

Table 11.1. Factors to consider in selecting an irrigation system

	Sprinkler system solid set	Surface flood systems			Microirrigation systems	
		Graded border	Level border	Furrow	Drip	Microsprinkler
Maximum slope (%)						
irrigation direction	none	0.2–2.0	0.0	1.0–2.0	none	none
cross-slope	none	0.2	0.2	6.0	none	none
Soil						
water intake (in/hr)						
minimum	0.05	0.30	0.10	0.10	0.02	0.02
maximum	3.00	2.00	2.00	3.00	none	none
erosion hazard	slight	moderate	slight	severe	none	slight
saline-alkali hazard	slight	moderate	slight	severe	moderate	moderate
Water						
(TDS)*	severe	slight	slight	moderate	slight	slight
suspended solids	moderate	none	none	none	severe	severe
rate of flow	low	moderate	moderate	moderate	low	low
dissolved iron (1 ppm or greater)	none	none	none	none	moderate to severe	moderate
Climate						
temperature controlled	yes	yes	yes	yes	no	no
wind affected	yes	no	no	no	no	yes
Irrigation efficiency (%)†	75–85	65–80	70–80	65–80	80–95	80–95

Notes:
*Total dissolved solids.
†Consumptive use ÷ applied water = irrigation efficiency, assuming good to excellent management and design. High efficiency in surface irrigation systems requires a tailwater return system.

for integrating these factors when choosing an irrigation system. The sections in this chapter provide general information on each type of irrigation system; for more information, consult the references at the end of the chapter.

Microirrigation Systems

Microirrigation systems (surface drip, subsurface drip, and microsprinklers) are increasingly being used in orchards. These systems wet only a portion of the orchard floor. Generally, it is recommended that 40 to 60% of the orchard floor be wetted for good tree performance. Although wetted soil volume is often a better indicator of adequate wetting by a microirrigation system, it is difficult to determine wetted soil volume without extensive soil moisture monitoring or excavating a backhoe pit.

Advantages of Microirrigation Systems

- A high degree of water application uniformity, often the highest degree of all irrigation systems available. This means less applied water is required to meet tree evapotranspiration demand (ET_c) than with most surface irrigation systems.
- Excellent control of the amount and timing of irrigation. Small, frequent irrigations can be applied to match the trees' water needs. Runoff is minimized because of the low application rates, and deep percolation losses can also be minimized if the correct amount of water is applied. Frequent irrigation provides an excellent balance between aeration and adequate soil moisture for optimal tree performance.
- Easy irrigation of irregular terrain.
- Weed growth is minimized, since only a portion of the orchard floor is wetted.
- Well suited to soils that have poor infiltration.
- Chemigation (injection of fertilizers and other chemicals) is easily accomplished using microirrigation systems. Chemigation is often simpler, more efficient, and safer than other methods of chemical application.
- Can provide limited frost protection.

Disadvantages of Microirrigation Systems

- High initial cost.
- Excellent management is needed to maintain the system, since clogging of the emitters by physical particles, organic materials, or chemical precipitates may occur.
- Irrigation water must be pressurized, which increases energy costs. The required pressures are generally less than those needed for sprinkler systems, but they are higher than those needed for surface irrigation.
- System must be run more frequently than surface or sprinkler irrigation systems.
- Cover crops cannot be grown year-round due to the localized nature of the water applications.
- System can be damaged by insects and animals.
- Systems with low application rates requiring long set times have been shown to be associated with increased incidence of brown rot in prune orchards.
- Provide little to no frost protection.

Components and Considerations

The various types of microirrigation systems—microsprinkler, surface drip, and subsurface drip systems—are made up of the same basic components. Figure 11.1 shows the components of a typical microirrigation system. Generally, a microirrigation system consists of

- pump
- flow meter
- mainlines and submains
- drip or microsprinkler lateral lines
- valves
- filter(s)
- injection equipment

For the most part, microirrigation systems differ only in the emitter spacing, the type of emission

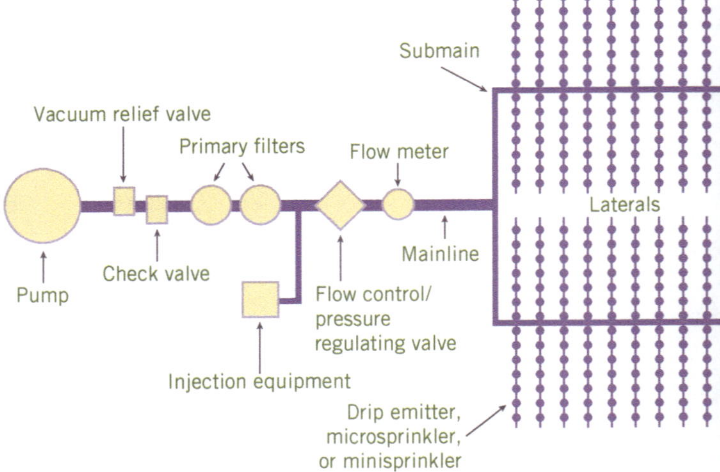

Figure 11.1 Components of a microirrigation system.

device used, and the size of the components. The type of emitter used affects the size of the other components. For instance, microsprinklers, due to their higher application rates, generally require larger filters, mainlines, and submains than do drip systems.

In a microirrigation system for trees, the emitter spacing and discharge rate depend primarily on the tree spacing and water needs of the trees. The emission devices must be capable of supplying each tree with enough water during the peak water use periods to satisfy the evapotranspiration (ET) requirement. The peak prune ET for an average year in the Tehama County area is 4.25 inches in a 15-day period, or 0.28 inches per day (see chapter 12, "Irrigation Scheduling and Tree Stress"). Frequent irrigation with microsprinklers can cause extended high humidity, which increases the incidence of brown rot in prunes. To combat this threat, higher-output microsprinklers are used, which allow a longer dry period between irrigations.

Following is a brief overview of the components and operation of a microirrigation system for trees.

Pumping plant

The pump and motor (or engine) selected must deliver the correct pressure and flow rate at the highest possible efficiency. The microirrigation system designer determines the flow rate and pressure to be delivered by the pump, and the pump supplier uses this information to select the most efficient pump for a given system.

Flow meters

A flow meter should be part of the system. Knowing the flow rate is necessary for determining the amount of water being applied, which, in turn, is critical to efficient irrigation and scheduling. A propeller meter, which displays either the flow rate and/or the total water applied, gives an accurate measurement.

Injection equipment

Microirrigation is well suited to injecting chemicals such as soluble fertilizer. Various types of injection equipment, differential pressure tanks (batch tanks), venturi devices, electrical or water-driven pumps, and solutionizer machines can also be used, depending on the chemical applied, the accuracy level needed, and the injection rate required.

Valves

Valves are the control mechanisms of microirrigation systems. Several types are common: control valves, air or vacuum relief valves (which allow air to escape when the system is turned on and to enter when the system is shut down), and check valves (which prevent undesirable flow reversal). Pressure-regulating valves help maintain a constant operating pressure in the system.

Filters

Selecting the appropriate filter requires consideration of water quality factors. Particulate matter (such as sand) in the water can be removed with vortex filters (frequently referred to as sand separators). Screen, disk, or sand media filters are also effective in removing particulate matter.

Organic matter such as algae or slime can be removed using screen, disk, or sand media filters. Since organic matter can quickly clog a screen or disk filter and is difficult to flush from the screen or disk elements, sand media filters are the usual choice for filtering surface waters containing algae and slime.

Screen and media filters must be periodically backwashed. A pressure drop across the filter indicates that backwashing is required. Backwashing can be accomplished either manually or automatically, with automatic backwashing taking place on a defined schedule or when the filter senses a predetermined pressure drop. The water used to backwash is frequently discharged out of the system.

Mainlines and submains

Main and submain pipes, usually made of PVC, deliver water to the lateral lines and emitters. The mains and submains must be sized carefully, with the cost of the pipe balanced against pressure losses caused by elevation differences across the orchard and friction as water moves through the pipe. A qualified microirrigation system designer should design the system.

Lateral lines

Emitters are attached to tubing, or lateral lines, usually made of polyethylene. The length and diameter of the lateral lines to be used also depend on economics, balancing the tubing cost against pressure loss along the lateral. If the lateral lines are too long or the wrong diameter, the emitters may discharge water at different rates, causing nonuniform irrigation.

Emitters

The many different types of microirrigation emitters available can be grouped generally into above- or belowground drippers and microsprinklers. Choosing which emitter to use depends not only on physical

issues such as soil type and water quality but also on the personal preferences of the manager.

Surface drip. Surface drip irrigation, along with microsprinkler irrigation, is one of the most commonly used microirrigation systems in tree crops. The drip emitters can either be punched-in to the drip tubing or formed integrally in the drip hose (in-line) at a specific spacing by the manufacturer. Both single and double lines per tree row are used. Double lines are usually chosen to achieve a greater wetted area in the orchard or to increase the application rate of the drip system. Increasing the wetted area is frequently an issue in orchards with sandy soils that do not move water laterally ("sub") well. Increasing the application rate of a drip system requires fewer operating hours to satisfy tree water needs. This gives the manager more time when the drip system is not operating to accommodate other orchard cultural practices such as spraying or mowing. A higher application rate may also allow the manager to more easily use off-peak power rates. Double-line drip systems are initially more expensive than single-line drip systems.

Subsurface drip. Orchard subsurface drip systems are most often installed as in-line drippers in hard drip tubing, due to their longevity. Drip tape products are seldom used in orchard applications. Both single-line and double-line subsurface drip systems have been used successfully. Subsurface drip systems are usually installed to a depth of 10 to 24 inches. Deeper depths are chosen to minimize surface wetting and rodent damage, but they make installation and repairs more difficult. Advantages of subsurface drip systems include reduced weed growth since the soil surface usually stays dry, reduced irrigation system damage during harvest or other cultural practices, and the ability to irrigate almost anytime, even right up to harvest. The major disadvantages of subsurface drip are the difficulty in detecting clogging, rodent damage, the hazard of root intrusion into the emitters, and crimping of the buried hose as primary roots expand. Root intrusion may be mitigated by using drip products with herbicide-impregnated emitters, but these products are more expensive.

Microsprinklers. The advantages of microsprinklers over drip systems include a larger wetted area, often a higher application rate, less susceptibility to particulate clogging since the flow path is larger, and easier visual inspection for clogging. While wetting a larger area may be a benefit for tree growth, it is a disadvantage due to increased weed growth. Another disadvantage of microsprinklers is that insects can enter or lay eggs in the microsprinkler orifice, causing them to clog. To prevent this, some microsprinklers have a "bug cap" that closes the orifice when the microsprinkler is not operating.

Operation and Maintenance

Microirrigation systems can apply irrigation water quite efficiently only if they are properly operated and maintained. Irrigation scheduling—determining when to irrigate and how much water to apply—is critical to operating the system efficiently. Effective irrigation scheduling requires knowing how much water the tree is using or has used since the last irrigation (the evapotranspiration, or ET, of the tree) and the application rate (how much water the irrigation system applies in a given period of time). Because determining the application rate of a microirrigation system can be confusing at times, a procedure is presented at the end of this section. Irrigation scheduling and the application rate determine how long the microirrigation system should be run.

A virtue of a microirrigation system is its ability to deliver a uniform amount of water to each location it serves so that water is applied evenly over the orchard. But, because of pressure differences throughout the system and variability in emitter manufacture, even new systems may not apply water evenly. A carefully designed system can use pressure regulators in mains, submains, laterals, or hoses (or pressure-compensating emitters) to overcome pressure differences and variation in emitter discharge rates.

While a well-designed system can deliver water with a high degree of uniformity, the system must be properly maintained to keep the application uniform. A principal cause of nonuniformity in microirrigation systems is emitter clogging by particulate or organic matter, lime precipitates, or iron precipitates.

Particulate (silt and sand particles) clogging can occur with both groundwater and surface water sources, but it can often be prevented by good filtration. Microirrigation emitter manufacturers frequently specify a filtration level for their emitters.

Biological clogging is caused by bacterial slimes, algae, and other materials that are either present in the irrigation water or growing in the irrigation system. Biological clogging is more frequent with surface water supplies than with groundwater, but groundwater can have biological clogging associated with iron bacterial slimes. Maintenance practices to minimize biological clogging include injecting a biocide such as chlorine and good filtration. Serious biological clogging may require continuous injection of 2 to 5 parts per million (ppm) chlorine, measured at the end of the lateral line. Less severe clogging may require only periodic chlorine treatments at 10 to 20 ppm every few weeks.

Chemical precipitate clogging can often be anticipated by an irrigation water analysis. For example, irrigation water with a pH of 7.5 or higher and a bicarbonate level of 2 milliequivalents per liter (meq/l) or greater has an increased hazard for calcium carbonate precipitate clogging. Waters with iron levels of 0.5 to 1.0 ppm or higher are at risk of iron precipitate clogging. Calcium carbonate precipitate clogging can be prevented or mitigated by lowering the water pH to 7.0 or less by injecting acid. Iron precipitate clogging is usually handled by storing the water in a reservoir in which the iron precipitate is allowed to settle prior to use (see Schwankl et al. 1998 and 2008).

Microirrigation systems should be flushed regularly until mainlines, submains, and lateral lines run clear. Fine particles passing the filters and other contaminants tend to accumulate at the end of the lateral lines, and flushing removes them before they clog emission devices. Begin flushing on a 3- to 4-week interval. Adjust the flushing frequency depending on how dirty the system is: if flushing a lateral line clean takes more than a minute or two, consider flushing more frequently. If flushing a lateral line happens quickly (less than a minute), the system can probably go longer between flushing events.

Distribution Uniformity

How uniformly or evenly water is applied is also important to efficient irrigation water management. In an irrigation system that applies water uniformly, all trees receive nearly the same amount of water during an irrigation event. Uniformity can be measured as distribution uniformity (DU), or it can be estimated if measurements have not been made. For microirrigation systems, distribution uniformity is often referred to as emission uniformity. A well-designed and maintained microirrigation system typically has a DU of 85 to 90%. Nonuniform water distribution occurs from three main causes: emitter manufacturing variability, irrigation system pressure changes due to friction losses and elevation changes, and emitter clogging.

A field evaluation of a microirrigation system, in which a sampling of emitter discharges is measured, is the preferred method of determining the distribution uniformity. In some areas, mobile irrigation evaluation teams are available to do field evaluations; in most locales, consultants can be found who will do such evaluations. If a field evaluation has not been done, estimate the orchard's distribution uniformity when determining the irrigation schedule.

Application Rate

The application rate of a microirrigation system can be determined by calculating the operating time based on the tree water use (ET) and the application rate of the emission devices. This method helps avoid confusion because information on irrigation scheduling and tree water use is usually measured in inches per day (in/day), while discharge from microirrigation emitters is measured in gallons per hour (gph). An alternative to calculating the operating time is to calculate the depth of water applied by monitoring a flow meter at the pump.

Calculating the operating time

- **Step 1:** Convert the total tree water use (ET) value from inches per day to gallons per day using the following formula (or see table 11.2) :

$$\text{Tree water use (gal/day)} = \text{Tree spacing (ft}^2\text{)} \times \text{Tree water use (in/day)} \times 0.623$$

Table 11.2. Prune water use (gal/day) for various tree spacings and tree water use (in/day)

Tree spacing (ft²)*	Tree water use (in/day)						
	0.05	0.1	0.15	0.2	0.25	0.3	0.35
100	3.1	6.2	9.4	12.5	15.6	18.7	21.8
150	4.7	9.4	14.0	18.7	23.4	28.1	32.7
200	6.2	12.5	18.7	24.9	31.2	37.4	43.6
250	7.8	15.6	23.4	31.2	39.0	46.8	54.5
300	9.4	18.7	28.1	37.4	46.8	56.1	65.5
350	10.9	21.8	32.7	43.6	54.5	65.5	76.4
400	12.5	24.9	37.4	49.9	62.3	74.8	87.3
450	14.0	28.1	42.1	56.1	70.1	84.2	98.2
500	15.6	31.2	46.8	62.3	77.9	93.5	109.1

Note: *Tree spacing (ft²) = row spacing (ft) × tree spacing in the row (ft).

For example, if tree spacing is 20 feet by 20 feet and tree water use is 0.25 inches per day:

Tree water use (gal/day) = (20 ft × 20 ft) × 0.25 (in/day) × 0.623 = 62.3 (or 63) gal/day

- **Step 2:** Determine the application rate per tree of the irrigation system in gallons per hour (gal/hr). For both drip emitters and microsprinklers, this requires determining the number of emission devices per tree and the discharge rate per emission device (gal/hr/emitter):

  Application rate (gal/hr) = Number of emission devices per tree × Discharge rate per device (gal/hr)

  Drip emitters: For example, if there are 5 drip emitters per tree and the discharge rate per emitter is 1 gal/hr:

  Application rate per tree (gal/hr) = 5 (emitters/tree) × 1 (gal/hr) = 5 gal/hr

  Microsprinklers: For example, if there is 1 microsprinkler per tree and its discharge rate is 12 gal/hr:

  Application rate per tree (gal/hr) = 1 (emitters/tree) × 12 (gal/hr) = 12 gal/hr

- **Step 3:** Determine the irrigation system operation time in hours per day. This requires using the tree water use determined in Step 1 and the application rate determined in Step 2 in the following formula (or see table 11.3):

  Net hours of operation per day = Tree water use (gal/day) ÷ Application rate (gal/hr)

  Drip emitters:

  Net hours of operation per day = 63 (gal/day) ÷ 5 (gal/hr) = 12.6 hr/day

  Microsprinklers:

  Net hours of operation per day = 63 (gal/day) ÷ 12 (gal/hr) = 5.3 hr/day

  Note that table 11.3 indicates these hours of operation for these examples.

- **Step 4:** Account for the microirrigation system uniformity, quantified as distribution uniformity or emission uniformity (same numerical value). Irrigation nonuniformity means that some areas of the orchard receive less water than others. To ensure that most of the orchard is adequately irrigated, additional water beyond that calculated in Step 1 must be applied. Therefore, the operation times calculated in Step 3 must be made longer to apply more water. This longer irrigation system operation time can be calculated as follows:

  Gross hours of operation per day = (Net hours of operation per day ÷ Distribution uniformity %) × 100

  Drip emitters: For example, the net hours of operation per day is 12.6 hr/day (from Step 3); the distribution uniformity can be assumed to be 85% in a well-designed, well-maintained system:

  Gross hours of operation per day = (12.6 hr/day ÷ 85) × 100 = 14.8 hr/day

  Microsprinklers: For example, the net hours of operation per day is 5.3 hr/day (from Step 3), and the distribution uniformity can be assumed to be 85% in a well-designed, well-maintained system:

  Gross hours of operation per day = (5.3 hr/day ÷ 85) × 100 = 6.2 hr/day

Calculating the depth of water applied

An alternative method of determining the application rate is to calculate the depth of water applied (in inches) by monitoring a flow meter (if available) at the pump. The following formula can be used:

Applied depth of water (in) = [Pump discharge (gpm) × Irrigation time per set (hr)] ÷ [449 × Acres irrigated per set]

For flow meters without an instantaneous readout in gallons per minute, use the flow meter's totalizing

Table 11.3. Hours of operation per day for various application rates (gal/hr) and tree water use (gal/day)

	Application rate (gal/hr)										
Tree water use (gal/day)	1	2	4	6	8	10	12	14	16	18	20
5	5.0	2.5	1.3								
10	10.0	5.0	2.5	1.7	1.3	1.0					
15	15.0	7.5	3.8	2.5	1.9	1.5	1.3	1.1			
25		12.5	6.3	4.2	3.1	2.5	2.1	1.8	1.6	1.4	1.3
50			12.5	8.3	6.3	5.0	4.2	3.6	3.1	2.8	2.5
75			18.8	12.5	9.4	7.5	6.3	5.4	4.7	4.2	3.8
100				16.7	12.5	10.0	8.3	7.1	6.3	5.6	5.0
125				20.8	15.6	12.5	10.4	8.9	7.8	6.9	6.3
150					18.8	15.0	12.5	10.7	9.4	8.3	7.5

readout. Keep track of the total flow change during a recorded period of time to determine the average flow rate in gallons per minute. This method can often be more accurate than using an instantaneous readout.

Frequency of Irrigation

Determining how often to irrigate with a microirrigation system must include consideration of microirrigation system capacity as well as soil and tree parameters. During peak ET periods, most drip irrigation systems are operated daily to meet ET demands. Microsprinkler systems, which generally have a higher application rate than do drip systems, are usually operated with multiple days between irrigations: 3 or 4 days between irrigations is common for microsprinkler systems during peak ET periods. The operation time of a microsprinkler irrigation system should be more than just a few hours. Short microsprinkler irrigation events wet only the soil surface, rather than the deeper root zone. Particularly in young orchards, this shallow water may be lost to evaporation and not be available for tree uptake. Longer irrigation events allow for deeper penetration of water. Soil moisture monitoring is helpful both before and after irrigation.

Sprinkler Irrigation Systems

Sprinkler irrigation can be an excellent method of irrigating tree crops. Solid-set sprinklers and hand-move sprinklers are the most common types of sprinkler systems used in prunes. When properly designed, sprinkler systems can apply water uniformly and can be efficient when properly managed.

Advantages of Sprinkler Systems

- They can apply water uniformly when well designed, ensuring that all trees receive nearly the same amount of water.
- They can be easily operated, lending themselves well to automation if desired.
- They are full-coverage irrigation systems. This takes advantage of the full orchard soil volume to store soil moisture and allows a cover crop to be grown throughout the season.
- They are the best of all irrigation systems in providing frost protection.
- They require less maintenance than do microirrigation systems.

Disadvantages of Sprinkler Systems

- Wetting the entire orchard floor requires more extensive weed control than with microirrigation systems.
- During orchard establishment, when trees are not at full canopy, irrigation efficiency may be reduced due to evaporation from bare soil surfaces.
- Hand-move sprinkler systems require significant labor.
- High-volume sprinkler irrigation systems have been shown to be associated with increased incidence of brown rot in prune orchards.
- Permanent-set sprinkler systems can be the most expensive of all irrigation systems used in orchards.

Operation of Sprinkler Irrigation Systems

To be efficient, sprinkler irrigation systems must be operated on the correct irrigation schedule. This means that the correct amount of water must be applied when the tree needs it. To accomplish this, information on tree water needs (see chapter 12, "Irrigation Scheduling and Tree Stress") and on the sprinkler application rate is required. Tree water needs, often referred to as tree evapotranspiration (ET), are most often provided in inches of water use per day (in/day) and are available from a variety of sources. The sprinkler application rate is determined in inches of applied water per hour (in/hr). Because tree ET and the sprinkler application rate are in similar units, it is easy to determine the hours of application required to replace the soil water used by the trees since the last irrigation.

Application rate

The sprinkler system application rate, usually measured in inches per hour (in/hr), is determined from the sprinkler nozzle size, the system's operating pressure, and the spacing of the sprinklers. The discharge rate (gallons per minute, gpm) of a sprinkler head is determined by the sprinkler nozzle size and the operating pressure at the sprinkler.

Discharge rate. Table 11.4 gives the sprinkler discharge for various nozzle sizes and operating pressures. The nozzle size is frequently engraved on the side of the nozzle, or it can be determined by testing the opening with drill bits of known size. The operating pressure can be determined by plac-

Table 11.4. Sprinkler discharge rate (gpm) for various nozzle sizes (in) and pressures (psi)

	Nozzle size (in)										
psi	3/32	7/64	1/8	9/64	5/32	11/64	3/16	13/64	7/32	15/64	1/4
20	1.17	1.60	2.09	2.65	3.26	3.92	4.69	5.51	6.37	7.32	8.34
25	1.31	1.78	2.34	2.96	3.64	4.38	5.25	6.16	7.13	8.19	9.32
30	1.44	1.95	2.56	3.26	4.01	4.83	5.75	6.80	7.86	8.97	10.21
35	1.55	2.11	2.77	3.50	4.31	5.18	6.21	7.30	8.43	9.69	11.03
40	1.66	2.26	2.96	3.74	4.61	5.54	6.64	7.80	9.02	10.35	11.79
45	1.76	2.39	3.13	3.99	4.91	5.91	7.03	8.30	9.60	10.99	12.50
50	1.85	2.52	3.30	4.18	5.15	6.19	7.41	8.71	10.10	11.58	13.18
55	1.94	2.64	3.46	4.37	5.39	6.48	7.77	9.12	10.50	12.15	13.82
60	2.03	2.76	3.62	4.50	5.65	6.80	8.12	9.56	11.05	12.68	14.44
65	2.11	2.88	3.77	4.76	5.87	7.06	8.45	9.92	11.45	13.21	15.03
70	2.19	2.99	3.91	4.96	6.10	7.34	8.78	10.32	11.95	13.70	15.59
75	2.27	3.09	4.05	5.12	6.30	7.58	9.08	10.66	12.32	14.19	16.14

ing a pitot tube (available at most irrigation supply stores) screwed to a pressure gauge into the water stream just outside the nozzle opening.

The sprinkler discharge rate can also be determined by using a short length (approximately 4 feet) of garden hose, a 5-gallon bucket, and a stopwatch. By placing the hose over the nozzle and directing the water into the bucket, then timing how long it takes water to fill the bucket, the discharge rate from the sprinkler can be determined.

Discharge rate (gpm) = Volume (gal) ÷ [Time to fill (sec) ÷ 60]

For example, if it takes 100 seconds to fill the 5-gallon bucket:

Discharge rate = 5 ÷ [100 ÷ 60] = 5 ÷ 1.7 = 2.9 gpm

Measurements should be taken at various locations within the orchard, whether the sprinkler operating pressure and table 11.4 or the bucket and stopwatch method are used. As pressure varies within the orchard, so will the sprinkler discharge. It will be a management decision whether to select irrigation set times to adequately irrigate the orchard section(s) receiving the least water or to irrigate based on the orchard's average sprinkler discharge rate.

Calculating the application rate. Calculating the application rate (in/hr) can be done once the discharge rate of the sprinkler (gpm) and the sprinkler spacing are known:

Application rate (in/hr) = [96.3 × Discharge rate (gpm)] ÷ [Sprinkler spacing along row (ft) × Sprinkler spacing across row (ft)]

For example, if the sprinkler discharge rate is 1.0 gpm and the sprinkler spacing is 48 feet along the tree row and 20 feet across the tree row:

Application rate (in/hr) = [96.3 × 1 (gpm)] ÷ [48 (ft) × 20 (ft)]
Application rate = 96.3 ÷ 960
Application rate = 0.1 in/hr

Table 11.5 can also be used to determine the sprinkler system application rate. The sprinkler area in square feet (ft^2) is determined by multiplying the sprinkler spacing along the row by the sprinkler spacing between rows: in the previous example, the sprinkler area would be 960 square feet. Looking in table 11.5 under 960 square feet and a sprinkler discharge rate of 1.0 gallons per minute gives the same sprinkler application rate of 0.1 in/hr.

Design considerations. When designing a sprinkler irrigation system, it is often difficult to choose an application rate. A major objective is to minimize runoff from the orchard. Ideally, water should soak in near where it lands. The infiltration rate of the soil is not easy to determine since it changes during an irrigation and may change across the season. During an irrigation event, the infiltration rate initially starts out at the highest rate and decreases during the irrigation until a relatively constant, final infiltration rate is reached (fig. 11.2). This final infiltration rate should be equal to or greater than the sprinkler application rate so that no runoff occurs. Often the best guidance for choosing a sprinkler application rate can be made by seeing what has worked on other orchards in the area with similar soil conditions, slope, management, and so on.

It is preferable to choose the application rate correctly at the design stage, but nozzle sizes can be retrofitted if needed. It is better to overestimate

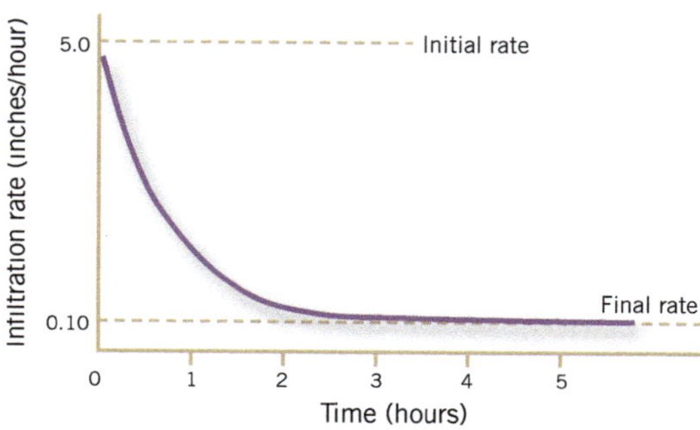

Figure 11.2 Typical infiltration rate curve showing a high infiltration rate at the beginning of the irrigation decreasing to a lower, constant final infiltration rate.

the application rate during design and decrease the nozzle sizes later since underground piping, pump sizing, and so on are based on the flow rate of the system. If larger sprinkler nozzles are retrofitted to the system, the piping and pump capacity will be too small, causing inadequate pressure in the system.

Determining application rate based on volume of water applied. An alternative method of determining the depth of applied water (inches) during an irrigation is to monitor the flow meter (if available) at the pump. As indicated earlier in this chapter, the following formula can then be used to estimate the applied water using flow meter readings:

Application rate (in/hr) = Pump discharge (gpm) ÷ [449 × Acres irrigated per set]

For flow meters without an instantaneous readout in gallons per minute, use the flow meter's totalizing readout. Keep track of the total flow change during a recorded period of time to determine the average flow rate in gallons per minute. This method can often be more accurate than using an instantaneous readout.

Determining irrigation set time

Once the sprinkler system application rate has been determined, the irrigation set time (hr) can be chosen in order to replace the soil water depleted since the last irrigation plus any irrigation inefficiencies. The soil water depletion can be estimated by evapotranspiration (ET) or by soil moisture monitoring (see chapter 12, "Irrigation Scheduling and Tree Stress"). Since no irrigation is 100% efficient, additional irrigation water must be applied to ensure adequate irrigation. Efficiencies of 75 to 85% are desirable for solid-set sprinkler systems.

The desired irrigation amount (in) is therefore

Desired irrigation amount (in) = Soil moisture depletion (in) ÷ Irrigation efficiency (%) × 100

The irrigation set time (hr) is

Irrigation set time (hr) = Desired irrigation amount (in) ÷ Sprinkler application rate (in/hr)

For example, if the interval since the last irrigation is 10 days and the estimated soil moisture depletion is 2.0 inches (from ET information), the sprinkler application rate is 0.1 in/hr, and the irrigation efficiency is 80%:

Desired irrigation amount (in) = 2 (in) ÷ 80 (%) × 100 = 2.5 in

and

Irrigation set time (hr) = 2.5 (in) ÷ 0.1 (in/hr) = 25 hr

Table 11.5. Average application rate for various sprinkler discharge rates (gpm) and areas of coverage (ft^2)

Sprinkler area (ft^2)	Sprinkler discharge rate (gpm)							
	0.75	**1.00**	**1.25**	**1.50**	**1.75**	**2.00**	**2.25**	**2.50**
500	0.14	0.19	0.24	0.29	0.34	0.39	0.43	0.48
550	0.13	0.18	0.22	0.26	0.31	0.35	0.39	0.44
600	0.12	0.16	0.20	0.24	0.28	0.32	0.36	0.40
650	0.11	0.15	0.19	0.22	0.26	0.30	0.33	0.37
700	0.10	0.14	0.17	0.21	0.24	0.28	0.31	0.34
750	0.10	0.13	0.16	0.19	0.22	0.26	0.29	0.32
800	0.09	0.12	0.15	0.18	0.21	0.24	0.27	0.30
850	0.08	0.11	0.14	0.17	0.20	0.23	0.25	0.28
900	0.08	0.11	0.13	0.16	0.19	0.21	0.24	0.27
950	0.08	0.10	0.13	0.15	0.18	0.20	0.23	0.25
1,000	0.07	0.10	0.12	0.14	0.17	0.19	0.22	0.24
1,050	0.07	0.09	0.11	0.14	0.16	0.18	0.21	0.23
1,100	0.07	0.09	0.11	0.13	0.15	0.18	0.20	0.22
1,150	0.06	0.08	0.10	0.13	0.15	0.17	0.19	0.21
1,200	0.06	0.08	0.10	0.12	0.14	0.16	0.18	0.20
1,250	0.06	0.08	0.10	0.12	0.13	0.15	0.17	0.19
1,300	0.06	0.07	0.09	0.11	0.13	0.15	0.17	0.19
1,350	0.05	0.07	0.09	0.11	0.12	0.14	0.16	0.18
1,400	0.05	0.07	0.09	0.10	0.12	0.14	0.15	0.17
1,450	0.05	0.07	0.08	0.10	0.12	0.13	0.15	0.17
1,500	0.05	0.06	0.08	0.10	0.11	0.13	0.14	0.16
1,550	0.05	0.06	0.08	0.09	0.11	0.12	0.14	0.16
1,600	0.05	0.06	0.08	0.09	0.11	0.12	0.14	0.15

Distribution uniformity and application rate

How uniformly or evenly water is applied is also important to good irrigation water management. In a sprinkler irrigation system that applies water uniformly, all trees receive the same amount of water during an irrigation event. No tree will get too much or too little water. Potential irrigation efficiency can be estimated by measuring the distribution uniformity of a sprinkler system. Distribution uniformity can be quantified by field-measured catch can tests. Mobile irrigation evaluation teams or private consultants can perform this test.

A single sprinkler head does not apply water evenly over the area it wets. Figure 11.3 shows the pattern of water distribution for a typical impact sprinkler head. Note that more water is applied to some areas than to others. In order to make the water application more uniform, the patterns from adjacent sprinklers must be overlapped (fig. 11.4). This overlapping of sprinkler patterns makes the water application more uniform, but it increases the application rate. To see this, examine table 11.5. For a particular sprinkler discharge rate, such as 1.5 gallons per minute, as the sprinkler area decreases (closer sprinkler spacing), the application rate increases. Thus, closing up the sprinkler spacing may increase distribution uniformity, but it can also increase the application rate to the point where runoff may become a problem. A second obvious disadvantage of closing up sprinkler spacing is the added cost of more sprinkler heads and the increased size of pipe supplying the sprinklers. Design of a good sprinkler system is therefore a balancing act of choosing an acceptable application rate and achieving an acceptable irrigation uniformity.

Surface Irrigation Systems

Surface irrigation systems (border flood irrigation and furrow irrigation), while the simplest irrigation systems with regard to hardware, are the most complicated irrigation systems to manage properly.

Advantages of Surface Irrigation Systems

- Surface irrigation systems have the lowest energy demands of all irrigation systems. If water is delivered from an irrigation district, little or no pumping energy is required. Groundwater sources require energy to obtain the water but little additional pressure is needed for the surface irrigation system.
- Maintenance of a surface irrigation system is minimal, although delivery and runoff ditches may need to be created and/or maintained.

Disadvantages of Surface Irrigation Systems

- Irrigating efficiently and uniformly is more difficult than with microirrigation or sprinkler systems.
- Tailwater runoff is common with surface irrigation systems and often requires management using tailwater return systems.
- Weed control can be difficult.
- Land must be relatively flat.
- Initial land preparation costs may be significant.
- The infiltration rate and other factors affecting surface irrigation performance change during the season. This complicates irrigation scheduling and makes it difficult to apply the desired irrigation amount.

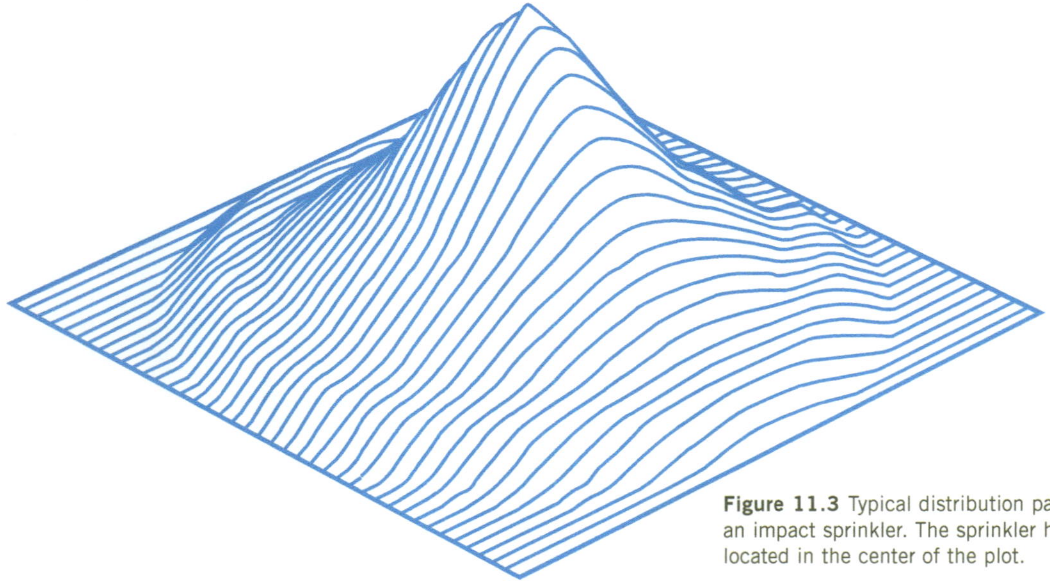

Figure 11.3 Typical distribution pattern of an impact sprinkler. The sprinkler head is located in the center of the plot.

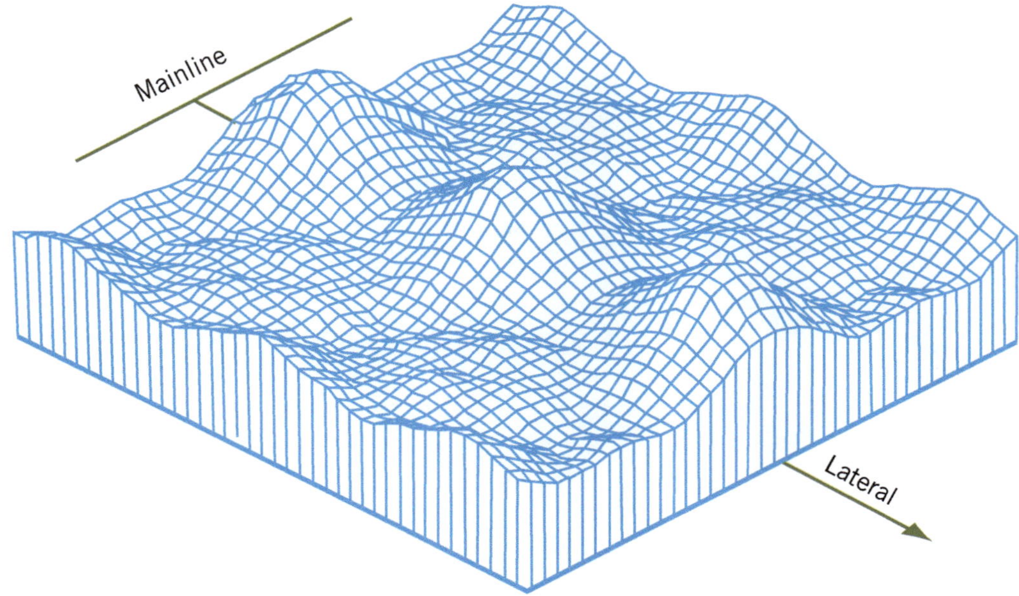

Figure 11.4 Water distribution pattern for overlapped impact sprinklers. Sprinkler heads are located along a lateral line.

Operation of Surface Irrigation Systems

The objective of surface, or flood, irrigation is to have water infiltrate for the same length of time at all parts of the field. This is difficult to accomplish because water begins to infiltrate at the head of the field as soon as irrigation begins, but it begins to infiltrate at the tail of the field only later, after traveling down the entire field. Thus, the head of the field almost always receives more water than the tail.

The exception is if water is allowed to "pile up" at the end of the field while the head of the field begins to dry; in this case, the part of the field approximately two-thirds to three-fourths of the distance down the field receives the least water. Many variations of border irrigation exist due to adjustments in border length, slope, and operation time. In general, it is advantageous to keep borders as short as practical to keep irrigation uniformity high. The tradeoffs for short borders are increased labor and pipeline costs.

An additional difficulty in managing surface irrigation systems is measuring the volume of water applied. If water is supplied from a pump, a flow meter such as a propeller meter can be installed in the outlet pipe. To obtain accurate measurements, follow the manufacturer's installation recommendations. If water is supplied from an open ditch, consult the local irrigation district to obtain an estimate of the flow rate to the orchard.

The following equation may be used to determine the average amount of water applied to an orchard through surface irrigation.

Flow (cfs) × Irrigation time (hr) = Area irrigated (ac) × Depth of water (in)

If the flow meter reads in gallons per minute (gpm) rather than in cubic feet per second (cfs), convert the reading: 1 cfs = 449 gpm.

The depth of water applied in the above equation should be the amount of water used by the orchard since the last irrigation; it can be estimated by orchard evapotranspiration (see chapter 12, "Irrigation Scheduling and Tree Stress"). Additional water should be applied to account for distribution nonuniformity, as discussed above. The efficiency of flood-irrigated orchards is generally lower than that of sprinkler-irrigated orchards. If no irrigation evaluation information is available for the orchard, an irrigation efficiency estimate of 65 to 75% is realistic.

References

Schwankl, L. J., B. R. Hanson, and T. L. Prichard. 1998. Micro-irrigation of trees and vines. Oakland: University of California Agriculture and Natural Resources Publication 3378.

———. 2008. Maintaining microirrigation systems. Oakland: University of California Agriculture and Natural Resources Publication 21637.

Schwankl, L. J., T. L. Prichard, B. R. Hanson, and I. Wellman. 2000. Cost of pressurized irrigation: Systems for tree crops. Oakland: University of California Agriculture and Natural Resources Publication 21585.

12 Irrigation Scheduling and Tree Stress

• Ken A. Shackel, Terry L. Prichard, and Lawrence J. Schwankl

Prune trees, like all other crop plants, require substantial amounts of water each season to grow and to produce a crop. Most of the water used by a prune tree is lost by evaporation from the leaves to the atmosphere; hence, the demand for water is largely determined by weather conditions such as sunlight intensity, air temperature, humidity, and wind speed. Essentially all of the water used by a prune tree is supplied by water stored in the soil following irrigation or rainfall. The quantity of water that is available to the tree is determined by the depth of soil occupied by the root system and soil physical properties such as sand, silt, and clay content, soil compaction, and so on. The atmospheric demand and the soil water supply determine what is called the water balance of the tree. One approach to irrigation management in prune orchards is to use this water balance to determine the frequency and amount of irrigation over the season. In this approach, the atmospheric water demand for the orchard as a whole is referred to as crop evapotranspiration (ET_c).

This chapter is divided into three sections. The first section is devoted to an explanation of ET_c and how it is used for irrigation management in prune orchards. The second section describes the soil factors that are important in determining the supply side of the water balance approach and how these factors may impact irrigation management. The third section describes the physiology of water stress in prune trees and how water stress is measured with the pressure chamber technique. This technique is important for the practice known as regulated deficit irrigation (RDI), in which mild to moderate tree water stress is used as a management tool to reduce fruit moisture content and increase fruit soluble solids, both of which are economically beneficial to prune production.

Using Evapotranspiration Estimates to Schedule Irrigations

Knowing the pattern of water use of a prune orchard over the season is important in developing a successful irrigation schedule. The water use of orchard crops can be estimated by combining evapotranspiration reference values (ET_o) derived from climatic data with a site-specific measure of crop development called the crop coefficient (K_c). Historical ET_o averages for an area can be used to predict crop water use over the season. Real-time ET_o (current season) daily values can be used to refine the historical ET_o estimates and determine current-year water requirements.

Evapotranspiration (ET)

Evapotranspiration consists of two components: evaporation and transpiration. Evaporation is water evaporated from the soil surface. Transpiration is water evaporated from plant leaves. The two terms are combined into one term because it is difficult to measure the two components separately. In the spring, as leaves expand and the area shaded by the tree canopy increases, evaporation from the soil surface decreases and transpiration from leaf surfaces increases. During and after an irrigation event, evaporation is higher until the soil surface dries over a few days' time.

Transpiration is water evaporating from plant leaves mainly through pores in the leaf called stomata. The rate of transpiration depends on the solar energy available for evaporating the leaf water, the movement of the air around the plant, and the relative dryness of the air. About 95% of the water taken up by plants is transpired. Transpiration is a necessary consequence of the need for plants to take up carbon dioxide in photosynthesis, but transpiration also cools the leaves, and the flow of transpiration water facilitates the transport of many nutrients from the roots to the leaf.

Solar radiation is the primary source of energy for evapotranspiration. Increased solar radiation increases the potential evapotranspiration. The amount of solar radiation depends on the time of day, time of year, latitude, and cloud cover. Maximum radiation in California occurs in June. Although solar radiation provides the energy needed for evapotranspiration, the evaporated water or water vapor must also flow away from the leaf and soil surface for evapotranspiration to occur. The rate at which water vapor flows away from a leaf depends on the water vapor content of the surrounding air and the air turbulence. The water vapor content of the surrounding air depends on temperature and humidity. High temperatures mean increased potential for transpiration. High humidity means that the potential for transpiration is less. The rate at which water vapor flows is also affected by turbulence caused by eddies and wind. Turbulent air transports water away from the leaves into the atmosphere. This is the reason leaves on even well-watered trees can show some wilting on very hot, dry, windy days, as the evaporated water is being pulled away from the leaves very rapidly. Day length and canopy size (measured as orchard floor shading at noon) also determine the water use. Maximum day length occurs in late June. Maximum crop water use is achieved when 60 to 70% of the orchard floor is shaded, measured at midday.

Estimating crop water use from climate data

Climate data such as net solar radiation, temperature, humidity, and wind speed are commonly used to calculate a reference crop evapotranspiration (ET_o). The reference crop evapotranspiration is that of an adequately irrigated grass crop clipped to 3 to 6 inches, completely covering the soil and not water deficient. The reference crop provides a standard measure of evapotranspiration that can be determined easily at various locations using climate data. The evapotranspiration for a particular crop is then calculated from:

$$ET_c = K_c \times ET_o$$

where

ET_c = crop evapotranspiration

K_c = crop coefficient

ET_o = evapotranspiration of the reference crop

Crop coefficients relate crop evapotranspiration to the reference crop evapotranspiration. The value of the crop coefficient depends on crop type and stage of growth. Crop coefficients are simply the ratio of actual measured crop ET_c to ET_o. Crop coefficients for a mature prune orchard over one growing season are shown in figure 12.1.

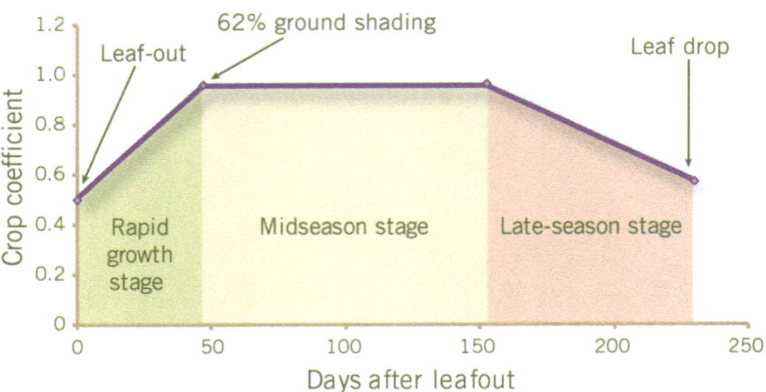

Figure 12.1 Crop coefficients for a mature prune orchard over the growing season, from leaf-out to leaf drop.

Prune water use begins at leaf-out and rapidly increases as the canopy develops until approximately 62% of ground shading. The maximum canopy coincides with an increase in evaporative demand as days lengthen and air temperatures increase. The canopy in mature orchards is fully developed by mid-April. Peak water use occurs in June and July. As the leaves become less functional, the K_c declines toward leaf drop, where K_c for the orchard becomes 0 if there is no cover crop. Water use may be 20% greater (not to exceed K_c = 1.15) with an actively growing cover crop or if substantial weed cover is present. Table 12.1 contains numerical K_c values for prunes in 2-week time periods over the season.

Table 12.1. Numerical crop coefficients (K_c) for prunes in 2-week periods

Period	K_c
3/16–31	0.45
4/1–15	0.62
4/16–30	0.84
5/1–15	0.96
5/16–31	0.96
6/1–15	0.96
6/16–30	0.96
7/1–15	0.96
7/16–31	0.96
8/1–15	0.95
8/16–31	0.92
9/1–15	0.84
9/16–30	0.78
10/1–15	0.69
10/16–31	0.57

Availability of ET_o information

Evapotranspiration information is often available in two forms: historical ET information and real-time weather data. Historical ET_o information is the crop water use estimated from long-term averages. For any particular day, the actual ET_o may vary from the historical average, but it has been found that the actual variability of weekly May to September reference crop evapotranspiration from year to year is small. Thus, differences between historical (average) ET_o values based on many years of data and actual values for any given year are small. Table 12.2 shows historical ET_o from three prune-growing areas of the state assembled from more than 20 years of detailed CIMIS station data. The sample irrigation worksheet in table 12.3 combines the ET_o and K_c for a 2-week period for prunes each month in Tehama County, California. The product of ET_o and K_c is the ETc for each period, totaling 41.83 inches of total water requirement.

Real-time ET estimates are available to most growers in California. Real-time ET_o is determined by collecting weather data and estimating ET_o. California uses the CIMIS (California Irrigation Management Information System) network of more than 100 active weather stations to collect the data and then estimate the ET_o almost anywhere in the state. This information is readily available from a number of sources. Two good Web-based sources are the UC Statewide Integrated Pest Management website, http://www.ipm.ucdavis.edu, and the California Department of Water Resources' CIMIS website, http://www.cimis.water.ca.gov. The pasture grass ET_o must then be converted to the ET_c of the crop by using a crop coefficient.

Scheduling irrigations using crop water use estimates, rainfall, and soil moisture use

Full-potential water use (ET_c) represents water demand, and this demand must be met by the water supply:

Total water supply = Effective rainfall + Soil moisture contribution + Net irrigation

For water supply, the stored soil moisture and effective rainfall are important factors in addition to irrigation, and each must be considered to construct an effective and efficient irrigation schedule.

The amount of soil moisture can be quantitatively determined using soil moisture measuring devices, as discussed later in this chapter. The two times when measurement is most critical to an irrigation schedule are at leaf-out and at the driest point (usually after harvest and just before postharvest irrigation). Failure to irrigate prune trees after harvest can yield small

Table 12.2. Historical biweekly ET_o during prune season

Period	Gerber CIMIS Station 8, Tehama County	Nicolaus CIMIS Station 30, Sutter County	Clovis CIMIS Station 39, Fresno County
3/16–31	2.15	2.03	2.05
4/1–15	2.38	2.28	2.45
4/16–30	2.69	2.64	2.82
5/1–15	3.22	3.01	3.32
5/16–31	3.77	3.51	3.91
6/1–15	4.05	3.65	4.05
6/16–30	4.20	3.93	4.23
7/1–15	4.28	3.95	4.32
7/16–31	4.26	3.96	4.42
8/1–15	3.80	3.53	3.93
8/16–31	3.70	3.43	3.82
9/1–15	3.06	2.81	3.10
9/16–30	2.72	2.33	2.50
10/1–15	2.30	1.95	2.07
10/16–31	1.75	1.53	1.57
Total	48.33	44.54	48.56

Table 12.3. Sample irrigation scheduling worksheet showing potential water use

Prune ET_o historical from Gerber CIMIS Station 8, Tehama County
Assumptions:
1. Leaf-out 3/15
2. Leaf drop 10/15

Date	A = Historical ET_o (in/period)	B = Crop coefficient* (K_c)	C = A x B: Potential water use (in)
3/16–31	2.15	0.45	0.97
4/1–15	2.38	0.62	1.48
4/16–30	2.69	0.84	2.26
5/1–15	3.22	0.96	3.09
5/16–31	3.77	0.96	3.62
6/1–15	4.05	0.96	3.89
6/16–30	4.20	0.96	4.03
7/1–15	4.28	0.96	4.11
7/16–31	4.26	0.96	4.09
8/1–15	3.80	0.95	3.61
8/16–31	3.70	0.92	3.40
9/1–15	3.06	0.84	2.57
9/16–30	2.72	0.78	2.12
10/1–15	2.30	0.69	1.59
10/16–31	1.75	0.57	1.00
Total			41.83

Note: *Crop coefficient is based on a midsummer minimum of 62% land surface shaded at midday.

fruit buds for the following year. The amount of stored water for a given soil is calculated as the available water-holding capacity (AWHC) times the depth of rooting, or more simply as the root zone available water. Nonstressed ET_c equals the effective rainfall plus stored soil water depletion until about 50% of the root zone available water is extracted. Scheduling should begin at leaf-out. If effective rainfall exceeds ET_c during the spring, the appropriate time to measure root zone moisture would be when the effective rainfall does not exceed ET_c. Irrigation should be scheduled to begin at that point. The sample irrigation scheduling worksheet in table 12.4 shows a water balance that slowly uses the stored soil water over the season. This schedule minimizes the leaching of nitrogen fertilizers, avoids stress on the tree, and leaves a drier root zone at the end of the season that can efficiently store winter rain water.

It is difficult at best to determine the effective rainfall portion of the total rainfall. During spring rains, about 50 to 70% of the total rainfall enters the soil to become effectively used by the orchard. Another rule of thumb is to sum the rainfall for a single event, however long that may be, subtract 0.25 inches (a common amount that just evaporates from wet soil), then take 80% of the remainder as effective rainfall.

In accounting for crop use of soil moisture when no irrigation or effective rainfall takes place during an irrigation period, the moisture extracted would equal the ET_c. Likewise, effective rainfall would decrease the soil moisture extraction. Examples of both these instances are shown in table 12.4, which uses the calculated ET_c from table 12.3 and adds the soil contribution and effective rainfall to yield the net irrigation requirement.

The net irrigation requirement is the amount that all portions of the field must receive to sustain full potential water use. Unfortunately, no irrigation system is 100% efficient, even microirrigation systems (see chapter 11, "Irrigation Systems"). The sample irrigation worksheet in table 12.5 uses a double-line drip system that has a distribution uniformity of 92%. Since the system is operated with no runoff and no deep percolation losses and evaporation is at a minimum, the distribution uniformity equals the irrigation efficiency. The gross irrigation volume must be increased to account for the remaining 8% nonuniformity, to ensure that even the driest parts of the field receive the minimum amount of water to supply ET_c (the net requirement calculated in table 12.4). This final total is the gross irrigation requirement.

Table 12.4. Sample irrigation scheduling worksheet showing water balance over a season

Prune ET_o Historical from Gerber CIMIS Station 8, Tehama County

Assumptions:

1. Leaf-out 3/15
2. Leaf-drop 10/15
3. Harvest 9/1–9/15
4. Root zone extracted soil moisture from 4/15 to leaf drop = 4.8 in

Date	C = A × B: Potential water use (in)	E = Soil contribution (in)	F = Effective rainfall* (in)	G = (C – E – F): Net irrigation requirement (in)
3/16–31	0.97	0	1.0	–0.03
4/1–15	1.48	0	1.75	–0.27
4/16–30	2.26	1.0	1.5	0.26
5/1–15	3.09	1.4	0.0	1.19
5/16–31	3.62	0.3	0.5	2.82
6/1–15	3.89	0.3	0	3.59
6/16–30	4.03	0.3	0	3.73
7/1–15	4.11	0.3	0	3.81
7/16–31	4.09	0.3	0	3.79
8/1–15	3.61	0.3	0	3.31
8/16–31	3.40	0.3	0	3.10
9/1–15	2.57	0.3	0	2.27
9/16–30	2.12	0.0	0	2.12
10/1–15	1.59	0.0	0	1.59
10/16–31	1.00	0.0	0	1.00
Total	41.83	4.8	4.75	32.28

Note: *Effective rainfall is calculated from actual rainfall.

Table 12.5 continues with a simple calculation that yields the duration of operation to meet the gross irrigation for the period. Required inputs include the tree spacing and irrigation application per tree.

Using real-time evapotranspiration to schedule irrigations

It is possible to more accurately estimate tree crop evapotranspiration using real-time evapotranspiration data provided by systems such as CIMIS. All that is necessary is to substitute real-time ET_o data into the ET_o column in table 12.3 when the data are available. Using this method, one can schedule forward in time using historical ET_o, then enter the real-time data to refine the irrigation schedule. If the historical ET_o underestimated real-time ET_o, the irrigation time on the next irrigation can be increased to match the difference. If the historical ET_o was an overestimate, irrigation can be reduced accordingly.

Soil-Based Irrigation Scheduling

Soil-based methods of irrigation scheduling assume that the soil moisture content largely determines the moisture status of the plant. If there is less water in the soil and the water is held at a greater tension, it will be more difficult for the plant's roots to take up water. The result is that the plant will be under greater stress. Thus, irrigation scheduling is done by monitoring the soil moisture and maintaining it within acceptable limits that do not adversely affect plant growth.

The use of soil moisture monitoring devices such as tensiometers, gypsum blocks, neutron probes, or other devices is an excellent way to manage irrigation scheduling. Soil moisture monitoring in microirrigated orchards is complicated by the importance of monitoring device placement. Monitoring devices should be placed so that they reflect the average changes in the emitters' wetted volume. A shrinking wetted volume or slowly drying root zone during the season indicates underirrigation. Overirrigation would increase the wetted volume during the season and cause loss of water to deep percolation. In flood- or sprinkler-irrigated orchards, monitoring the soil moisture before and after irrigation provides additional information on selecting the duration of irrigation as well as the depth to which irrigation water will penetrate.

The most basic method of monitoring soil moisture is to collect soil samples from various depths (e.g., 1-foot increments) within the root zone and either estimate soil moisture using the "feel method" (see the USDA NRCS Estimating Soil Moisture by Feel and Appearance website, http://www.mt.nrcs.usda.gov) or measure it directly by weighing, oven drying, and reweighing. Using a soil auger, sampling tube, or even a shovel requires significant time and effort to sample throughout the root zone. If weight data are used to calculate a volumetric soil moisture value (e.g., inches per feet of soil moisture), an undisturbed soil sample must be collected, a difficult task. In order to make monitoring of soil moisture easier and quicker than collecting soil samples, a number of soil moisture monitoring tools are available. Some of these soil moisture monitoring tools are described below.

Tensiometers

A tensiometer is a device for measuring soil water tension, a measure of the surface tension force holding water in the soil. A tensiometer (fig. 12.2) consists of a cylindrical pipe about 1 inch in diameter with a porous ceramic cup attached to one end and a vacuum gauge attached to the other.

Table 12.5. Sample irrigation scheduling worksheet for a double-line drip system with a distribution uniformity of 92%

Prune ET_o historical from Gerber CIMIS Station 8, Tehama County

Assumptions:

1. Leaf-out 3/15
2. Leaf-drop 10/15
3. Tree spacing 20 × 20 feet

Date	G = (C − E − F): Net irrigation requirement (in)	H = Emission uniformity* (%)	I = G/H: Gross irrigation amount (in)	J = Tree spacing (ft²)	K = (I × J × .623): Gallons per tree/ period (gal/period)	L = Average application rate (gph/tree)	M = (K/L): Hours of predicted irrigation time (hr/period)
3/16–31	–0.03	92	–0.04	400	–9.0	8	–1.1
4/1–15	–0.27	92	–0.30	400	–74.4	8	–9.3
4/16–30	0.26	92	0.29	400	71.0	8	8.9
5/1–15	1.19	92	1.30	400	322.9	8	40.4
5/16–31	2.82	92	3.07	400	764.3	8	95.5
6/1–15	3.59	92	3.90	400	972.2	8	121.5
6/16–30	3.73	92	4.05	400	1,009.8	8	126.2
7/1–15	3.81	92	4.14	400	1,032.9	8	129.1
7/16–31	3.79	92	4.12	400	1,027.3	8	128.4
8/1–15	3.31	92	3.60	400	895.9	8	112.0
8/16–31	3.10	92	3.37	400	839.9	8	105.0
9/1–15	2.27	92	2.47	400	615.6	8	76.9
9/16–30	2.12	92	2.31	400	574.5	8	71.8
10/1–15	1.59	92	1.73	400	430.2	8	53.8
10/16–31	1.00	92	1.09	400	270.9	8	33.9
Total	32.28		35.09		8,744.0		1,093.0

Note: *Irrigation efficiency is assumed to be equal to the distribution uniformity of the drip system.

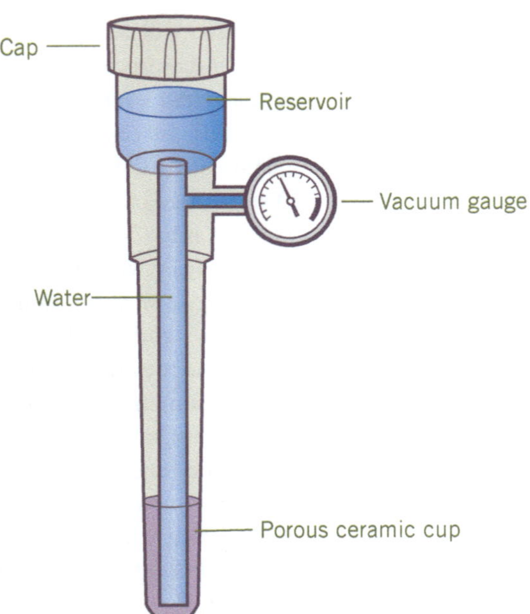

Figure 12.2 Tensiometer.

Tensiometers indirectly measure soil moisture. They are installed for the entire season or longer and provide readings at the same location over an extended period of time. Tensiometer readings are easily interpreted and indicate the soil water conditions experienced by the trees' roots. Soil salinity does not affect the readings.

In an unsaturated soil, soil water tension (frequently referred to as a suction or matric potential) falls below atmospheric pressure. As wet soil dries, this soil water suction increases, causing water to flow out of the tensiometer through the porous cup. The small pores in the cup prevent air from entering the tensiometer if the tension is within the range of the vacuum gauge. This outflow of water creates a vacuum inside the tensiometer and increases the reading on the vacuum gauge. If the soil is rewetted by irrigation, water will be drawn back into the tensiometer, reducing the vacuum inside, and the reading on the tensiometer gauge will decrease.

The vacuum gauge measures the suction in centibars (cb: 100 cb = 1 bar = 1 atm), with a range of 0 to 100. A reading of 0 cb indicates a saturated soil in which plant roots will probably suffer from poor aeration. A reading of 10 to 25 cb reflects a soil at field capacity. The lower reading is for sandy soils at field capacity, and the higher reading is for finer-textured soils. Readings of 70 to 80 cb indicate a dry soil. Tensiometers do not read above about 85 cb, as this is about the point at which air begins to enter the porous tip and the vacuum is lost.

Tensiometers do not provide information on the amount of water depleted from the soil unless they have been calibrated for the particular soil type. They therefore indicate when to irrigate, but not how much to irrigate.

Although tensiometers are used most frequently for monitoring soil moisture, they can also be incorporated into automated irrigation systems. Tensiometers with solenoids can be used to control an irrigation system, and tensiometers equipped with transducers can be used with computerized irrigation systems.

Installation

To measure soil water tension, insert the porous cup end of the tensiometer at the depth desired through a pilot hole, made with a soil probe, in the soil. The porous cup should be soaked in water for several hours before installation, filled with water, and then allowed to dry temporarily to confirm that the tensiometer can develop a vacuum. Water should be placed in the bottom of the pilot hole and allowed to soak in. The tensiometer's porous tip can then be pushed into the wet soil in the bottom of the pilot hole, ensuring good contact between the porous tip and the soil. After installation, fill the tensiometer with water and allow it to equilibrate with the soil water for about 24 hours. A tensiometer should be installed in the zone of greatest root density, at about one-quarter to one-third of the maximum root depth. A tensiometer at this depth can be used to schedule irrigations. For sprinkler- or flood-irrigated orchards, a tensiometer reading of approximately 60 cb indicates a need for irrigation in the near future. For microirrigated orchards, the frequent irrigations should keep the soil in the emitter's wetted volume near field capacity, and the tensiometer should read near the 15 to 35 cb range. Refinements based on soil sampling will be required to adjust for site-specific conditions.

A second tensiometer should also be installed in the bottom one-third of the root depth to assure that the moisture is maintained to an adequate depth. Usually, the shallower tensiometer will begin to show an increased reading (indicating drier soil) before the deeper tensiometer begins to change. Trees tend to take up water from the shallower depths first. Some water managers using tensiometers in flood- or sprinkler-irrigated orchards use the beginning of moisture extraction from the deeper depths, indicated by a change in the lower tensiometer's readings, as an indicator of when to irrigate. This practice seems to work well as long as irrigation water can penetrate to the lower depths to replace the soil moisture used. If the tensiometer reading at the lower root depth remains unchanged following an irrigation or continues to rise during the growing season, irrigation applications may be insufficient, and frequency or duration of irrigation should be increased.

Tensiometers should be placed in the tree row to minimize damage by equipment. In addition, placement of tensiometers relative to microirrigation emitters, especially drippers, is critical to their use. If the tensiometer is too close to the drip emitter, it will constantly read wet; if it is too far from the emitter, the tensiometer will dry out and not be rewetted. The tensiometer should be placed near the edge of the emitter's wetted volume so that changes in the wetted area, reflecting over- or underirrigation, can be detected (see chapter 11, "Irrigation Systems").

The number of tensiometer stations required depends on the irrigation system and on soil uniformity and management. For areas up to 40 acres, at least two stations should be established. Stations should be located in areas representative of overall moisture status, with separate stations for problem areas or for areas with different soil conditions. In sprinkler irrigation, tensiometers should not be placed in areas in which the sprinkler pattern is "shaded" by a tree.

Maintenance

Tensiometers must be properly maintained. This requires periodically refilling the pipe with water and replacing porous cups that have cracked or become clogged with salts. If the soil becomes too dry (tensiometer readings greater than 85cb), the porous cup will break tension and air will enter the tensiometer. A cracked cup or a leaking vacuum gauge also prevents a vacuum from developing in the tensiometer and causes the instrument to always read zero. In locations where temperatures fall below freezing, tensiometers should be protected or removed from the field during cold periods. Finally, other than testing before installation, the porous cup of a tensiometer filled with water should not be exposed to the atmosphere for long periods of time. Such exposure causes evaporation of water from the cup's surface, which in turn may cause salt buildup and clogging of the cup.

Electrical Resistance Blocks

Electrical resistance blocks, commonly referred to as gypsum blocks, are an inexpensive and simple soil moisture measurement tool. They are useful for timing irrigations but do not provide information regarding the amount of irrigation water required. Electrical resistance blocks are installed permanently at a site and have the advantages of allowing their user to quickly monitor soil moisture at the same location throughout the season. Resistance blocks monitor soil moisture indirectly by measuring the electrical resistance between two electrodes attached to a small cast block of gypsum buried in the soil. Some electrical resistance blocks have electrodes mounted in fiberglass or in plastic- or metal-encased blocks containing a sand-gypsum mixture. Wires, available in various lengths, lead from the blocks to the soil surface. The electrical resistance is read with a portable conductance meter. Measuring conductance (1 ÷ resistance) requires alternating current (AC), which the resistance block meter produces from its direct current (DC) battery. Using a simple resistance or ohmmeter to read the electrical resistance blocks will not give stable or reliable readings. Electrical resistance blocks and meters vary in cost by manufacturer, with blocks costing from $6 to $25 each and meters costing approximately $250. Since the meter is portable, only a single meter is required to monitor all the installation sites. At a normal site, a block may last 2 to 3 years, but it may require annual replacement in areas that have frequent irrigation or a high water table.

One manufacturer produces a block and meter that read soil water tension directly in centibars, while most other manufacturers provide charts or tables to correlate the meter readings to centibar values. Soil water tension information indicates the timing of irrigation but does not indicate the amount of irrigation water required. Indirectly, the irrigation amount can be arrived at through a trial-and-error procedure of monitoring the resistance blocks before and after irrigations. Blocks whose readings indicate wetter conditions following irrigation reflect irrigation water reaching their depth. If a block does not reflect wetter soil conditions after irrigation, the irrigation was insufficient to reach the block's depth. Observing the changes in resistance block readings after a number of irrigations allows an irrigator to gauge the irrigation to match the soil moisture deficit.

Data loggers, which can be used with soil moisture blocks, are available to record moisture block readings on a frequent and automated schedule. This "continuous" measurement of soil moisture can help the irrigator improve irrigation water management while minimizing the labor associated with reading the moisture blocks.

Installation

Gypsum blocks should be placed at several soil depths. At a minimum, blocks should be placed in the top one-third and bottom one-third of the root zone. Consultants who install soil moisture

monitoring sites using resistance blocks often place blocks at 1-foot intervals throughout the root zone. Soil moisture conditions are monitored only in the soil closely surrounding a resistance block. Wires from the blocks are brought to the soil surface to facilitate periodic connection to the portable meter for soil moisture measurements. The matric potential of the blocks is assumed to be in equilibrium with the surrounding soil, so the blocks act much like the surrounding soil, taking up and releasing water as the soil wets and dries. The electrical conductance between the electrodes varies according to the water content of the block. The higher the water content, the higher the conductance and the lower the electrical resistance.

The placement of resistance block stations in an orchard depends primarily on the soils. Generally, monitoring stations should be chosen to be representative of what is happening in the surrounding area. At a minimum, two stations per 40 acres are recommended, with additional stations in problem areas or in areas with different soil conditions. Since gypsum is soluble, gypsum blocks slowly dissolve. The life of a block can be extended 1 or 2 years by adding a small quantity of gypsum to the backfill soil at the time of installation. A small quantity of lime ($CaCO_3$) may likewise be useful in an acid soil to prolong the life of a block.

Most electrical resistance blocks monitor soil moisture more effectively in the drier range of soil water tensions (in excess of 33 cb). Consequently, blocks are more useful in the medium- to heavy-textured soils (loams and clays), which retain more available water in the higher soil water tension range. Sands and coarse-textured soils tend to release much of their water at low tensions, where the blocks are not as accurate or responsive. Blocks are also affected by highly saline-alkali soil conditions and may therefore not be appropriate for this use.

In prunes, moisture resistance blocks can often be installed in the tree row to protect them from traffic or cultivation damage. They should be installed within the drip line of a tree. When sprinkler irrigation is used, avoid placing the resistance blocks in the drier "shadow" area caused by the tree blocking the sprinkler's spray.

At a station, resistance blocks to be placed at various depths can either be sequentially stacked in the same hole or placed in separate holes. Placing blocks in separate holes has the advantage of locating the block in the soil environment closest to an undisturbed condition, but it has the disadvantage of requiring more work. Both techniques have been used successfully.

Prior to installation, test each block by soaking it in water and connecting it to its companion conductance meter. The meter should indicate a wet moisture block. The installation hole should be only slightly larger than the block and can be made using a soil probe or small auger. The hole should be to the depth at which the block will be installed. Before inserting the block, add about 2 ounces of water to the hole and allow it to soak into the bottom of the hole.

To place the block in the hole, run the wire lead through an appropriate length of ½-inch-diameter PVC pipe and put tension on the wire to hold the block on the end of the pipe. One manufacturer's block has a special collar that fits into the end of ½-inch PVC pipe to facilitate installation. Lower the block into the hole and push it firmly into the moist soil in the bottom of the hole. Remove the PVC pipe and fill the hole for several inches with the removed soil. Pack the soil firmly, preferably using a wooden rod such as a broom handle so as to avoid damaging the blocks or wires. Continue refilling the hole in stages, making sure that the soil is firmly packed at each stage in an effort to simulate undisturbed conditions. Refilling the hole properly is important to ensure that water and roots do not penetrate the hole more easily than the surrounding soil.

When installing blocks in trafficked or cultivated areas, trench or bury the lead wires to prevent damage. Identify the blocks by knotting the wire, color coding, or tagging. If possible, protect the bare ends of the wire leads since they tend to become corroded. Scraping the wire leads with a knife blade or fingernail file prior to attaching the meter will provide a better connection.

How to use the information from the blocks

Resistance block information is best used by evaluating changes in block readings. In flood- or sprinkler-irrigated orchards, a block that shows surrounding soil becoming drier between irrigations is a good indicator that water uptake by roots is occurring at that depth. Most frequently, trees preferentially take up water from the shallower depths, drying shallow soil before taking up water from deeper soil. Resistance blocks placed in shallower soil therefore show a change first. Some block users use this transition of water uptake from drier shallow soil to the deeper soil, along with a particular meter reading of the shallow block, as a signal to irrigate. These signals to irrigate are often arrived at over time by watching and comparing the tree response with block readings.

As with tensiometers, placement of moisture blocks relative to microirrigation emitters, especially drippers, is critical to their use. If a moisture block is too close to the drip emitter, it will constantly read wet, while if placed too far from the emitter, it will dry out and not be rewetted. Moisture blocks should be placed on the edge of the emitter's wetted volume so that changes in the wetted area, reflecting over- or underirrigation, can be detected.

As mentioned earlier, the blocks do not provide information on how much to irrigate, only on when to irrigate. Irrigation rates can be estimated through a process of trial and error. Reading the blocks just prior to irrigation and then reading them 2 to 3 days after irrigation (when drainage has nearly stopped) can provide information on the depths to which irrigation water penetrated. If a block's readings remained unchanged (assuming it is not already reading completely wet), the water did not reach that depth. To wet the soil to that depth, a longer irrigation would be needed.

If the block(s) in the deeper soils never change readings, the period between irrigations may be lengthened; another possibility is that there is no root activity at that soil depth. A lack of root activity can be due to saturated soil conditions (caused by overirrigation or a high water table), disease, or nematodes. In addition, if the deeper blocks continually read wet, it is impossible to tell whether the soil is being overirrigated or is losing water to deep percolation.

If the accuracy of the readings may be in question, often indicated by readings remaining unchanged when they should be changing, checking a nearby location's soil moisture using a soil auger and the "feel" method is an excellent first check.

Neutron Soil Moisture Meters

A neutron soil moisture meter (neutron probe) is a device for measuring soil moisture consisting of an electronic gauge, a connecting cable, and a probe cylinder containing a nuclear source and a detector tube (fig. 12.3).

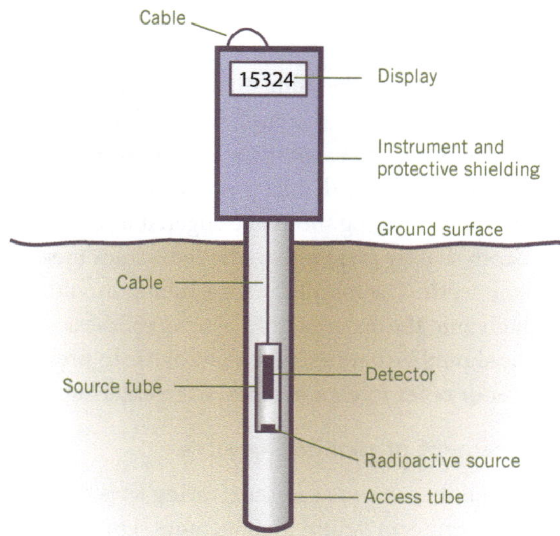

Figure 12.3 Neutron soil moisture gauge.

The nuclear source is an Americium 241/ Beryllium capsule that emits high-energy neutrons. When lowered down an access tube into the soil, the high-speed neutrons collide with surrounding soil particles and lose energy, creating low-energy, or thermal, neutrons. Some of these neutrons are reflected back to the probe cylinder and are counted by the thermal neutron detector tube. These counts are transmitted through the cable back to the gauge and displayed to the user. The neutron probe is therefore a sophisticated hydrogen counter. Since water is the primary source of hydrogen atoms in the soil, the neutron probe indirectly measures the soil moisture level.

Neutron probe measurements are taken by lowering the probe cylinder into the access tube to the desired depth. The neutron probe measures a large (volleyball-size) sample. This is a larger sample than most other soil moisture sensors measure, an advantage of using the neutron probe. The probe cylinder can be lowered to the same depth each time, allowing repeated measurements at the same location. The neutron probe allows the user to obtain very accurate soil moisture measurements in a matter of minutes.

The neutron probe has some disadvantages, however. Soil moisture measurements at shallow depths (6 inches or less) are often unreliable because neutrons escape through the soil surface. Initial calibration of the neutron probe at a site can be time consuming. Since the neutron probe uses a nuclear source, the operator must be trained in its handling, storage, and use. A license to operate the gauge is required, as are periodic inspections. The initial cost of the neutron probe is also quite high, approximately $4,000.

Installation

Establishing a neutron probe monitoring site entails installing an access tube to the desired depth. Access tubing can be aluminum pipe, PVC pipe, or electrical metal tubing (EMT). Since the probe readings are affected by the access tubing material, the material used for the tubing should be consistent throughout the orchard. Aluminum has the least effect on the readings, while PVC pipe will cause lower readings because the chlorine in the PVC absorbs "slow" neutrons.

Access tubing should be approximately 2 inches in diameter. The inside diameter of the access tube

should be nearly the same diameter as the probe cylinder. In areas with a high water table, the bottom of the access tubing should be sealed.

A hole the same diameter as the outside diameter of the access tubing should be augered into the soil to a depth slightly greater than the maximum measurement depth. The installed access tubing should fit snugly into the augered hole. The top of the access tube should be covered when not in use to prevent water or other foreign material from entering.

Placement of monitoring sites

The number of required monitoring sites depends on soil uniformity and management. For areas up to 40 acres, at least two sites should be established. Stations should be in areas representative of overall moisture conditions, with separate sites established for problem areas or for areas with different soil or management conditions.

The maximum depth at which the access tube is installed should be at least the depth of the active root zone. For established prune trees with no restricting soil layers or high water table, measuring to a depth of 5 feet will monitor most water usage by the tree. Installing to a depth slightly greater is recommended for monitoring deep percolation losses below the root zone.

The access tube should be placed in the tree row. In flood- or sprinkler-irrigated orchards, the access tube should be installed within the drip line of the tree, but not directly next to the trunk. In sprinkler-irrigated orchards, the access tube should not be placed in locations where trees "shade" the sprinkler distribution pattern. In microirrigated orchards, access tubes should be placed near the edge of the emitters' wetted soil volume. This allows monitoring changes in soil wetted volume caused by over- or underirrigation.

What do the readings mean?

Neutron probe readings are simply counts of returning neutrons, so a calibration must be done to translate the readings to soil moisture measurements. For timing irrigations, it may be sufficient to simply define the neutron probe reading at which the soil is "full" (approximately at field capacity) and to identify the reading at which irrigation should begin (refill point). A more thorough calibration, in which neutron probe readings are compared with actual soil moisture as determined by volumetric soil sampling, allow the user to determine not only the timing of irrigations but also the required irrigation amounts. The latter calibration technique is time-consuming but need be done only once for a given monitoring site. Sites with different soil types must be calibrated separately. Once calibrated, the neutron probe is very accurate.

Calibrations should relate soil water content to a relative count, which is the actual gauge count divided by a standard count. This procedure is recommended because the actual count can change with time due to deterioration of the radioactive source and because of electronic drift. The standard count is made by placing the probe at least 2 feet above the soil and 2 feet away from any source of hydrogen ions, such as a human body. The count is made with the source tube locked in the gauge.

Maintenance

Neutron probes require little maintenance beyond checking to ensure that the probe is operating properly and that the access tubes are free of water or foreign material. If repairs are required, they can be quite costly and must be done by the gauge manufacturer.

Dielectric Constant Soil Measuring Devices

A relatively new technology for soil moisture measurement makes use of the dielectric constant, a physical property of materials. The dielectric constant of a material impacts the travel time of a high-frequency electromagnetic pulse passing through the material. The dielectric constant of soil and water are significantly different (soil = 3 to 7; water = 80). Two soil moisture measurement techniques take advantage of these differences in dielectric constant. Time domain reflectometry (TDR) measures the travel time of an electromagnetic wave along two or more waveguides of known length; frequency domain reflectometry (FDR), also known as capacitance measurements, measures the soil capacitance of a high-frequency radio wave pulsed through the instrument and soil. These devices generally require that sensors be permanently installed in the soil.

The most common soil-installed TDR sensor has been a pair of parallel stainless steel rods spaced a known distance apart, usually about 5 centimeters. The length of the rods determines the monitored soil volume: the device measures a cylinder the length of the probes with a diameter slightly greater than the spacing of the rods. Monitoring various depths thus requires installation of a number of pairs of parallel rods, each pair of different lengths. One manufacturer of TDR sensors has combined the parallel rods with remote diode shorting into a single segmented probe that can measure various depths.

A number of FDR sensors are currently available. They either require the permanent installation of a sensor or sensors or installation of a PVC access tube down which a probe is lowered; in one case, a portable sensor can be pushed into the soil as a soil probe and allow a measurement to be taken.

If TDR or FDR devices are to be used, growers should work closely with a consultant or other professional familiar with their use. There are numerous stages, ranging from installation to interpretation of the data, where their expertise may prove valuable.

Advantages of dielectric constant devices

A major advantage of most TDR and FDR devices is that they can be automated to take continuous measurements. This allows the user to have an accurate picture of how soil moisture is changing with time, such as how soil moisture changes during soil moisture uptake by the tree or as a result of an irrigation. The electromagnetic wave–generating component of TDR and FDR devices can monitor and store information from multiple sensors using a multiplexing system and a data logger. These capabilities are unique to TDR and FDR devices. TDR and FDR devices can also be accurate once calibrated to the site.

Disadvantages of dielectric constant devices

TDR and FDR devices have significant disadvantages when used for field applications. First, both types monitor a relatively small volume of soil, measuring only a slight distance from the sensors. For TDR devices using pairs of parallel metal rods, pairs of different lengths must be installed if soil moisture readings for various depths are desired. Excellent contact between the soil and the TDR or FDR sensor must be maintained. Some of the greatest difficulties in using the devices have been caused by poor contact following installation; it is often difficult to attain good contact, particularly in gravely or heavy soils. The devices are also difficult to calibrate to achieve the desired accuracy. Finally, the cost of TDR and FDR devices is high. The systems, including the wave generators, multiplexer, and data storage units, often cost $7,000 or more. A portable FDR probe costs approximately $600.

Placement of monitoring sites

A minimum of two sites per 40 acres is recommended; however, as with other soil moisture measurement devices, soil, irrigation, tree, and management variability often determine the number of sites necessary in an orchard. The multiple-station monitoring capability of TDR and FDR devices, with their necessary wiring connections, limits the distance between monitoring sites. At each site, it is useful to take measurements at 1-foot depth intervals to monitor root uptake and irrigation refill and penetration.

Monitoring sites should be placed to obtain the information desired. If the information is to be used for scheduling irrigations for the orchard, a location typical of the surrounding area should be chosen. Choosing a site in the tree row within the drip line of the tree and a few feet from the trunk is preferred. If a problem area of the orchard is to be investigated, siting should be chosen accordingly.

In microirrigated orchards, as with the other soil moisture monitoring devices, dielectric constant devices should be installed near the edge of the wetted soil volume to monitor any changes in it. These changes often reflect over- or underirrigation.

Plant-Based Irrigation Scheduling, Water Stress, and Regulated Deficit Irrigation

The water that is lost by each leaf of a prune tree (transpiration, the "T" of "ET," fig. 12.4A) is continuously being replaced by water that the leaf

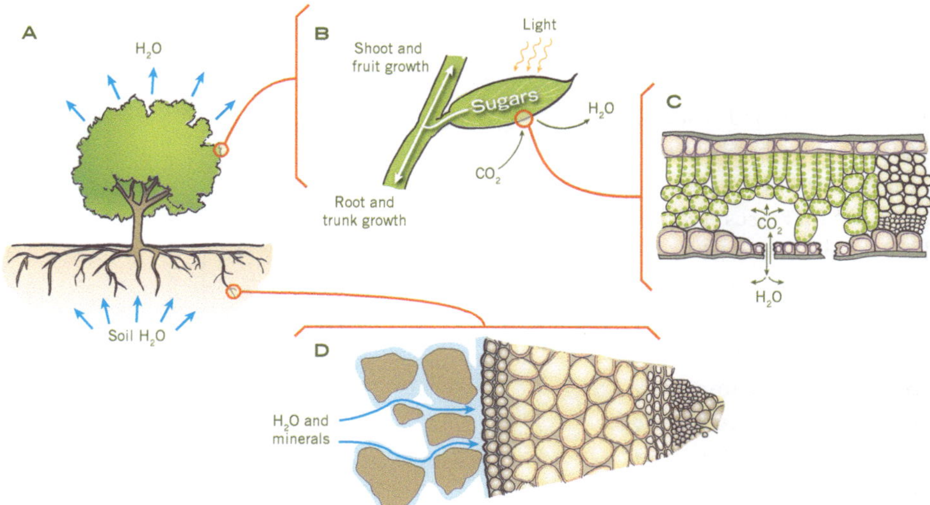

Figure 12.4 Water flow through the soil-plant-atmosphere continuum (SPAC). Water evaporates through the same stomatal pores in the leaf (C) that allow the uptake of atmospheric CO_2 for the production of sugars in the process of photosynthesis (B). The water leaving the tree canopy (A) must be continuously replaced by water uptake from the soil (D).

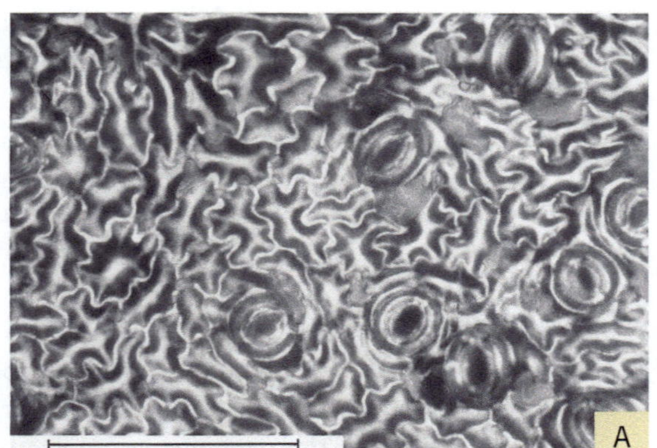

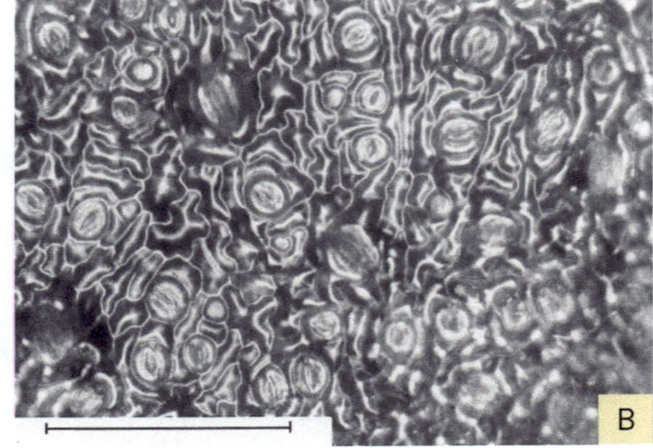

Figure 12.5 Apple leaf surfaces observed under the microscope, showing open (A) and closed (B) stomatal pores. The bar on both micrographs equals 100 µm.

pulls from its stem, much like a wick or a siphon pulls water from a reservoir. The stem, in turn, pulls water from the branch, and so on through the plant to the roots, which ultimately pull water from the soil (fig. 12.4D). The need for water is driven by the need to convert atmospheric carbon dioxide into sugars by the process of photosynthesis (fig. 12.4B). All the sugars that build the prune fruit during its development and are present in the fruit at harvest are obtained by photosynthesis. In order to allow carbon dioxide into the leaf, specialized cells on the leaf surface called stomatal guard cells (fig. 12.5) must press open a gap called the stomatal pore. This gap allows carbon dioxide uptake, but it also allows water loss (see fig. 12.4C), initiating the transpiration process. The stomatal pore can be thought of as a main door to the photosynthetic factory in the leaf, and this door must remain open for the factory to receive its most important raw material (carbon dioxide). When prune trees are under water stress, stomata close (fig. 12.5B), and photosynthesis is reduced.

Water Stress

Most of the transport of water within a plant is through the xylem, the woody tissue composed of long, hollow cells. These cells have very strong walls and are designed to withstand the tension that develops as water is being pulled from the soil. As the soil dries, it becomes harder for the plant to extract water from the soil, and the tension within the plant increases. As the tension increases, the degree of water stress within all parts of the plant increases. We can measure this stress using a pressure chamber, commonly called a pressure bomb. A leaf must first be covered to prevent moisture loss, and, depending on the type of measurement desired, either removed from the plant immediately or after a period of time (see below) and placed in a chamber with a small part of the leaf stem (the petiole) sticking out of the chamber through a seal (fig. 12.6). The chamber is slowly pressurized until the water begins to flow (upwell) out of the petiole xylem. This pressure is called the endpoint pressure (see fig. 12.6). The endpoint pressure equals the tension on the water in the leaf: a high pressure value means a high tension value in the leaf and a high degree of water stress in the tree. The units of pressure most commonly used are the bar (1 bar = 14.5 psi) and the megapascal (1 MPa = 10 bars). Because tension is the opposite of pressure, and pressure is expressed as a positive value, tensions are expressed as negative values (e.g., –1 MPa or –10 bar). An easy way to remember this is

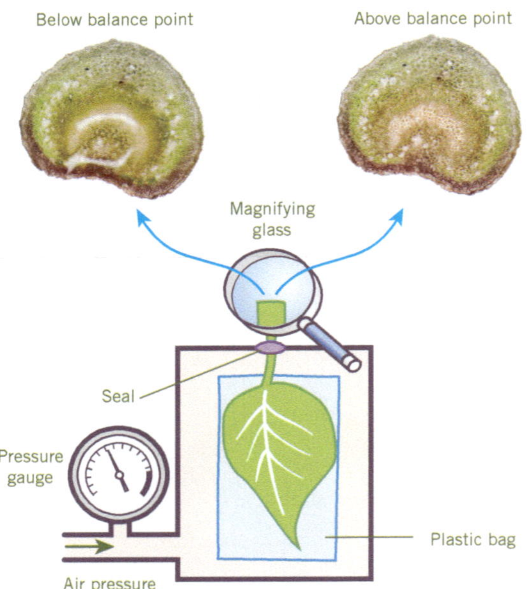

Figure 12.6 Schematic of a leaf inside a pressure chamber and photos of the appearance of the cut end of the petiole when it is dry (left, below the balance point, no water upwelling) and wet (right, above the balance point, water upwelling from the tissue).

to think of water stress as a deficit: the more the stress, the more the plant is experiencing a deficit of water. The scientific name given to this deficit is the water potential of the plant.

Stem Water Potential

For prune trees, many growers and researchers have adopted the use of midday stem water potential (MSWP) as a measure of water stress. In this method, a plastic or foil envelope is used to cover a fully expanded (mature) lower canopy shaded leaf that is close to the trunk or a main scaffold. The recommended minimum time between covering and measuring in the pressure chamber is 10 minutes, but longer times (hours or days) are acceptable as long as the leaf remains dry and undamaged. The leaf must be dry, in shade at the time of covering, and should remain in shade until the time of sampling. Choice of leaf (spur, shoot, nitrogen status, proximity to fruit, etc.) is not important, as long as the leaf is reasonably healthy and small enough to fit comfortably in the bag (flat is best, but folding is acceptable as long as the leaf does not break) and the petiole is a good size for sealing in the pressure chamber. Leaves that are slightly damaged mechanically or by insect feeding may be more difficult to read in the pressure chamber because of bubbling, but they can still give valid readings.

The correct sampling time is midday, when stress is at its peak, usually during a few hours after solar noon, between 1:00 p.m. and 3:00 p.m. (daylight savings time). Leaves should remain covered during the entire sampling and measurement procedure. Although the envelope will prevent leaf water loss after picking, the time between picking the leaf and completing the measurement should be minimized. It should be possible to complete the measurement in less than 1 minute. It is most convenient to pick the leaf from the plant by gently snapping the leaf off at its connection to the spur or shoot, then recutting the leaf petiole with a sharp razor. A dull razor or ordinary knife will usually give a rough-cut surface, which makes the endpoint (see fig. 12.6) more difficult to observe. The leaf petiole is inserted through the seal and the seal tightened. It is important to minimize the length of petiole that extends beyond the top of the seal to a few millimeters. The amount of petiole removed during recutting has no influence on the measurement, nor does the degree of seal tightening, as long as the petiole remains intact and the endpoint can be clearly observed.

With the leaf inside the chamber, the measurement is made by increasing the pressure in the chamber until water begins to come out of the petiole's cut surface. Usually, the pressure at which water appears is very definite, so it does not make much difference in the pressure reading to wait a little too long for a lot of water or stop a little short if there is evidence that water is just beginning to come to the surface. When using a good-quality hand lens (7×), the water coming out of the petiole cut surface looks like an upwelling of water from a porous surface (see fig. 12.6). The best endpoint is one where a small increase in pressure (such as 0.2 bar) causes a noticeable increase in the quantity of water at the cut end, and where a small decrease in pressure causes the water to disappear quickly back into the petiole. The rate of pressure increase itself does not influence the measurement, unless it is so fast that the endpoint is not noticed until after the pressure has passed the true balancing pressure (overshoot). The best method is to increase the pressure slowly, especially when close to the endpoint pressure. Two or more leaves on the same tree should give almost identical readings (i.e., within about 0.2 bar), so for beginners it is good practice to sample more than one leaf per tree to check for reproducibility of measurement. With experience, only one leaf per tree is necessary. Remeasuring the same leaf should get nearly the same value. This is done at the first endpoint by reducing the pressure enough that water disappears into the petiole, then increasing the pressure slowly until the endpoint appears again. Different trees can give different readings, however, and these reflect real differences in tree water potential, so it may be important to keep track of each tree separately.

For fully irrigated prune trees (i.e., no limitation in soil water), regardless of tree age, midday stem

Table 12.6. Midday stem water potential (bars) to expect for fully irrigated prune trees under different conditions of air temperature and relative humidity

Temperature (°F)	Air relative humidity (RH, %)						
	10	20	30	40	50	60	70
70	−6.8	−6.5	−6.2	−5.9	−5.6	−5.3	−5.0
75	−7.3	−7.0	−6.6	−6.2	−5.9	−5.5	−5.2
80	−7.9	−7.5	−7.0	−6.6	−6.2	−5.8	−5.4
85	−8.5	−8.1	−7.6	−7.1	−6.6	−6.1	−5.6
90	−9.3	−8.7	−8.2	−7.6	−7.0	−6.4	−5.8
95	−10.2	−9.5	−8.8	−8.2	−7.5	−6.8	−6.1
100	−11.2	−10.4	−9.6	−8.8	−8.0	−7.2	−6.5
105	−12.3	−11.4	−10.5	−9.6	−8.7	−7.8	−6.8
110	−13.6	−12.6	−11.5	−10.4	−9.4	−8.3	−7.3
115	−15.1	−13.9	−12.6	−11.4	−10.2	−9.0	−7.8

water potential will depend on the local weather conditions, as shown in table 12.6. For young prune trees, where maximum vegetative growth may be desirable, the average midday stem water potential value should be close to the baseline. If the average is more negative than the baseline value at the temperature and relative humidity conditions when the measurements are made, the trees are under water stress, and vegetative growth will probably be reduced. If trees remain more negative than the baseline even though the soil is wet, this may indicate either that the roots are not healthy and root water uptake is being impaired or that water movement through the xylem is being blocked.

Regulated Deficit Irrigation

For mature prune trees, a moderate level of water stress during the growing season has been shown to be beneficial in that it reduces fruit drying ratio, excessive shoot growth, and, in some cases, the need for irrigation by 50% or more. The purposeful withholding of irrigation for a horticultural benefit is called regulated deficit irrigation (RDI). For prune trees, the current best estimate for the desired level of water stress at different times in the season, regardless of weather conditions, is given in table 12.7.

For full-coverage irrigation systems such as flood or sprinkler, RDI values can be used as irrigation thresholds: wait until the average of the monitored trees reaches these values before irrigating. For higher-frequency irrigation systems such as drip or microsprinkler, irrigation should be started once the trees reach the target, but the frequency or duration (or both) should be adjusted during the season to keep the trees close to the target value for that time of year. For either full-coverage or higher-frequency irrigation, allowing the trees to reach a stress of –15 bars or more during late June and early July can be dangerous because fruit cracking can occur when tree water stress is suddenly reduced by irrigation. Early irrigation cutoff has not been associated with increased preharvest fruit drop, although sudden increases in tree water stress near harvest, which may occur after irrigation cutoff in shallow soil, have been associated with increased levels of preharvest fruit drop.

Since each orchard is unique, only experience will tell the grower what it takes to achieve these targets. The pressure chamber should be used to manage irrigation by fine-tuning current irrigation practices. For instance, if trees show no stress immediately after a flood irrigation and show only slight stress just before the next, it should be possible to lengthen the irrigation interval, assuming that poor water penetration does not prevent adequate refilling of the soil profile. In addition to having horticultural benefits on fruit quality, reducing the amount of applied irrigation water conserves water and reduces the flow of nutrients and other pollutants into the groundwater.

Experimental results using RDI in prunes

Three irrigation treatments were applied to drip-irrigated prune trees in Butte County, California, during 1994 to 1996 in an 18-year-old commercial French prune (*Prunus domestic* L. cv. 'French') orchard near Gridley (Lampinen et al. 2001). The soil was classified as a Gridley clay loam, and trees were planted at an 18-by-18-foot spacing on Marianna 2624 rootstock. The control irrigation treatment applied approximately 100% of crop ET (ET_c), and two RDI levels were adjusted as necessary to achieve target SWP levels of –15 bars (moderate stress, as in table 12.8) or –20 bars (severe stress) by the end of September. Figure 12.7 shows the monthly average SWP for all 3 years combined.

It is clear that for this orchard, application of 100% ET_c (based on the crop coefficients given in table 12.3 and real-time ET_o) did not maintain the control treatment trees at baseline values during

Table 12.7. Suggested target levels of midday stem water potential (bars) during the growing season in prunes

Period in season	Month						
	Mar	Apr	May	June	July	Aug	Sep
early	–6	–8	–9	–10	–12	–13	–14
mid	–7	–8	–9	–11	–12	–13	–15
late	–7	–9	–10	–11	–12	–14	–15

Table 12.8. Three-year average production data for prunes under full irrigation (control) and at moderate and severe levels of RDI, with least significant difference (LSD) at the 5% level of significance

Irrigation treatment	Applied water (%ETc)	Spring bloom (flowers/cm branch)	Total number fruit dropped/ac (× 1,000)	Fresh fruit yield (tons/ac)	Dry fruit yield (tons/ac)	Number fruit/ac at harvest (× 1,000)	Dry fruit count/lb	Number of undersize fruit (%)	Fruit hydration ratio	Side cracks (%)	End cracks (%)	Growth in trunk cross-sectional area (cm²)	Fresh pruning mass (lb/tree)
control	108%	0.161	700	18.1	6.07	5,020	61	7.04	2.99	1.17	1.37	10.53	27
moderate RDI	57%	0.166	560	17.5	6.03	5,090	62	6.80	2.90	0.78	0.93	9.68	24
severe RDI	38%	0.161	490	16.2	5.73	4,940	68	8.32	2.83	0.75	1.67	8.86	23
LSD (5%)		0.023	100	1.77	0.54	680	3	2.49	0.07	0.38	0.67	1.30	4.5

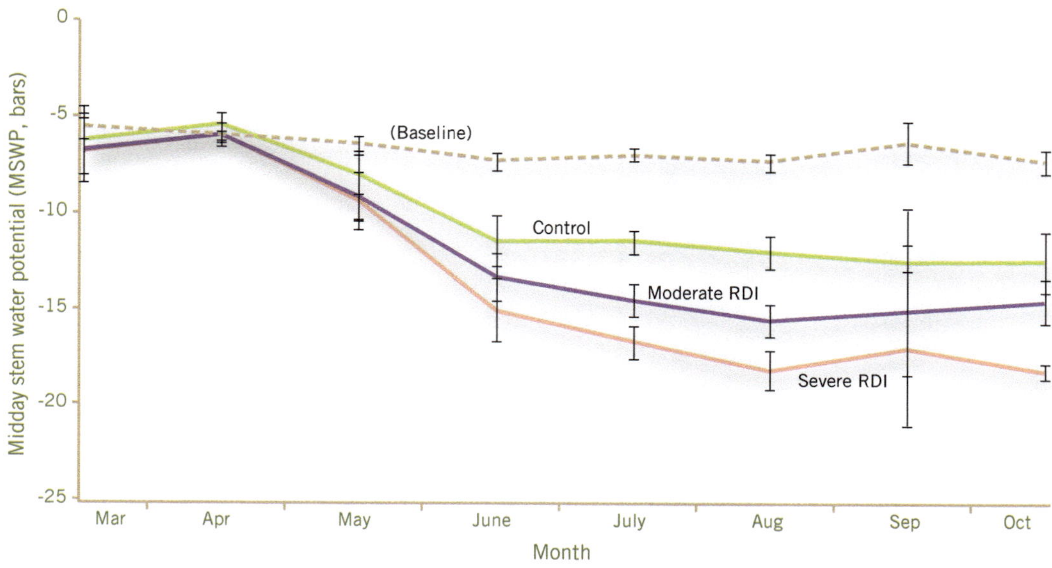

Figure 12.7 Monthly average values of stem water potential for 3 years of an irrigation experiment in prunes. Values for the three treatments (control, moderate RDI, and severe RDI), as well as the baseline SWP (see table 12.6) expected under fully irrigated (nonstressed) conditions.

most of the summer. There are many possible reasons for this: for instance, the drip irrigation system did not wet the entire soil volume occupied by roots, and hence there were roots in soil that became progressively drier as the season progressed. However, comparing the three treatments, the control had the highest SWP, the moderate RDI achieved the target value of −15 bars by mid-September (table 12.7), and the more severe RDI produced the lowest values of SWP from June to October (fig. 12.7).

Over the 3 years of the study, there was no indication that RDI was having a long-term detrimental effect on the trees; table 12.8 shows the combined years' water application and prune production data. The only statistically significant effects of the moderate RDI compared with the control treatment were decreases in fruit drop and fruit hydration ratio, both of which are economically beneficial. Long periods of irrigation deprivation followed by flood irrigation have been associated with fruit cracking in prunes, but cracking was not found to be a problem in this RDI study. The average reduction in applied water was also substantial (about 50%), which may be economically significant, depending on water and pumping costs. These data show that prune production is relatively tolerant of stress, and that moderate levels of stress may be economically beneficial.

References

Lampinen, B. D, K. A. Shackel, S. M. Southwick, and W. H. Olson. 2001. Deficit irrigation strategies using midday stem water potential in prune. Irrigation Science 20:47–54.

13 Orchard Planning, Design, and Planting

• G. Steven Sibbett, John P. Edstrom, and Wilbur O. Reil

The grower who is establishing a prune orchard must consider a number of factors if the trees are to mature successfully and produce regular crops of high-quality prunes. In addition, returns should exceed harvesting and processing costs as soon after planting as possible. This chapter discusses the factors that affect the success of a new prune planting.

Orchard Planning

Site Selection

Selecting a site is a critical step in establishing a new prune orchard. The main considerations are soil, water availability, water quality, and climate.

Soil

When assessing a possible site for an orchard, the important physical properties of the soil to consider are depth and texture (see chapter 8, "Soil Evaluation and Physical Modification"). Prune trees develop a moderately deep root system, with some roots growing 4 feet deep or deeper. For this reason, prune trees grow best on deep, unstratified, medium-textured, loamy soils. A shallow hardpan or other stratified layer in the soil profile or a high water table that causes saturated soil conditions can restrict the depth to which the trees can root, thereby limiting tree size and vigor. Soil profile modification practices, such as ripping or slip plowing, can in some cases alleviate restrictions imposed by hardpan or stratification. A backhoe can be a useful tool for evaluating soil depth and profile.

Prunes are often planted on less than perfect soils, the best soils being used for other more soil-sensitive tree crops. Such less-than-perfect soils might include heavier, clayey soils, those that are quite sandy, or shallow soils. Smaller tree size can be expected when planting in these types of soils, so closer tree spacing can be used to maximize productivity (see the section "Tree spacing," below).

The most important chemicals in soil for prune production are soluble salts, including calcium (Ca^{2+}), sodium (Na^{+}), magnesium (Mg^{2+}), chloride (Cl^{-}), sulfate (SO_4^{2-}), bicarbonate (HCO_3^{-}) and carbonates such as exchangeable sodium and boron (B) (see chapter 7, "Managing Salinity in Soils and Water"). Prune trees need some boron and soluble salts for growth, but they will die if large amounts are present. Too much exchangeable sodium in a soil is not only toxic but also affects soil structure and water penetration. High levels of chloride can cause leaf burn. Soil pH is another important factor to consider. Prunes grow best in soils with a pH from 5.5 through 7.5.

To determine the chemical properties of the soil in the area under consideration, take soil samples from different sections of the site at 1-foot increments to a depth of 6 to 7 feet. Ask an agricultural laboratory's guidance in taking soil samples for chemical analyses. The local University of California Cooperative Extension office does not perform soil analysis but does offer help in interpreting such analyses.

Another valuable source of information is the soil survey for the county or area in which the possible site is located. The soil survey consists of a map showing the types of soils present, their distribution, and their acreage. It also describes each soil type, including information about drainage, flooding, alkali (exchangeable sodium) content, and similar details that can indicate whether a soil is suitable for growing a particular crop. If a soil survey is not available for distribution, look at one at your local Cooperative Extension office or the local U.S. Department of Agriculture (USDA) Natural Resource and Conservation Service (NRCS) office. Most soil surveys are now on the NRCS website, http://www.websoilsurvey.nrcs. California county maps are also available at the UC Kearney Agricultural Center's Geographic Information

Systems Facility website, http://www.gis.uckac.edu.

Soils previously planted with tree or vine crops should be assessed for nematode populations, particularly for root lesion nematode *(Pratylenchus vulnus)* and ring nematode *(Criconemoides xenoplax)* (see chapter 19, "Plant-Parasitic Nematodes"). Soils containing root lesion nematode should be fumigated before planting prune trees in them, or they should be planted to a crop that is not a host to these nematodes.

Water

Many regions of California are semiarid. For this reason, plan on supplying supplemental water to the orchard (see chapter 11, "Irrigation Systems"). Although in a few small districts growers still dry-farm prunes, the quantity they produce is low. Investigate the possibility of obtaining supplemental water from surface delivery systems or wells.

For optimal production, mature prune trees must extract 38 to 40 inches of water per year, depending on the growing district. Supplemental irrigation must provide the part of the water requirement that rainfall does not supply. When irrigating, an additional amount of water (from 5 to 30%, depending on the irrigation system) must be added to allow for runoff, evaporation, and the other inefficiencies that exist in all irrigation systems. Be sure to select an area where an adequate supply of quality water is available. Prune trees cannot tolerate water that contains an excess of boron, chloride, carbonate, bicarbonate, or other salts.

Climate

Temperature extremes can severely affect prune production. Once growth begins in spring, blossoms and prune shoots are easily killed by freezing temperatures. Low-lying areas in an orchard may accumulate cold air, increasing the potential for frost damage.

Once growth begins, prunes prefer minimal rainfall. Excessive cold, wet conditions during and shortly after bloom encourage the brown rot fungus. The physiological disorder lacy scab damages young fruit soon after bloom if conditions are continually wet. In summer, rainfall or heavy dews prior to harvest encourage both brown rot and leaf rust fungi (see chapter 18, "Diseases and Physiological Disorders"). Brown rot damages the fruit, and prune rust defoliates the tree. Both reduce fruit quality. Sites with warm, dry growing conditions during the season are preferred for prunes. Hot, dry, windy conditions during bloom may reduce fruit set.

Design

When planning a new prune orchard, tree spacing and design decisions are fundamental to its ultimate profitability. The prune grower's objective is to plan and design an orchard that attains economic productivity as soon as possible after planting, then develops and maintains optimal productivity of high-quality fruit for as long as possible. A good orchard design with optimally spaced trees also provides for efficient equipment operation and cultural practices that minimize costs.

Tree spacing

When a tree's fruit-bearing wood completely occupies its periphery and orchard trees are spaced at the maximum density to still allow sufficient light penetration to maintain each tree's productivity, the orchard is said to be at its optimal bearing potential. Before an orchard reaches its optimal bearing potential, an acre's productivity is directly related to the number of producing trees on it. The prune orchardist must consider both early and long-term production when making initial tree spacing decisions. The following are important considerations.

Sunlight. Sunlight is the most important factor affecting tree growth and production of high-quality fruit. Closely spaced prune trees that are allowed to crowd each other lose productivity as fruitwood shades and dies out, and any prunes produced in shaded conditions are small because of low sugar content, causing a very poor drying ratio. Such prunes are expensive to produce and are of low value and reduced overall quality (see table 13.1). Tree spacing must provide adequate light to all parts of the tree at the least management cost throughout the orchard's life. Excessive crowding at tree maturity due to unmanaged high tree density for early production is extremely costly in the long run.

Tree vigor. Inherent tree vigor determines the potential size of a tree, but trees are also influenced by site and cultural factors. The warm, long growing season of the Central Valley prune-growing districts promotes maximum vigor. Trees planted in the cooler coastal districts are less vigorous. Prunes growing in well-drained fine sandy loam to loam soils will be the most vigorous,

Table 13.1. Canopy location effects on fruit set, fruit weight at reference date, fresh and dry fruit weight at harvest, and drying ratio in a high-density prune orchard

Canopy location	Fruit set %	Fruit weight at ref. date	Fresh (g)	Fruit weight at harvest Dry (g)	Dry ratio
north top (sun)	45.0	4.15	18.3	5.3	3.4
south top (sun)	53.3	4.30	19.3	5.8	3.3
north bottom (shade)	37.5	3.68	16.0	4.2	3.8
south bottom (shade)	44.2	3.86	16.8	4.7	3.6

while trees growing in heavy or shallow soils are not as vigorous and can be spaced closer together to ensure optimal productivity. Prunes can be grown on peach or plum rootstock (see chapter 6, "Rootstocks"). Plum-rooted trees are generally less vigorous than those growing on peach. Peach-rooted trees can be excessively vigorous when placed in good, light soils. Although attention to cultural practices is essential to production of high-quality fruit, good culture promotes tree vigor, requiring more attention to tree crowding.

Tree spacing options

Conventional spacing. Conventional spacing remains popular with prune growers. Conventional spacing for prune trees ranges from 16 to 22 feet between trees and 18 to 22 feet between rows (151 to 90 trees per acre, respectively). Although these spacing options are productive, they sacrifice some early production per acre. Some orchardists plant a row crop in between conventionally spaced trees to generate income and tax deductions before the prunes come into economic bearing. This strategy is to be viewed with caution, as attention to the intercrop can often stunt the prune trees, adding a further delay to bearing.

Conventional spacing, including filler trees. To offset early production loss in conventionally spaced trees, interplanting with temporary "filler" trees is an option. Filler trees are destined to be reduced in size and removed altogether as they crowd conventionally spaced trees. Unlike some other tree crops, this practice has not been popular with prune growers, likely due to the size of the equipment required for harvest and the difficulty in removing the trees.

High-density planting. High-density plantings, where in-row spacing is 9 to 14 feet and row spacing is 18 to 22 feet (268 to 141 trees per acre, respectively), are becoming popular because of their early heavy productivity per acre before maximum tree size is reached. Such trees are permanently placed in the orchard, and the rows are often managed as a hedgerow. Tree removal is not considered. Although productivity can be exceptional, high-density orchards must be very well managed to be profitable. Cultural practices such as pruning, fertilization, and thinning are usually more expensive on a per-acre basis and must be carefully adapted to avoid quantity and quality loss. Tree height must be reduced to prevent shading out of the lower canopy. Conventional harvest machinery must also be adapted to the closer in-row spacing. Production and quality from one high-density planting is presented in table 13.2 to demonstrate potential the detrimental effects of alternate bearing, fruit size, and drying ratio. In windy locations, place tree rows parallel to the prevailing wind.

Table 13.2. High-density yield at designated tree age for grower treatment, 1984–1993*

	1984 4th leaf	1985 5th leaf	1986 6th leaf	1987 7th leaf	1988 8th leaf	1989 9th leaf	1990 10th leaf	1991 11th leaf	1992§ 12th leaf	1993 13th leaf
Dry yield (ton/ac)	2.77	2.88	3.81	7.95	5.53	9.18	6.74	2†	8.12	5.17
count/lb	37	58	87	96	61	78	63	6.5†	67	79
drying ratio	20	2.79	3.46	3.04	3.16	3.26	2.23	3†	2.97	3.78
% undersize	—	—	—	33†	6	18	3	5†	8	14

Notes:
*Trees spaced 9 feet apart in row with 16 feet between rows (303 trees/ac).
†Estimated.
§For 1992, the dry yield and count per pound were determined after the undersized prunes were removed.

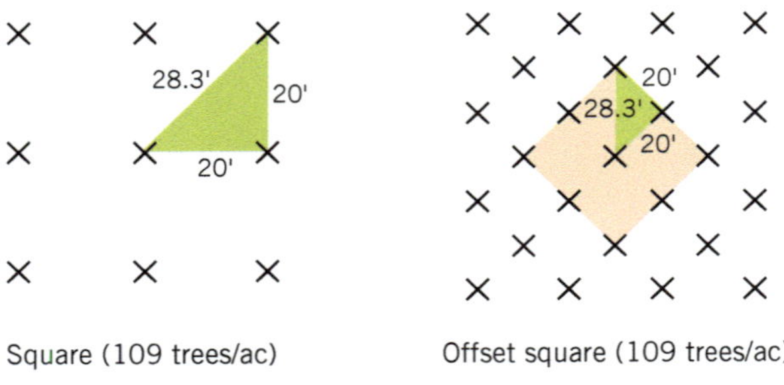

Figure 13.1 The four most commonly used orchard designs for prunes.

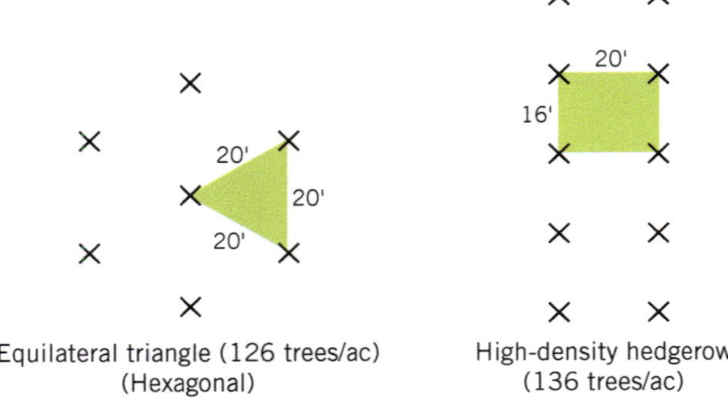

Orchard Design

The best planting design must provide for maximum tree canopy exposure to sunlight for the planned number of trees as well as efficient orchard operations. An efficient design considers the size and operation of orchard equipment, the irrigation system, and whether a filler tree strategy will be used. The following four basic design options are available to prune growers (fig. 13.1). The number of trees per acre at varied spacing for each orchard design is given in table 13.3.

- **Square.** The square is a popular and common design because it is easy to lay out and can accommodate a filler tree strategy. Trees are placed at equal distances down and between tree rows. Orchard operations can be easily conducted in two directions in this design.
- **Offset square.** The offset square is similar to the square except that the trees are offset one-half tree space in every other row. An offset square design facilitates spray coverage. Where equipment, such as pruning towers, can work on two rows at a time an offset square can be more efficiently managed than the square because pruning towers can "swing" from tree to tree rather than continually having to back up to prune an adjoining tree. There is no improved light access with an offset square design, as is often believed. At any given tree spacing the square and offset square have the same number of trees per acre and the same square footage per tree allotted for sun access.
- **Equilateral triangle.** For the equilateral triangle (also known as a hexagonal or diamond) design, trees are spaced the same distance apart when measured in any direction from any other tree; a line connecting any three adjoining trees forms an equilateral triangle. The equilateral design is the most efficient use of the orchard space at any given tree distance, as it allows approximately 15%

Table 13.3. Number of trees per acre for square or offset square, equilateral triangle, and hedgerow orchard designs

Distance between trees (ft)	Square or offset square	Equilateral triangle
16	170	195
18	134	155
19	120	138
20	109	126
21	99	114
22	90	104
	Hedgerow	
Distance between trees (ft)	Distance between rows (ft)	Trees per acre
9	16	303
18	10	242
18	12	202
18	14	173
20	10	218
20	12	182
20	14	156

more trees per acre than a square or offset square design. The design is adaptable only to permanently placed trees, however. Filler trees cannot be efficiently used and removed in this design.

- **High-density hedgerow.** The high-density hedgerow design places trees closer in the row than between the rows. This design is usually managed as a hedge, with the row, not the tree, being the producing unit in the orchard. Hedgerows shade rapidly down the tree row and must be managed very carefully if alternate bearing and small fruit are to be minimized. Hedgerows should be planted in a north-to-south direction for best light distribution.

Planting the Trees

Delivery and Storage

June-budded prune trees are 1 year old when they are dug from the nursery. They are removed bare root in late December or early January for delivery and planting prior to bud break in spring (fig. 13.2), and they are delivered to the grower as soon as possible after digging. If trees are not planted immediately on receipt, they should be heeled in to prevent root desiccation, that is, they should be placed with their roots in a trench in a well-drained area; the roots are moistened and covered with soil to prevent drying out and freezing (fig. 13.3). Heeled-in trees can be successfully stored in this condition until warming temperatures initiate tree growth. If later planting is anticipated, the trees should be placed in moist wood shavings or sawdust in cold storage (36° to 40°F) to prevent growth until planting.

Digging the Tree Holes

Tree holes should be dug once the orchard has been marked and just before planting. For most prune trees, tree holes are hand-dug 12 to 18 inches wide and deep enough to accommodate the root system. Prune root systems are relatively small compared with the root systems of other trees such as walnut, so large augers are rarely used. Deep tree holes are not desirable because excessive settling occurs after planting, and soil can cover the graft union. Tree holes should be dug when the soil is not excessively wet. If tree holes are dug when soils are wet, the sides become glazed with the shovel. Glazed sides resist water movement and root growth and should be roughened prior to planting.

Planting the Trees

The trees must be planted as soon as possible in winter. As winter progresses, root growth begins before top growth. Nursery trees that have new root support before top growth begins will be more vigorous than those whose top growth begins before root growth occurs, as with later plantings. Nursery trees in cold storage can be planted through June. Such late planting is successful, but top growth the first season is minimal. Before planting, the tree's roots should be pruned to further accommodate the tree hole, and any

Figure 13.2 Bare-root nursery trees just delivered from the nursery. *Photo:* G. S. Sibbett.

Figure 13.3 Bare-root nursery trees heeled in to prevent desiccation of the root system prior to planting. *Photo:* G. S. Sibbett.

Figure 13.4 Bare-root nursery tree with roots pruned. *Photo:* G. S. Sibbett.

damaged root tissue should be removed (fig. 13.4). Place the new tree so the roots are well spread and distributed in the tree hole, not crammed, kinked, circling, or bent upward. Lift the tree to the point where its nursery soil line matches that of the orchard floor, then fill with granular soil and tamp it firmly around the root system. Do not place fertilizers into the tree hole at planting.

Postplanting Care

If soils are moist at planting, irrigation is not necessary until 4 to 6 inches of new growth has developed. If soils are dry at planting, they should be irrigated, or "tanked" (basin irrigated), to further firm the soil around the root system. They should be fertilized only after growth is well along and not at planting. A 50-50 mix of white latex paint mix and water should be applied at planting to the entire top to prevent sunburn during the first growing season. A safe preemergent herbicide can be applied pre- or postplanting down the tree row to prevent weed competition for moisture and nutrients. Begin irrigating when tree growth begins.

14 Pruning and Tree Training

• William H. Krueger, Theodore M. DeJong, and William H. Olson

Physiological Principles of Pruning and Training

Tree training and pruning should be approached with an understanding of tree functions, realizing that trees have their own internal controls that adapt them for survival in their native habitat. It is necessary to train and prune trees to work with those natural growth tendencies to optimize cropping efficiency rather than species survival.

Fruit trees are trained and pruned for several reasons. Early pruning, commonly called training, is done to obtain vigorous, mechanically strong trees with good branch angles that are properly shaped for convenient orchard management (mechanical harvest, pest control, etc.). Mature trees should be pruned to increase the uniformity of light distribution within the tree canopies so that leaf photosynthetic activity can be optimized throughout the tree and fruitwood can be maintained in as much of the tree as possible. Pruning can also help regulate annual crop production and improve fruit size.

Principles of Tree Physiology and Growth

Regardless of the objectives that are being considered in developing an effective pruning strategy, it is necessary to understand some of the basic principles of plant growth and physiology. The tree can be thought of as a framework designed to efficiently display solar energy collectors (leaves). The objective with regard to solar energy collection is to maximize leaf exposure to light while maintaining optimal uniformity of light exposure between different leaves on the same tree. The light energy that is captured is transformed into chemical energy through photosynthesis. While the energy is transported around the plant in the form of carbohydrates (sugars), fruit and fruitwood located in close proximity to where photosynthesis occurs have the greatest access to available photosynthates and therefore are the principal sites of fruit crop productivity. Areas of a tree canopy that are heavily shaded will be less productive than well-exposed portions of the same canopy. Similarly, young trees fully exposed to light, with adequate supplies of water and nutrients and good growing conditions, will grow in relation to active leaf surface area. Any reduction in surface area causes a net reduction in tree growth. Therefore, it is important to shape, form, and maintain trees with as little reduction in leaf area as practically possible. In order to do this, one must be familiar with some basic facts of vegetative growth.

Vegetative Growth

Primary and secondary growth

Aboveground vegetative growth consists of extension (primary growth) of shoot or branch length, and diameter growth (secondary growth) that increases branch or trunk girth. Primary growth is initiated in the shoot meristems (lateral or terminal buds), where regions of rapid cell division and differentiation give rise to the tissue types that are later found in leaves, stems, and fruits. These physiologically active shoot meristems are also the source of several natural growth-influencing compounds called plant hormones.

Three types of primary shoot growth occur in prune tree canopies. Most shoots arise from terminal or lateral vegetative buds that develop in one season and grow out in the next. These types of shoots are called proleptic shoots. Vegetative buds that grow into proleptic shoots generally contain several preformed nodes (up to 12) that simply expand in early spring. These shoots often stop growing shortly after bud break and form short shoots or spurs. This is referred to as preformed growth. A subset

of proleptic shoots continues to form additional nodes and extension growth after the preformed nodes fill out; this is called neoformed growth because the growth is the result of newly formed nodes developed in the same season as the axial growth occurs. Thus, some proleptic shoots consist of preformed growth, while others consist of both preformed and neoformed portions. Most proleptic shoot growth is complete by the end of spring, and these are the primary shoots that are responsible for developing flower buds and bearing the next year's crop.

In very vigorous growing conditions, some lateral buds may grow out in the same season as they are formed; in this case they are called syleptic shoots. Syleptic shoots are most often found on very vigorous shoots called water sprouts. Most water sprouts develop from epicormic buds or meristems that form lateral to the positions where there was formerly a lateral vegetative bud. Thus, they most often arise from stems that are 2 years old or older. Epicormic meristems are often stimulated to grow by a traumatic event that occurs distal (farther up the stem) from the meristem, such as a pruning cut. Heavy pruning into older wood tends to increase the amount of water sprout growth in a tree. Unlike proleptic shoot growth, epicormic (water sprout) shoot growth often does not stop in the spring but continues to grow vigorously as long as environmental conditions are favorable for growth, usually into the fall.

Secondary growth originates in two lateral sheaths of rapidly dividing cells called the vascular and cork cambia. The vascular cambium is the more important of the two. It is responsible for the development of wood (xylem) on the inside of the tree and sugar and/or nutrient conduction tissue (phloem) to the outside. The cork cambium is a layer of actively dividing cells that forms to the outside of the phloem each year and produces a protective layer known as cork. The cork oak *(Quercus suber)* is an example of a tree that has a particularly prolific cork cambium that is responsible for the production of commercial cork. Most commercial fruit trees have minimal development of the cork cambium.

On a daily basis, both primary and secondary growth are highly responsive to temperature. Temperature is closely linked to respiration, the process by which energy is liberated from carbohydrates to do cellular work (in this case, growth). Daily patterns of growth are moderated by various environmental conditions such as water, nutrient, and salinity stresses. Most extension shoot growth actually occurs in the afternoon when temperatures are high, the water status of the tree is recovering from a midday low, and carbohydrates are abundant from daily photosynthesis.

On a seasonal basis, vegetative growth is minimal during the dormant season. In temperate deciduous crops, growth is reactivated in the spring after exposure to periods of cold, followed by postchill heat units. In mature prune trees, growth is greatest (nearly exponential) from bud break until late spring. Proleptic shoot growth, completed in spring, is referred to as the "grand period of growth." It is responsible for more than 90% of the total extension growth that occurs in a given year in mature trees. In young, vigorous trees, a significant amount of extension growth can occur later in the summer, because there is more neoformed growth in these trees. Young trees are usually pruned harder relative to their size, and this induces more epicormic shoot growth.

Unlike primary growth, secondary growth occurs more or less uniformly throughout the growing season from bud break to leaf fall, provided environmental conditions are adequate and there is no excessive competition for carbohydrates from other active processes such as fruit development and primary growth. Some research indicates that under natural circumstances there may be mechanisms for coordinating some growth processes to avoid competition for carbohydrates at particular times during the season. For example, flushes of root growth may occur before the early flush of primary shoot growth and after the final stages of fruit development.

Vegetative shoot growth

Vegetative shoot growth is influenced by four factors: genetic potential, morphological origin, endogenous regulation, and environmental stimulation. The genetic potential sets the maximum limit to the growth rate of a given organ. We have little ability to know when a leaf, shoot, or root is growing at its maximum genetic potential, but it is well documented that some cultivars have different genetic capacities for growth than others.

As discussed above, morphological origin (i.e., prolepsis, sylepsis, or epicormic) appears to influence whether the shoot terminates growth early in the growing season or continues to grow later into the summer, but this may just be a function of proleptic shoots being more susceptible to endogenous regulation than are syleptic and epicormic shoots.

Endogenous regulation of growth is important when considering pruning practices. Growth is regulated within the plant by hormonal control and metabolite regulation. In metabolite regulation, growth is limited if metabolites such as carbohydrates and nutrients needed for growth are in short supply at any given growing point. This type of regulation is thought to occur through competition for metabolites among different organs that have the potential for simultaneous growth. For instance, rapid shoot growth may limit root growth, or rapid fruit growth may limit shoot growth because of a lack of metabolites to simultaneously support growth of both organs.

Hormonal control of growth is very complicated and not well understood. There is evidence that hormones produced in root tips influence shoot growth, and vice versa. There is also some evidence that hormones produced in developing seed in fruit influence the growth of various organs. However, one basic phenomenon is well documented, predictable, and central to the understanding of pruning responses. This is the general phenomenon of apical dominance, that is, the control of growth and development of lower buds or shoots by more distal buds or shoots. In trees, there are three major manifestations of apical dominance: correlative inhibition, apical control, and shoot epinasty. Correlative inhibition is the suppression of lateral buds by a vigorously growing apical meristem during the current season's growth. Correlative inhibition is thought to be caused by the production of hormones (mostly auxins) by the rapidly growing terminal meristem that suppresses the growth of the lateral buds. Whenever the terminal meristem is damaged or removed during the active growing season, one to three lateral buds closest to the shoot apex will begin to grow.

Apical control is the tendency for terminal and upper lateral shoots to depress the growth of lower, more basal (subordinate) shoots. This phenomenon is most readily apparent on shoots growing on the previous year's wood. Usually the terminal bud and the upper one to three lateral buds will grow and produce relatively vigorous shoots. The growth of shoots arising from lateral buds on that 1-year-old branch will be progressively less the farther they are from the vigorous shoots near the apex. Often these short shoots near the base of the 1-year-old wood are referred to as spurs. Their growth is controlled (limited) by hormones arising from the more active growth near the terminal end of the branch. Mature prune trees tend to set more fruit on spurs than on longer extension shoots, and therefore the development of these spurs is important for cropping.

Shoot epinasty is the tendency for actively growing upper shoots to influence the branch angle of lower shoots (usually making them wider). This is primarily manifested on new shoots arising from a 1-year-old branch. The upper, more vigorous shoots tend to have narrow angles of attachment. Branch angles of new shoots tend to become wider or flatter the farther down they are the 1-year-old branch. Hormonal control mechanisms for this phenomenon are thought to be similar to those for apical control.

If the apex of a shoot is removed by pruning, the natural apical dominance of the shoot is disrupted. If the pruning is done in the dormant season, the control of the apical shoots over the more basal shoots is rapidly reestablished, but usually the upper two or three shoots arising below the pruning cut will be nearly upright and equal in vigor and become codominant.

The final phenomenon controlling vegetative growth is environmental stimulation. If a given shoot has the genetic potential for a certain growth rate and the shoot is in a position on the branch such that it is not being endogenously controlled by other plant parts, the environmental conditions still must be favorable for rapid growth to occur. Growth is very sensitive to temperature because growth involves tissue respiration and enzyme activity. The temperature optimum for shoot growth of most temperate fruit crops is between 77° and 86°F. Growth is also sensitive to water stress because turgor (internal water pressure) is needed to drive cell expansion and metabolism. The direction of growth is also sensitive to light (phototropism) and gravity (geotropism), but these are not too important for our consideration because light and gravity seldom elicit other than normal phototropic or geotropic responses in orchards. Most of the so-called growth responses to light seen in orchards are probably more related to carbohydrate availability than a hormonal phototropic response.

Responses to Pruning

Once we have a rudimentary understanding of the timing and regulation of vegetative growth, we can begin understanding and even predicting responses to certain types of pruning. There are two basic types of pruning cuts: heading cuts and thinning cuts. With heading cuts, a terminal portion of a shoot or branch is removed and a basal

portion remains attached to the tree. With thinning cuts, entire shoots or branches are removed at the base. The different responses to heading and thinning cuts are illustrated by figure 14.1 and the accompanying table 14.1. In this figure, there are two idealized dormant shoots. For the purposes of this illustration, the upper two shoots on both branches are the same length; the next two are smaller than the first pair but equal to each other, and so on. The dashed lines across the shoots indicate pruning cuts. In the branch on the left, three shoots are eliminated by thinning cuts. In the branch on the right, the same length of wood is removed by six heading cuts such that the terminal half of the length of each shoot is removed. The invigorating response to pruning is predicted to be much greater with the heading cuts than the thinning cuts for the following reasons:

- Heading cuts remove all of the terminal buds, and therefore all of the natural growth-controlling mechanisms related to apical dominance are temporarily eliminated.
- Heading cuts remove more than 50% of the lateral buds because the number of buds per shoot length is usually greater on the terminal half of the shoot than the basal half. Therefore, the number of potential growing points on the headed branch is less than on the thinned branch. This should stimulate more vigorous growth from remaining buds.
- Heading cuts remove proportionally less mass of wood and stored carbohydrates than the thinning cuts because the basal half of the shoots are greater in diameter and often contain two to three times as much carbohydrate per unit dry weight as the terminal half. This means that the remaining lateral buds on a headed branch would have access to proportionately more carbohydrates than the same buds on the thinned branch.

Although this illustration is idealized, it demonstrates the fundamental reasons for differences in growth responses to the two different types of pruning cuts. An important consequence of these different pruning responses is that in the year after pruning there will probably be more spur development on the thinned branch than on the headed branch, but the overall length of new shoot growth will be greater on the headed branch. It is up to growers to apply these principles in their own situations. For instance, the long-pruning and short-pruning methods for early training of prune trees are based on these principles. In situations favorable for tree growth, long pruning with minimal use of heading cuts can promote rapid spur development on vigorous trees, resulting in increased early fruit production. On the other hand, in less favorable sites, early fruitwood development may limit the shoot growth necessary to rapidly fill the allotted space. Heading cuts can be used to invigorate vegetative growth and delay cropping until adequate tree size is obtained.

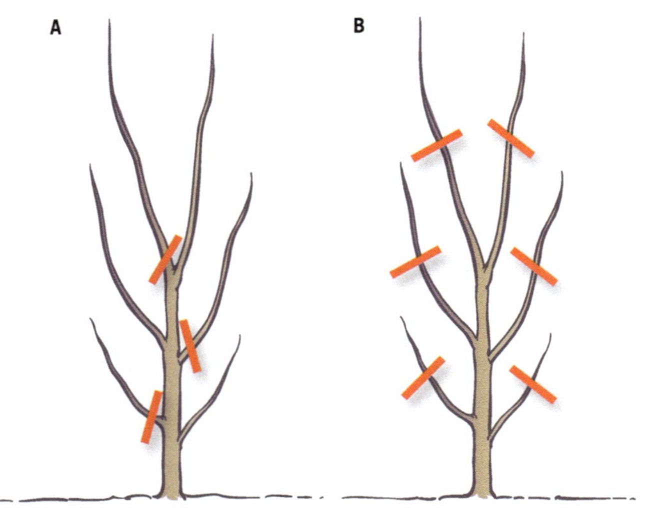

Figure 14.1 Thinning versus heading cuts.

Pruning responses also depend on the age of the wood that is being removed. Removing large limbs or heading older branches often stimulates the growth of epicormic shoots (water sprouts), which can be positive if one is trying to replace old branches or invigorate the canopy. In vigorous trees, such cuts will generate excessive new growth, and this can develop into a futile cycle of repeated pruning that reduces crop yield. In this case, the grower would be better off to stop or reduce pruning for a year since epicormic shoots will not grow in mature trees unless there are heavy pruning cuts or limb breakage.

Pruning responses are also highly dependent on timing. Pruning in the dormant season is the most invigorating because leaf area is not removed

Table 14.1. Relative effects of thinning versus heading cuts

	Thinning	Heading
terminal bud removal	50%	100%
lateral bud removal	~ 50%	> 50%
mass of wood removed	~ 50%	< 50%
loss of stored carbohydrate in wood	~ 50%	< 50%
invigoration effect	small	large
relative spur formation the next year	more	less

at the time of pruning. Dormant pruning increases the relative amount of carbohydrates and nutrients available to the potential growing points left on the tree. Pruning in spring and early summer, after the trees have expended substantial amounts of energy developing leaves and shoots but before the tree receives any return on its investment in those leaves, can be particularly debilitating. However, if the pruning is done in early spring during the "grand period of growth," it will generally stimulate new growth in the area of the pruning cuts. After summer dormancy begins to take hold in mid-July, responses to summer pruning will decrease until leaf fall. In mature trees, pruning in August or September often produces very little stimulation for new growth in the same year. Responses to heading cuts are also minimal during this time because the small amount of shoot growth they stimulate late in the season may just reestablish the apical dominance and control effects before the tree goes dormant for the winter.

With the knowledge of the factors regulating which shoots grow and a sense of the timing of that growth, the grower can obtain a desired pruning response. There is no one best way to prune. Cultivars vary greatly in growth and fruiting habit, vigor, and so on. Improved French is the primary cultivar in the industry. If new cultivars become important, these differences must be taken into account. Orchard sites vary greatly in soil type, nutrients, climate, and so on; good orchard management requires the ability to adjust pruning strategies to fit the goals and situation of the individual orchard.

Pruning and Training Young Trees

The objective of pruning and training young trees is to bring them into production as soon as possible while developing an upright tree structure capable of sustained production of heavy crops. Pruning reduces tree size and delays the onset of production, but it can increase the structural strength of the tree and set the tree up for good sunlight distribution within the canopy.

Pruning Equipment

Pruning shears are used for making most of the cuts associated with training young trees; a collapsible pocket saw can be used for making an occasional larger cut. A ladder may be necessary for making heading cuts in the top of the trees in the second and third dormant pruning. Pruning shears should be kept sharp. When making cuts, the hook of the shears should be placed on the top of the limb to be removed with the blade of the shear closest to the trunk or scaffold branch. The limb should be cut by upward motion of the blade. This will avoid splitting or breaking the limb. The cuts should be made just outside the collar of where the limb originates from the parent limb (fig. 14.2). This makes the smallest wound possible and encourages faster healing.

Figure 14.2 Proper limb removal. *Photo:* M. L. Poe.

Heading Versus Thinning Cuts

In a heading cut, a terminal portion of a limb is removed at some point on the limb. As discussed above, heading cuts remove the apical dominance exerted by the growing tip and cause strong upright vigorous growth near the cut. The more severe the cut, the more pronounced the response. Growth originating nearest the cut will be the most upright and vigorous. Growth originating farther from the heading cut will be less vigorous and have flatter angles of attachment. Heading cuts invigorate trees and promote branching at a desired location. With thinning cuts, the entire limb is removed from the parent limb. This is invigorating, but less so than with heading cuts. Thinning cuts are used to direct growth, thin the canopy, manage light within the canopy, control cropping, and control tree height.

Time of Pruning

Most pruning for training young trees is done during the dormant season because the invigorating response is usually desired and trees will be less subject to sunburn than if they were pruned in the summer.

Prunes are typically trained to an open center system with three or four primary scaffolds spaced approximately equally around the tree. The center

of the tree is kept relatively open to allow light to filter down through the canopy, providing adequate light to maintain healthy wood and production of high-quality fruit in the lower part of the canopy.

At planting

When trees are dug from the nursery, most of the roots are left behind. At planting, head trees at 32 to 42 inches and remove all side branches to rebalance the top with the root system and promote vigorous growth for primary scaffold development. Trees should be headed high enough to allow primaries to develop with good separation between scaffolds; the lowest scaffold should be high enough to allow shaker attachment to the trunk (usually at least 2 feet). Heading the tree higher will result in more opportunities for vertical spacing of the scaffolds and a longer area for shaker attachment. The canopies of higher-headed trees develop more leverage and may be more subject to blowing over, especially if they are planted on a high berm perpendicular to prevailing winds.

First growing season

It may not be necessary to prune at all during the first growing season. Retain as much of the leaf surface as possible to maximize tree growth. If desired, direct growth into the permanent limbs by pinching back growth that is in undesirable locations or that competes with the limbs selected to be scaffolds. By just pinching back the growing tips, the growth of these limbs is slowed and directed into the permanent limbs while a majority of the foliage is left to develop the trunk and roots. Undesirable growth can be completely removed during the first dormant season.

Good cultural practices during the first growing season include adequate irrigation, fertility, and pest control (especially weeds), which ensure good growth and yield more choices for scaffolds at the first dormant pruning.

First dormant season

During the first dormant pruning, select permanent primary scaffolds (fig. 14.3). Choose three or four scaffolds with an upward and outward orientation spaced approximately equally around the tree. These scaffolds are the beginning of the open-centered prune tree. Significant vertical spread of the scaffolds on the trunks is desirable to ensure the best chance of a strong branch attachment and reduced future breakage. One or two inches of separation may seem adequate at initial selection, but when the trunk has developed in a few years they will look as though they all originate from the

Figure 14.3 First dormant pruning: primary scaffold selection and heading length: long (A), intermediate (B), and short (C). *Photos:* M. L. Poe.

same place. Avoid selecting limbs with included bark (fig. 14.4). A bark ridge should be visible on the upper side of the limb where it attaches to the trunk (fig. 14.5). If in doubt, pull the limb down gently. If the branch separates from the trunk on the top of the attachment, remove it. Ideally, scaffolds should have an angle of attachment of approximately 45 degrees from the tree trunk. This upright architecture transfers weight to the trunk and roots, allowing the primary scaffold to support heavy loads. Branches that are too upright can be spread with tree spreaders or by twisting two pruned branches together and wedging them between two limbs. Primary limbs will open up some when the trees start to bear. If primary limbs are too flat they may lose vigor and dominance and be susceptible to sunburn. The strongest primary scaffold should be on the north side of the tree, if possible. A strong vigorous limb on the north side helps keep the tree from being dominated by growth on the sunny south side. It is undesirable to have two lower scaffolds opposite each other on the trunk with another scaffold above. The upper scaffold will be stunted, or "choked out," by the lower scaffolds. Select two primaries if there are not at least three good primary scaffold candidates; this is better than choosing poorly positioned or weak scaffolds. The two-scaffold tree will likely grow very vigorously the second season, giving many choices for limb selection the following year.

After selecting the primary scaffold branches, decide how long to head the primaries. Heading length, which can vary from 12 to 18 inches from the trunk (short) to completely unheaded (long), has a dramatic effect on early production (Krueger and Heath 2000) (table 14.2). Normally, the tallest primary or the branch desired to be the tallest is left slightly taller to maintain apical dominance. If the primaries are not balanced, the weaker primary should be left longer and the stronger one(s) headed back to bring the tree into balance.

Short pruning (strong heading of primary scaffolds) creates vigorous growth at the expense of spur and fruitwood development, delaying the onset of production. It will, however, afford more precise control of the tree structure because many shoots will develop near the cut and there will be many choices for secondary scaffold branches. Excess vigorous shoot growth near the pruning cuts will have to be thinned out during subsequent pruning. Short pruning develops stiff primaries that will probably not require tying and are less likely to bend or break with future cropping.

Long pruning (no heading of primary scaffolds) encourages earlier production. Long-pruned primaries may be prone to bending or breaking because of early fruit set on smaller-caliper limbs and will probably require tying. If the primaries bend over, they must be retrained. Long pruning should not be used on anything other than vigorously growing trees because lower-vigor trees may not have the necessary growth to fill the allotted space or develop adequate secondary growth up and down the limb.

Figure 14.4 Included bark. *Photo:* M. L. Poe.

Figure 14.5 Bark ridge. *Photo:* M. L. Poe.

Table 14.2. Effect of differential heading length on dry fruit yield

Treatment*	3rd leaf†	4th leaf	5th leaf	Accumulated dry yield (lb/acre)
headed at 30 in	136 a	727 a	3,350 a	4,214 a
lightly tipped	324 ab	1,903 b	4,838 ab	7,065 b
unheaded	521 bc	2,769 b	6,863 c	10,153 c

Notes:

*Treatments differed in heading length only. New growth was headed at the same lengths as described in the treatments for the first through fourth dormant prunings. No differential pruning was done from the fourth dormant season on. By the sixth leaf, no yield differences were measured.

†Values followed by different letters are significantly different at the 5% level using Fischer's method.

Intermediate pruning (lightly tipping the primary scaffolds, removing the last 6 to 12 inches of new growth) is a compromise strategy. As tree growth slows in the summer, flower buds often develop on the tips of the shoots. If these flowers set fruit, the fruit weight can pull the limbs out of position. Intermediate heading removes the end of the shoot tips with the flower buds. This type of pruning has many of the advantages of long pruning without most of the disadvantages. The response to the lighter heading cuts is usually not strong enough to inhibit the development of fruiting wood, and early yields can be close to that of the long pruning. If early fruit set is heavy, tying may be necessary.

Regardless of the heading length, small, flat, noncompetitive wood can be left for early production. There will be much less of this type of wood on the shorter-pruned trees.

Second dormant pruning

During the second dormant season, select secondary branches originating from the primaries (fig. 14.6). Usually two or three upright, vigorous shoots per primary are selected. The secondaries should have an upward and outward orientation and be spaced at least 12 inches apart on the primary. The primary can be cut to an outside secondary to open up the tree if desired. Select the other secondaries on the primary below the upper secondary. If the secondary forks into two approximately equal branches, remove one. Remove any other vigorous upright branches that compete with the selected secondaries. Also remove lower lateral branches that could interfere with harvest equipment. Flat, thin, noncompetitive branches can be left to develop fruitwood.

It will be necessary to decide how long to leave the secondaries after they have been selected. As with the primaries, leaving the secondaries longer encourages heavier early production (Weakley and Yeager 1986). Most of the advantages and disadvantages related to length of heading with the primaries also apply to the secondaries. As with primaries, light tipping of secondaries provides most of the advantages of long pruning without the disadvantages.

Third dormant pruning

After the third growing season, employ the same strategies as with the earlier pruning. Select upright tertiary limbs on the secondaries. Remove upright, vigorous, competitive limbs that compete with the selected limbs; also remove lower lateral limbs that may interfere with harvest machinery. The goal is to allow the upper periphery of the tree to fill with secondary and tertiary branches and fruitwood while maintaining a somewhat open center that allows sunlight to penetrate. After the third dormant

Figure 14.6 Second dormant pruning: secondary scaffold selection and heading length: long (A), short (B), and intermediate (C). *Photos:* M. L. Poe.

pruning, the training phase is essentially complete.

If heading or tipping the upright shoots in the top of the tree is desired in the second or third dormant pruning, it is often accomplished mechanically with rotary or sickle bar pruners. While this is less precise than hand pruning, it often eliminates the need for ladders and will be quicker and less expensive.

Tree blow-over

An obvious concern related to pruning strategies that bring trees into early production is a greater risk of blowing over before the roots have become anchored enough to withstand the forces exerted by wind on the larger canopy ("sail") created by less pruning. To take advantage of the early production offered by some of these pruning strategies, consider other practices affecting tree anchorage such as the following.

- Roostock selection: taprooted seedling rootstock may provide better anchorage.
- Planting depth: on well-drained soils, trees can be planted deeper than they were in the nursery without covering the graft union.
- Avoid irrigation prior to expected windy periods.
- Reduce canopy leverage:
  - o Avoid planting on high berms. perpendicular to prevalent winds.
  - o Avoid high heading heights where "blow-over" is a concern.
  - o Prune or defoliate with zinc sulfate prior to the onset of fall or winter storms.

Tying

It may be necessary to tie or rope the upper tree canopy to keep trees from opening up too much or bending over if pruning strategies are employed that encourage early production. Tying may be necessary as early as the second growing season and may need to be continued through the third or fourth season. It should be done in the spring before the weight of shoot growth and the developing crop cause the limbs to open up too much or bend over. Tying is accomplished by encircling the upper canopy with twine or tree rope as high as possible in the tree canopy that allows for support of the major upright limbs of most concern (fig. 14.7). It is not necessary to include the entire canopy within the circle created by the rope. Concentrate on limbs that must remain upright to maintain good tree structure. Flatter limbs can be left out of the circle and allowed to bend over and develop into fruitwood.

Figure 14.7 Tied tree, second dormant season. *Photo:* M. L. Poe.

Pruning Mature, Bearing Trees

The primary goals of pruning mature, bearing prune trees are to manage the tree canopy to optimize light distribution to fruiting branches, strengthen the tree structure so the tree can carry the crop, facilitate cultural practices, optimize crop load and fruit size, and control tree size. Pruning practices vary because of differences in initial tree training, tree vigor and acceptable tree size, desired cropping levels, and fruit size expectations. Mature tree pruning should reinforce the tree form that was established during the training stages and ensure that major structural branches maintain their dominance if they are sound or are replaced if they are weak or broken. In addition, mature tree pruning should manage the amount, quality, and position of fruiting shoots and spurs for optimal cropping. Pruning mature trees relies on the physiological principles outlined at the beginning of this chapter and an appropriate mix of heading

and thinning cuts, depending on the immediate objective. Pruning is needed to maintain good light penetration down to the fruiting wood to ensure that high-quality flower buds are formed for the subsequent year and good-quality fruit can be produced in the current year.

To manage tree structure, use thinning cuts near the branch origin to remove unwanted vigorous shoots that threaten to crowd the tree canopy and obscure the overall tree structure. To create vigorous new branches to replace old, weak, or broken scaffolds, head an older branch to stimulate epicormic shoot growth. In a healthy tree, especially in well-lit areas, such a cut will likely stimulate several new upright shoots. Subsequently, the optimal shoot or shoots for replacing the scaffold should be selected by removing all other shoots using thinning cuts.

Pruning for crop load involves using thinning cuts to remove excessive fruiting wood when stimulating more shoot growth is not desirable. Make heading cuts into older fruiting wood when renewal of fruiting wood is desired. Prune trees bear much of their fruit on spurs and short shoots. Spurs may be only a few inches long and can remain viable for more than 8 years. The terminal part of the spur that grew in the previous season produces flowers and bears fruit. The ability of spurs to produce large, quality fruit declines with spur age. In addition, it is usually necessary to prune to remove excessive fruiting potential. Thus, depending on the vigor, about 20 to 40% of fruitwood around the tree should be removed annually by leaving young shoots and removing stems with old spurs. Since prune trees tend to bear heavily in alternate years, growers should adjust their intensity of pruning according to the expected crop load; for example, prune lightly following a heavy crop load. One study suggests that it may be possible to eliminate pruning in alternate years without negatively affecting cropping. This practice did not yield much savings because trees that were not pruned in one year required more pruning in the subsequent season (Southwick et al. 1989a).

Mature tree pruning is usually done during the winter due to the ease of seeing the tree structure without leaves and the desired stimulatory effect of dormant pruning. Increasingly, however, growers are pruning earlier in the fall while some leaves are still on the tree because of labor availability after harvest and economic incentives to assess pruning costs against current-season crop income. This practice may slightly reduce the invigorating effect because early leaf removal may decrease storage of carbohydrates that prepare the tree for spring growth. This may be desirable for excessively vigorous trees but undesirable when the orchard is weak.

Mechanical Pruning

In recent years, there has been a tendency for growers to use mechanical pruning aids such as tree toppers after harvest to reduce the height of trees, increase light penetration into the canopy during the fall, and reduce the amount and cost of hand pruning. While this practice can have positive effects on light penetration into the tree canopy right after topping, it can also have negative effects on tree structure and long-term management. During a mechanical topping (or hedging) operation, all the pruning cuts are heading cuts. Headed shoots proliferate new shoots, and the canopy can become crowded and the fruitwood shaded.

Annual hand pruning with ladders and loppers has long been recommended for mature prunes because of the selective nature of the pruning that cannot be matched by mechanical pruning. However, increased cost and reduced availability of labor has increased interest in cheaper, less labor-intensive methods for accomplishing what has traditionally been done with hand pruning. Research with mechanical hedging and topping has shown that, by itself, mechanical pruning severe enough to achieve the same fruit and size and quality possible with hand pruning resulted in reduced production that was not completely compensated by higher fruit value and reduced pruning costs (Southwick et al. 1989b).

Several options are now available for reducing pruning costs, including doing a less-detailed pruning with pneumatic shears or pole saws from the ground. This can be done alone or in combination with mechanical topping. A 4-year trial conducted in Glenn County and completed in 2009 investigated mechanical pruning with various strategies and timings (Krueger et al. 2009) (table 14.3). The trial showed that mechanical pruning, in combination with a generally less-detailed annual hand pruning from the ground using pole saws, pole loppers, and long-handled loppers, along with crop control strategies such as mechanical thinning when warranted by crop load, resulted in equal or better accumulated production and value than the standard ladder and lopper treatment. The treatments without ladders had an annual pruning cost of 30 to 50% less than the standard treatment. Initially, some of the mechanically pruned

Table 14.3. Yield summary for mechanical pruning in combination with ground hand pruning

	Mechanical pruning treatments				Dry yield/ac (% of standard)					Cumulative value/ac (% of standard)
	Year				Year					
Treatment	1	2	3	4	1	2	3	4	Cumulative	
1	D T				166 a	97 cd	80 c	129 ab	113 bc	117 a
2	D T	S RT		S RT	145 ab	110 abcd	112 b	131 ab	121 ab	121 ab
3	Std	Std	Std	Std	100 c	100 bcd	100 bc	100 ab	100 c	100 b
4	D V	S V		S V	136 abc	114 abc	147 a	121 ab	127 a	129 a
5	D V (eastside)		PH RT		160 ab	117 a	81 c	133 ab	121 ab	121 a
6	S V		PH T	PH T	166 a	114 abc	100 bc	131 ab	124 ab	122 a
7	D RT	S V		PH RT	169 a	116 a	109 bc	137 bc	129 a	127 a
8	D RT	S V		PH RT	158 ab	100 cd	113 b	123 bc	118 ab	118 a
9	D MH	S MH		S MH	125 bc	114 ab	115 b	116 ab	116 ab	121 a

Note: All treatments, with the exception of the standard pruning treatment, received annual hand pruning from the ground with pole saws and/or pole loppers and long-handled loppers. Numbers followed by different letters are statistically different at the 5% level using Fisher's method.

KEY
D = dormant
MH = Mowhawk, notches mechanically cut into the shoulders of the canopy
PH = postharvest
RT = rooftop
S = summer
Std = standard or ladder and lopper
T = flat top
V = V cut in the center of the tree

treatments were applied during the dormant season. Due to an excessively vigorous response and results from other researchers (Olson 2007), these treatments were shifted to immediately after harvest. In general, this time proved to be more satisfactory, with a less-vigorous response. An initial concern was that the summer mechanically pruned treatments would lose too much yield due the partially developed fruit being pruned off in June, but over the course of the trial this did not prove to be a valid concern. The summer mechanically pruned treatments had yields and value equal or better than the other treatments.

References

Asai, W. K., J. P. Edstrom, and J. H. Connell. 1996. Training young trees. In W. Micke, ed., Almond production manual. Oakland: University of California Agriculture and Natural Resources Publication 3364. 121–124.

Krueger, W. H. and Z. Heath. 2000. Effect of differential training regimes on early production and gross revenues in French prunes. Prune Research Reports 2000:25–27.

Krueger, W., F. Niederholzer, E. Nielsen, and C. Garcia. 2009. Investigation of pruning strategies for dried plums including hand, mechanical and combinations. Prune Research Reports 2009:54–58.

Olson, B. 2007. Mechanical topping: When is the best time to top. Prune Research Reports 2007:56–60.

Southwick, S. 1993. Pruning and training young prune trees. DVD. Oakland: University of California Agriculture and Natural Resources Publication 6541D. 32 min.

Southwick, S., J. Yeager, M. Norton, and J. Osgood. 1989a. Alternate year pruning trials. Prune Research Reports 1989:73–79.

Southwick, S., W. H. Krueger, and J. Yeager. 1989b. Comparison of hand versus mechanical pruning in a high-density French prune orchard. Prune Research Reports 1989:64–72.

Weakley, C., and J. Yeager. 1986. Prune training trial. Prune Research Reports 1986:25–27.

15 Crop Control

• Richard P. Buchner, Stephen M. Southwick, William H. Krueger, James T. Yeager, and Cyndi K. Gilles

Fruit size, quality, and yield are the major factors affecting crop value. The goal in prune production is to produce high yields of high-sugar fruit while targeting specific fruit sizes to maximize crop value. Crop load, or, more specifically, the number of fruit per tree, primarily influences fruit size, sugar accumulation, and drying ratio. Excessive crop loads usually cause high yields of lower-value prunes, lowering the monetary return to the grower (fig. 15.1). Conversely, if crop loads are too light, fruit size and sugar accumulation are favored but yields are too low to achieve high value.

The prune payment schedule (table 15.1) represents an average California price schedule and is intended to illustrate how pricing changes by size and year. For example, 72/76 represents a dehydrated fruit count of 72 to 76 dry prunes per pound. Dried prunes in this group had an average 2007 grower payment of $1,397 dollars per ton. Notice that small prunes and prunes in the manufacturing grades are much less valuable. Figure 15.2 illustrates the monetary

Table 15.1. Average prune payment schedule by dry fruit size in dollars per ton, 2005–2007

Size	2005*	2006	2007
30/40	1,836	1,663	1,675
41/45	1,836	1,663	1,675
46/51	1,836	1,663	1,675
52/56	1,831	1,658	1,670
57/61	1,784	1,619	1,631
62/66	1,737	1,581	1,591
67/71	1,638	1,498	1,509
72/76	1,538	1,387	1,397
77/81	1,377	1,193	1,202
82/86	1,240	1,082	1,060
87/91	1,191	1,040	1,001
92/96	1,141	970	848
97/101	1,104	942	739
102/111	472	375	300
112/121	422	350	45
122/Sm	345	325	0
MFG† A (62/lgr)	666	515	468
MFG B (67/76)	631	480	433
MFG C (77/91)	571	420	373
MFG D (92/smr)	510	250	0

Notes:
**Payments* represent an average of Sunsweet and Prune Bargaining Association (PBA) price schedules.
†*MFG* = manufacturing grades.

Figure 15.1 Heavy crop set. *Photo:* M. L. Poe.

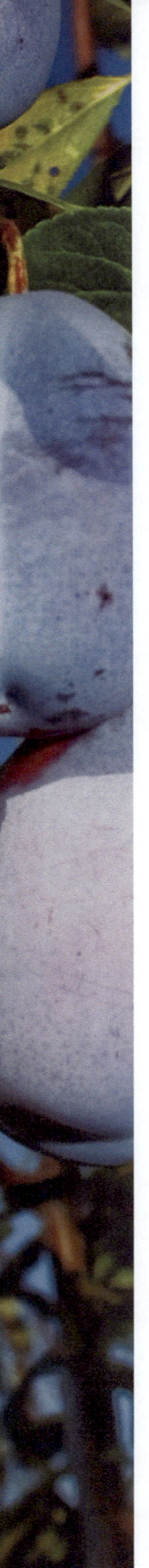

return per acre versus the crop load. The overall goal is to operate at the top of the curve, where fruit value and tonnage result in the highest return per acre.

Factors affecting orchard horticulture can affect fruit production and sizing potential. For example, dormant pruning removes flower buds that can reduce crop load, depending on fruit set and bloom. Pruning also affects light distribution and the development of flower buds. Adequate soil moisture is necessary for rapid fruit growth; water-stressed trees do not function properly, and growth and tree performance are compromised. Water management also influences fruit cracking and blue prune drop. Attention to adequate tree nutrition is critical for good sizing potential. Potassium deficiency, commonly associated with small fruit, causes leaf scorch and reduces leaf area and photosynthesis. Less photosynthesis reduces carbohydrate production, which in turn reduces fruit growth and soluble solids accumulation and increases drying ratio.

Prune crop control provides many benefits in prune production. Crop load is usually adjusted to

- improve fruit size and sugar accumulation
- manage or eliminate limb breakage (fig. 15.3)
- decrease the tendency for alternate bearing
- reduce potential for potassium limb dieback
- decrease tree stress and improve tree health and vigor

Many techniques are available to manage crop load for French prune production. Hand-thinning fruit is one possibility, but the cost is too expensive for the California prune industry. Dormant pruning is the first step in reducing crop load, but it may have to be supplemented with mechanical fruit thinning to achieve desired number of fruit per tree. Chemical thinning has not been widely adopted in California, mostly because of the risk of over thinning in a year of light crop set; more research is necessary to substantiate this technique. Mechanical thinning is the most-used technique.

Girdling Trunks or Scaffolds

Girdling, or removing a ring of bark tissue all the way down to the xylem, has been shown to improve fruit size in several stone fruit varieties. Sibbett and Day (1994) experimented with scaffold girdling on French prune and found that girdling did not improve fruit size, but it did increase preharvest drop, reducing dry yield. As a result, the University of California does not recommend trunk or limb girdling for California French prune production.

Chemical Thinning

Yoshikawa et al. (1992) experimented with applying surfactants and fertilizers for blossom thinning. All treatments reduced fruit set compared with untreated controls; however, zinc sulfate, urea, and Silwet (surfactant) were phytotoxic. Wilthin (monocarbmide dihydrogen sulfate) reduced fruit set compared with the untreated control, and no phytotoxicity was observed (Yoshikawa et al. 1993). Southwick et al. (1994) studied the effect of early June sprays of gibberellin on bloom and fruit size the following year. Sprays reduced flowering, but increased fruit size was not consistently measured. Armothin (AKZD) applied at 50% full bloom reduced fruit set, reduced undersized prunes, and increased fruit size (Southwick et al. 1995). The cytokinin CPPU (2-chloro-4-pyridyl-N-phenylurea) was investigated by Glozer and Niederholzer (2005). Preharvest applications on French prune showed promise in delaying

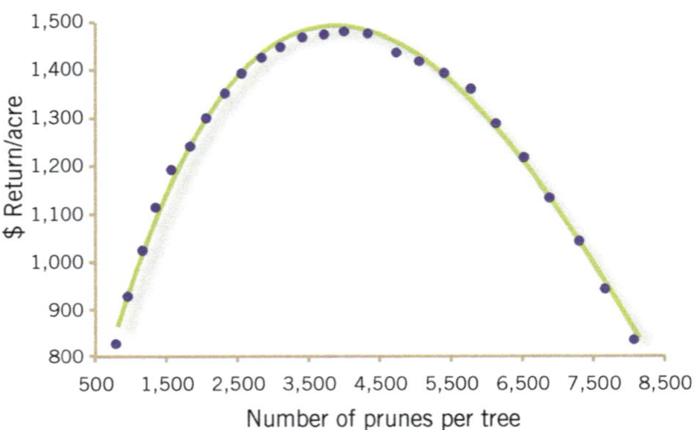

Figure 15.2 Crop load expressed in number of prunes per tree versus return per acre. *Source:* Thompson 1991.

Figure 15.3 Trees breaking from excessive fruit load. *Photo:* R. P. Buchner.

maturity, reducing fruit drop, and increasing fruit size, while maintaining fruit firmness. Sibbett and others (Martin et al. 1975; Sibbett and Martin 1982) applied 50, 100, and 150 ppm rates of Ethephon (2-chloroethyl phosphonic acid) as fruits attained a 0.31 inch seed length. All three rates thinned fruit and improved fruit size without phytotoxicity when applied to unstressed trees. Although chemical thinning shows some promise, uncertainty in performance and the inability to predict final fruit load at bloom has hindered commercial adoption in California prune production. In addition, effective chemical thinning compounds are not currently approved for use in California. Additional research is required before chemical thinning becomes a viable crop load management tool.

Mechanical Thinning

Mechanical fruit removal using harvest shakers with modified weights is the most common method for decreasing crop load in California prune production. Mechanical thinning is generally reliable, and growers can evaluate actual crop load before making thinning decisions. Mechanical fruit removal can cause minor prune skin damage and is not suggested by the University of California for fresh market prunes sometimes sold as sugar plums.

Determining Reference Date

Success with mechanical fruit removal requires knowledge of the orchard sizing potential and an estimate of fruit size at harvest. Each orchard or block within an orchard produces a certain average size of dried prunes. Some orchards consistently produce large fruit, and others produce only medium to small fruit. The ability of an orchard to size fruit with good sugar accumulation is referred to as the sizing potential for that orchard. Orchards have different sizing potentials because of location, soils, tree age or health, climate, cropping history, and management. The key to successful crop load management is to know an orchard's history and its ability to size a specific number of fruit per tree.

UC Farm Advisors, specialists, and faculty developed a reference date technique to predict final fruit size at harvest (Carnill et al. 1974). The count per pound at a reference date showed good correlation to dry fruit size at harvest. The reference date is set when 80 to 90% of a fruit sample shows visible endosperm development. Endosperm development can be monitored by randomly selecting 20 fruit from 20 trees (a 400-fruit sample). Prunes are cut lengthwise through the seed. The endosperm develops in the seed at the blossom end of the prune. If present, it will look like a distinct clear glob of cells at the blossom end of the seed. The endosperm can be confirmed by lifting it out using a knife point (fig. 15.4).

Figure 15.4 Endosperm. *Photo:* M. L. Poe.

The reference date is the most reliable stage of development to predict fruit size. This date is valuable to prune growers because green fruit count per pound at the reference date is a reasonably good indicator of final fruit size at harvest (table 15.2). In California, the reference date usually occurs about the first week in May, so fruit sampling starts in late April (possibly a little sooner in the state's southern prune districts). The

Table 15.2. Prune reference date size, and average harvest dry fruit size

Reference size, green (count/lb)	Harvest size (dry) (count/lb)		
	Orchard sizing potential		
	Average	Good	Excellent
50	32	31	30
55	36	34	32
60	39	37	35
65	42	40	38
70	46	43	41
75	49	45	43
80	53	48	46
85	56	51	48
90	60	54	51
95	63	57	54
100	67	60	56
105	70	63	59
110	74	66	61
115	77	68	63
120	81	71	66
125	84	74	68
130	88	77	70
135	92	79	73
140	95	82	75

Source: Carnill et al. 1974.

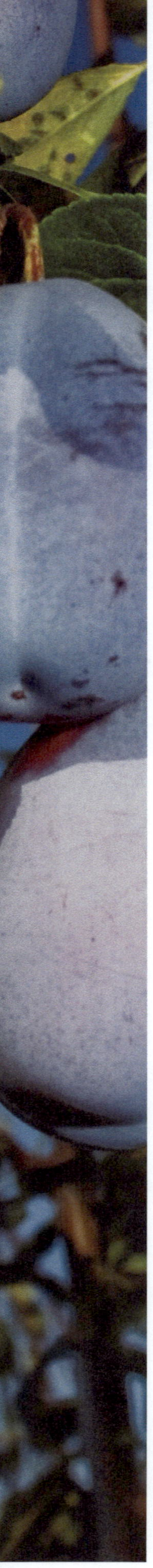

endosperm develops about 1 week after the seed tip begins to harden. Thus, tip hardening can be used to help determine the reference date.

Experience suggests that the reference date occurs too late for it to be used to achieve the most benefit from mechanical thinning. Removing excess fruit as early as possible provides more benefit and decreases the chance of removing too many large fruit. As a practical matter, mechanical thinning starts when shakers can effectively remove excess fruit. When fruit is roughly ½ inch in diameter, test trees can be shaken to observe fruit removal and to fine-tune the timing for thinning. Mechanical thinning could occur as much as 2 weeks before the reference date, depending on orchard location.

Estimating Harvested Fruit Size

Once the reference date is determined from a test sample, the sample can also be used to calculate the number of fresh prunes per pound: simply count the number of prunes required to weigh 1 pound. Several counts can be done to improve accuracy. Using the green sample count per pound at the reference date, refer to table 15.2 to find the corresponding harvest size. For example, if the reference date count is 115 prunes per pound, the predicted average harvest size would be 77 dry count per pound for an orchard with average sizing potential, 68 dry count per pound for an orchard with good sizing potential, and 63 dry count per pound for an orchard with excellent sizing potential. If samples and sample sizes accurately represent the orchard, this method is fairly accurate. More important, it gives growers early warning if crop loads are excessive and harvest size is in jeopardy. If small sizes are predicted, mechanical thinning should be considered to improve fruit size and profitability. This point cannot be overemphasized because fruit size and sugar accumulation are directly related to crop load, and both have an enormous impact on crop value.

Mechanical Thinning Procedure

Growing large prunes is largely a numbers game. There are various ways to remove excessive numbers of fruit (by pole, shaking individual limbs, etc.). The most typical technique is to mechanically shake whole trees for timed durations to remove specific numbers of fruit. If properly done, mechanical thinning has a long history of success. Concerns include removing bark from trees, damaging fruit if the crop is sold for the fresh market, removing large, high-value prunes, and exposing the crop to brown rot from shaken fruit on the orchard floor. In general, the earlier fruit are removed, the greater the sizing benefit. For successful mechanical thinning, follow these five steps:

- Determine the desired target number of prunes per tree at harvest.
- Determine the number of prunes per tree prior to thinning.
- Determine how many prunes must be removed to achieve the target size.
- Estimate prune drop due to natural causes.
- Use a timed mechanical shake to remove excess prunes.

Target number of prunes per tree

The sizing potential of each orchard is different and largely relies on orchard history. Harvest tonnage in California is usually reported as dry tons of fruit per acre (a dry ton is the fruit weight after dehydration). The dry count is the number of dehydrated fruit per pound. If an orchard has historically produced 4 dry tons of 75-dry-count prunes per pound and the grower wishes to produce 75-count prunes, the target tonnage would be 4 dry tons. A more realistic average target might be 3 dry tons to achieve 75-or-lower-count fruit with a margin of safety. Once the target tonnage is identified, calculate the number of prunes needed per tree using the following formula:

Target number of fruit per tree = (Dry pounds per acre × dry count per pound) ÷ Number of trees per acre

For example, assume that an orchard planted at 20 feet by 20 feet can produce 3 dry tons (6,000 lb) per acre of 75-count prunes. The calculation indicates that 4,128 fruit per tree are needed to achieve the target dry size:

Target number of fruit per tree =
(6,000 lb/ac × 75 dry count/lb) ÷ 109 = 4,128 fruit

Number of prunes per tree prior to thinning

The previous procedure identified the target number of prunes needed per tree. The next step is to measure the actual crop load per tree in the orchard.

- Select typical trees that accurately represent the crop load in the orchard. Using multiple trees improves accuracy; selecting three representative trees per orchard is a good compromise of time and accuracy.

- Place tarps under the selected trees and shake off as much crop as possible using a mechanical shaker. Mechanical shaking seldom removes all of the crop. Any prunes remaining on the tree can be stripped by hand. If hand stripping is not practical, estimate the number of prunes remaining on the tree; they will be added to the weighed count. If machine shaking is not practical, fruit can be hand-picked and dropped on the tarp.
- Collect shaken prunes and remove leaves, twigs, and yellow or shriveled fruit—anything that would not contribute to an accurate fruit weight. After cleaning, weigh the entire green fruit sample. If the tarp is part of the weight, subtract its weight from the sample.
- Select three 1-pound fruit samples from the weighed whole-tree sample. Using a scale, count how many prunes are required to weigh 1 pound. This is the green count per pound. Another way to determine this is to weigh samples and count how many prunes are in the sample weight. The objective is to develop a relationship between count and weight that can be used to calculate the total number of prunes from the total weight shaken from the test tree.
- Multiply the total fruit weight by the count per pound to calculate the number of prunes per tree.

$$\text{Total number of prunes} = \text{Fresh crop weight (lb)} \times \text{Number of prunes per pound}$$

- For example, shaking and hand removal yielded 130 pounds of fresh fruit and 75 green prunes per pound; the result is 9,750 total prunes.

$$\text{Total number of prunes} = 130 \text{ lb} \times 75 \text{ prunes/lb} = 9{,}750 \text{ prunes}$$

- Finally, if all the prunes were not stripped from the tree, visually estimate how many prunes are still on the tree. Assuming that 500 prunes remain on the tree, adding 500 to the calculated number results in a total of 10,250 prunes per tree.

How many prunes to remove to achieve target size

To calculate how many prunes to remove, simply subtract the target number of prunes per tree from the total number of prunes per tree. Using the above examples, in which the target number was 4,128 and the total number was 10,250, we must remove 6,122 prunes to achieve our goal of 3 tons of 75-count-per-pound prunes.

Estimated prune drop due to natural causes

The next consideration is to account for any natural fruit drop following thinning. Fitch et al. (1973) reported a 40% natural fruit drop between thinning and harvest. Other researchers (Buchner et al. 1998; Krueger and Nielsen 1998) measured natural fruit drop in the 10% to 20% range. In the final analysis, decisions as to how much natural fruit drop to expect will have to be made on an orchard-by-orchard basis. For orchards with an excessive drop history, 40% might be a good choice, but for most orchards a natural fruit drop of 10 to 20% is more realistic.

Selecting a 20% natural drop means that 20% more prunes must be left on the tree to achieve the target harvest crop load. In the example above, dividing 4,128 prunes by 0.80 gives a corrected target crop load at thinning of 5,160 prunes per tree.

The results from the whole-tree test shake indicated 10,250 prunes per tree. Subtracting the corrected target crop load (5,160 prunes) from this total number of prunes per tree indicates that 5,090 prunes must be removed per tree to achieve the target tonnage and fruit size.

Accurately estimating natural fruit drop is important because it affects the final fruit load. Overestimating the fruit drop risks a larger crop load with smaller fruit. Underestimating fruit

Step 1.	________	Total pounds of prunes per tree from the test tree.
Step 2.	________	Counted number of prunes per pound.
Step 3.	________	To calculate the number of prunes in the test shake, multiply values for step 1 and 2.
Step 4.	________	Estimate the number of prunes remaining on the tree.
Step 5.	________	To calculate the total number of prunes on the test tree, add steps 3 and 4.
Step 6.	________	Target number of prunes per tree to achieve harvest fruit size goals.
Step 7a.	________	Adjust for a 10% natural fruit drop, divide line 6 by 0.90.
Step 7b.	________	Adjust for a 20% natural fruit drop, divide line 6 by 0.80.
Step 7c.	________	Adjust for a 30% natural fruit drop, divide line 6 by 0.70.
Step 7d.	________	Adjust for a 40% natural fruit drop, divide line 6 by 0.60.
Step 8.	________	To calculate the number of prunes to remove, subtract the appropriate step 7 from step 5.

Figure 15.5 Data sheet for organizing fruit removal information and for calculating how many prunes to remove. For metric calculations, substitute kilograms for pounds.

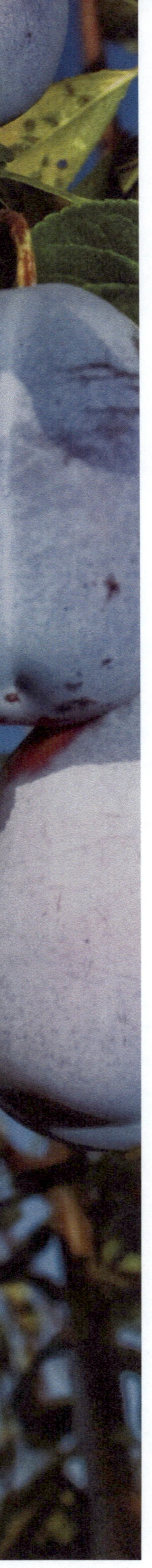

drop yields a smaller final crop load with larger fruit; marketable fruit will have been lost. Figure 15.5 provides a simple data sheet to manage and calculate crop thinning.

Timed shakes to remove excess prunes

The final step is to shake individual trees for specific intervals to discover how many seconds of shaking will be necessary to remove excess prunes. Place tarps under a test tree and apply a 1-second shake. Collect the fallen prunes and calculate the number of prunes removed using the technique described above. If needed, increase the shaking time until the required number of prunes are removed. A typical shake is usually 2 to 4 seconds at full throttle. If a shake of 6 or 7 seconds does not remove the required number of prunes, stop shaking. Multiple shakes or excessively long shakes can cause unacceptable skin damage and possible fruit drop as the crop matures.

Summary

Mechanical shaker thinning is the most reliable technique currently available for controlling crop load after bloom. Success depends on carefully developing shake times and applying them uniformly throughout the orchard. Plan on investing a day to shake sample trees, calculate fruit load, and calculate the number of prunes to remove to achieve the target size and tonnage. Mechanical thinning can damage fruit. Minor skin damage has not been a problem for dried prunes, but it could be for fresh market prunes. For that reason, mechanical thinning is not recommended for fresh market prunes. Growers have been concerned that larger fruit on the upper outside of the canopy may be preferentially removed instead of smaller fruit on the lower inside part of the canopy. However, numerous tests have confirmed the benefits of mechanical thinning compared with other methods of thinning or leaving the trees unthinned.

References

Buchner, R., M. Gilles, and K. Graham. 1998. 1997 Fruit drop following mechanical thinning. Prune Research Reports 1998:8–11.

Carnill, G. L., O. Lilleland, D. E. Ramos, L. B. Fitch, G. S. Sibbett, and J. W. Osgood. 1974. Physiological reference date for prunes, a means of determining thinning needs. International Horticultural Congress, Sept. 10–18, Warsaw, Poland.

Fitch, L. B., D. E. Ramos, J. T. Yeager, and G. S. Sibbett. 1973. Mechanically thinning French prunes: A progress report. University of California Agricultural Extension MA-68.

Glozer, K. G., and F. J. Niederholzer. 2005. Effects of CPPU on maturity delay, fruit firmness and fruit drop in French prune. Prune Research Reports 2005:18–21.

Krueger, W. H. and H. Nielsen. 1998. Determination of fruit drop from thinning until harvest in French prunes. Prune Research Reports 1998:7.

Martin, G. C., L. B. Fitch, G. S. Sibbett, G. L. Carnill, and D. E. Ramos. 1975. Thinning French prune (*Prunus domestica* L.) with (2 chloroethyl) phosphonic acid. Journal of the American Society for Horticultural Science 100(1): 90–93.

Sibbett, G. S., and K. R. Day. 1994. Effects of girdling French prunes on yield, quality and pre-harvest Fruit Drop. Prune Research Reports 1994:34–36.

Sibbett, G. S., and G. C. Martin. 1982. Cumulative effects of ethephon as a fruit thinner on French prune (*Prunus domestica* L.). HortScience 17(4): 665–666.

Southwick, S. M., J. T. Yeager, and W. H. Krueger. 1994. Effect of gibberellin on flowering in French prune. Prune Research Reports 1994:29–33.

Southwick, S. M., J. T. Yeager, and K. G. Weis. 1995. Chemical thinning on French prune with Annothin in 1995. Prune Research Reports 1995:17–19.

Thompson, G. 1991. Shaker thinning trial, Arbuckle, California. Yuba City: Prune Bargaining Association. 12.

Yoshikawa, F. T., S. M. Southwick, and J. T. Yeager. 1992. Effects of French prune blossom thinning with surfactants and fertilizers. Prune Research Reports 1992:40–42.

———. 1993. Effects of Wilthin blossom thinning in French prunes. Prune Research Reports 1993:40–41.

16 Nutrition and Fertilization

• Franz J. A. Niederholzer, Richard P. Buchner, and Stephen M. Southwick

The overall goals of orchard nutrient management are to maximize sustained economic production while minimizing or eliminating environmental contamination. This chapter provides background information on essential plant nutrients, tree physiology, and soil characteristics critical to achieving these goals.

General Plant Nutrition

The majority of plants, including prunes to the best of our knowledge, require 14 essential mineral nutrients to grow and produce a viable crop (table 16.1). Each nutrient performs at least one essential, nonreplaceable function in the plant. Some nutrients used in the physical structure or salt balance of the plant are required in larger amounts than are others and are referred to as macronutrients. Nutrients primarily involved in catalyzing chemical reactions in the plant are used in relatively smaller amounts and are called micronutrients. An example of the relative differences in nutrient contents of prune trees appears in table 16.2. The relative weight differences illustrate why nitrogen, potassium, calcium, and magnesium are considered macronutrients and boron, zinc, manganese, and copper are considered micronutrients.

Plants absorb nutrients through roots and leaves. The primary nutrient source for plant absorption is the soil, specifically the soil water (soil solution) around roots. Nutrients dissolved in the soil solution are available for root uptake. In the soil, nutrients are present as negatively or positively charged ions called anions or cations, respectively (see table 16.1). Nutrients present in the soil in undissolved form, combined in complex organic molecules, or strongly fixed to soil particles are not readily available for root absorption. Thus, there can be a significant difference between the amount of a nutrient in the soil and the amount available of that nutrient to plants. Plants can also absorb nutrients through leaf surfaces. As in root absorption,

Table 16.1. Essential plant elements and the form(s) of each most commonly absorbed from soil solution

Nutrient	Atomic symbol	Ionic form absorbed by plants	Names of nutrient forms
boron	B	H_3BO_3 $H_4BO_3^-$	boric acid, borate
calcium	Ca	Ca^{2+}	calcium
chlorine	Cl	Cl^-	chloride
copper	Cu	Cu^{2+}	cuprous copper
iron	Fe	Fe^{2+}	ferrous iron
magnesium	Mg	Mg^{2+}	magnesium
manganese	Mn	Mn^{2+}	manganese
molybdenum	Mo	MnO_4^-	molybdate
nickel	Ni	Ni^{2+}	nickel
nitrogen	N	NO_3^- NH_4^+	nitrate; ammonium
phosphorus	P	$H_2PO_4^-$ HPO_4^{2-}	orthophosphates
potassium	K	K^+	potassium
sulfur	S	SO_4^{2-}	sulfate
zinc	Zn	Zn^{2+}	zinc

Table 16.2. Selected nutrient contents of a 12-year-old, heavily cropping French prune tree growing in a commercial orchard and excavated just prior to fruit maturity

Nutrient	Fruit nutrient content (lb)	Leaf nutrient content (lb)	Remainder of tree* nutrient content (lb)	Total tree nutrient content (lb)	Relative weight difference compared with copper	% nutrient in leaves and fruit	% nutrient in fruit
nitrogen	0.55	0.37	0.61	1.53	1,628	60	36
potassium	0.80	0.22	0.12	1.14	1,213	88	70
calcium	0.08	0.31	0.70	1.09	1,159	36	7
magnesium	0.05	0.11	0.08	0.24	255	67	21
boron	0.00180	0.00070	0.00220	0.00470	5	53	38
zinc	0.00090	0.00050	0.00160	0.00300	3	47	30
manganese	0.00018	0.00027	0.00055	0.00110	1	50	16
copper	0.00038	0.00010	0.00046	0.00094	1	51	40

Source: Weinbaum et al. 1994.
Notes:
Trees were planted at a spacing of 16 feet by 16 feet and produced almost 5 dry tons of fruit per acre in the year of this study.
*Shoots (without leaves), branches, trunk, and roots.

nutrients absorbed into leaves must be dissolved in water for absorption to occur.

Essential elements can be beneficial or harmful to plants. Plant roots absorb nutrients either by active (energy-consuming) or passive (non-energy-consuming) processes. Plant absorption of certain nutrients, such as nitrogen, is self-regulated so that net nutrient uptake via the roots avoids excessive absorption of potentially toxic amounts. However, because plants can regulate uptake rates of certain nutrients such as nitrogen, care must be taken to match plant nitrogen needs with fertilizer application rates to avoid wasteful and expensive overfertilization. For other nutrients, especially micronutrients, plants do not have self-regulating systems and can accumulate toxic amounts of those nutrients. Thus, excessive availability of certain essential plant nutrients at the root surface can potentially damage tree health. Leaves do not appear to possess the capacity for self-regulation of nutrient absorption, so excessive concentrations of foliar fertilization can cause leaf, and potentially tree, damage.

How Soil Affects Plant Nutrition

Soil can be the source as well as a sink for plant nutrients. Nutrients are held or released from the soil solution depending on local soil physical and chemical conditions. Weathering of soil minerals releases nutrients, as does the decomposition of soil organic matter. The chemically active portion of soils is primarily found in the very fine particle sizes with significant surface area—clays and soil organic matter. Sand and silt particles generally affect soil physical properties such as air and water permeability, but they influence soil nutrient status only indirectly.

Colloidal soil particles (clays and organic matter) are primarily negatively charged and can hold positively charged nutrients (cations). The potential for a soil to hold positively charged nutrients is measured by the cation exchange capacity (CEC). Soils with higher organic matter or clay content have a higher CEC than do sandy soils, which usually have low organic matter content. When nutrient cations (e.g., ammonium or potassium) are absorbed by roots from the soil solution, other cations are released from the soil cation exchange sites to buffer the soil solution concentration of that nutrient. Because soils generally contain fewer positively charged binding sites, negatively charged nutrients (nitrate, chloride, sulfate, etc.) are generally more vulnerable to leaching with excess soil water (irrigation or rainfall) below the active root zone if their availability in the soil is not synchronized with plant needs.

Plant nutrient availability is primarily affected by soil physical, chemical, and biological conditions. Some soil conditions affect the absolute amounts of a nutrient in the soil, while other conditions affect the availability of a plentiful nutrient to the plant.

Soil chemical and physical conditions affect plant-available nutrients in several ways. Soil parent material can contain higher or lower amounts of certain nutrients. For example, soils formed from sedimentary rocks may contain higher amounts of certain potentially toxic nutrients (sodium, chloride, and boron) than do soils formed from granitic parent material under the same climatic conditions. Soil texture, determined largely by parent material in the alluvial valleys where prunes are often grown, influences the soil CEC and the amount of nutrients the soil can hold. Finer-textured soils generally contain more clay (and more soil organic matter), which has more negative charge to hold more nutrients and prevent them from being leached from the root zone than do coarse-textured (sandy) soils. (Sometimes soils can hold nutrients too tightly and limit nutrient availability.) Soil pH affects the availability of many plant nutrients (fig. 16.1), especially micronutrients. For example, while high levels of iron may be present in soils, the amount of iron dissolved in the soil solution and available for root absorption may be very low due to high soil pH. Soil temperature affects root function and nutrient absorption: cooler temperatures reduce root activity. Saturated soil conditions during the growing season can produce elevated soil bicarbonate levels that limit plant-available iron.

Soil biological conditions affect plant-available nutrient levels largely through changes in the soil organic matter levels and quality. Soil organic matter is made up of plant residue in various stages of decomposition as well as the macro- and microorganisms that use this plant matter for food and energy. Under well-aerated, warm soil conditions, by-products from organisms feeding on soil organic matter increase levels of plant-available nitrogen, phosphorus, sulfur, potassium, zinc, and copper, as well as soil humus, which contains binding sites for cations and anions. Soils with increased soil organic matter hold more positively charged nutrients (have a higher CEC) and water than do soils with smaller amounts of organic matter. Regular cultivation reduces soil organic matter, while growing cover crops, applying compost or manure, and practicing no-till farming tend to increase soil organic matter. In the warm climate of California, increases of soil organic matter content to more than 1% are very rare. However, this apparently small increase can significantly improve soil fertility.

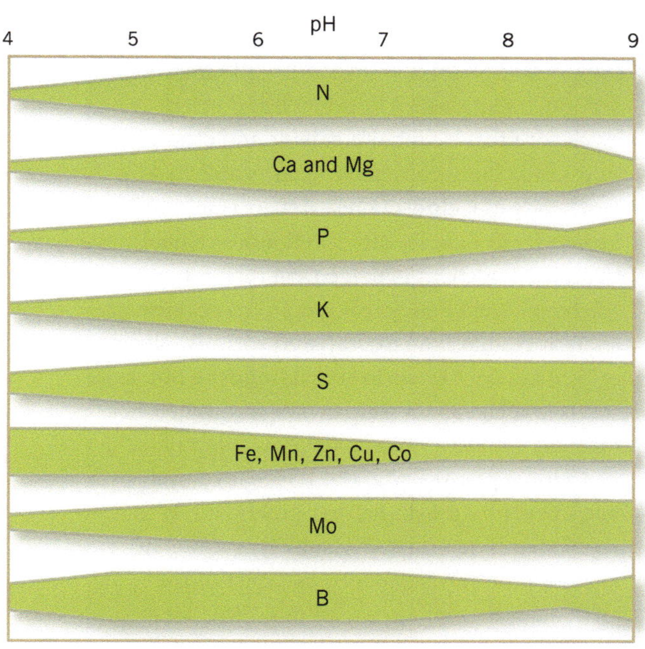

Figure 16.1 Effect of soil pH on relative availability of plant nutrients. A broad bar indicates high relative availability, while a narrow bar indicates low availability.

Assessing Orchard Nutrient Status

Prune tree nutrient sources include existing soil nutrients and irrigation water as well as added fertilizers or amendments. Depending on current orchard conditions, addition of fertilizer or soil amendments may not be needed.

Efficient prune orchard management is best achieved by regularly tracking orchard nutrient status (tree nutrition and soil conditions). Fertilizer or amendments can then be applied as needed. Blindly applying inputs can waste money and damage the environment. For example, excessive nitrogen fertilization can increase pruning costs, the risk of brown rot infection, and the risk of ground and surface water pollution. Evaluating orchard nutritional status consists of four practices: soil analysis, tree leaf tissue analysis, irrigation water analysis, and visual assessment of tree growth and appearance. A balanced approach to monitoring orchard nutrient status should consist of a combination of these practices, as each provides an important perspective on orchard health.

When evaluating an orchard for soil and plant nutrient status, planning is invaluable in improving the accuracy of sampling and the value of the work. Spatial variability is common in many orchards in California, so aerial photos along with careful ground checking and mapping provide some direction in deciding where to sample and how many samples to take to represent the orchard or to answer management questions.

Finally, soil and plant leaf sampling is often viewed as dull, simple work. Nothing is farther from the truth. The results of soil and leaf analysis often inform input decisions that cost hundreds of dollars per acre. The more accurate the sampling job, the better the chances that input investments will provide good returns.

Soil Analysis

Soil analysis can provide important information to prune orchard managers, but the practice has limitations. Soil analysis is an excellent way to establish soil pH and determine whether toxic levels of certain nutrients, particularly sodium, chloride, and boron, are present. However, there often is little relationship between soil analysis results for a given essential nutrient and the potential for plant response to fertilizer containing that nutrient (Marschner 1986). Because of this, critical levels for soil analysis results have not been developed by the University of California for prune production. Soil analysis is an effective practice to evaluate the environmental conditions in an orchard, which is especially important during orchard establishment, but it does not always accurately test levels of plant nutrient availability.

How soil sampling is done is determined by what information is needed. If root zone soil pH, nutrient status, salinity, or presence of potentially toxic ions (e.g., chloride or boron) are the target, samples should be taken that characterize the majority of the rooting zone of the tree, which differs from orchard to orchard but is usually 0 to 2 or 3 feet. Deeper sampling may be required, particularly if the soils are deep and well drained and the orchard is surface irrigated (not irrigated by low-pressure systems). If the irrigation water is very low in salinity or is sodic (has a high sodium adsorption ratio, SAR), it may have detrimental physical effects on the surface soil, causing crusting and reduced water infiltration. If this is suspected, taking soil samples in the top few inches (0 to 3) could give an indication of the exchangeable sodium percentage (the amount of sodium that is adsorbed on the soil relative to calcium and magnesium). It is important to note that irrigation water and fertilizer inputs may not be uniform around a tree, so sampling should differentiate between managed (fertilized or watered) and unmanaged soil.

The next decision to make is how many sampling holes will be needed to represent the root zone of a tree whose roots extend out laterally in the soil at least as far as the outer edge of the tree canopy. In general, a minimum of 10 to 20 sample holes, ½ to 3 inches in diameter, per sampling area should be taken. Multiple samples from the same depth should be bulked and carefully mixed in a clean bucket in order to obtain a representative sample for that depth (see the section "Preplant Soil Sampling" in chapter 7, "Managing Salinity in Soils and Water"). In general, sampling from 0 to 6 inches, 6 to 12 inches, 12 to 24 inches, and perhaps 24 to 36 inches from a uniform portion of the orchard represents a very thorough soil sampling of prune root zone.

The area to be sampled depends on prior knowledge of the orchard and the reason for sampling. For a simple comparison between the soil conditions under healthy and unhealthy trees, multiple sampling sites from under only a handful of trees should be sufficient. For an analysis of the general soil condition of the entire orchard, fewer sampling sites from a wider area should be taken.

Specific information on orchard nutrition obtained from soil analysis is limited; the following are general guidelines for soils for tree crop production. Soil pH values between 6.0 and 7.5 are generally considered acceptable. Essential mineral nutrient availability is generally reduced under higher or lower soil pH values (see fig. 16.1). Saturation paste soil salinity values greater than 1.5 dS/m are generally reported to reduce crop yield in plums and almonds, but data specifically for prunes are lacking.

Plant Tissue Analysis

Plant tissue analysis is generally more useful than soil analysis for assessing tree nutritional status for two main reasons. First, leaf nutrient levels more accurately represent the actual plant nutrient status than do data from localized soil sampling sites around the tree because soil sampling does not account for whether soil nutrients are available to the plant. Second, unlike with soil sampling, accurate relationships between leaf nutrient levels for most essential nutrients and crop yield or overall plant health have been established by University of California researchers (table 16.3). Leaf analysis data are also often more valuable in orchard nutrient management than relying solely on visual deficiency symptoms because leaf analysis results can be compared with established critical nutrient levels to reveal trends in nutrient status that can affect yield or tree health before a nutrient deficiency causes obvious symptoms.

Critical prune leaf plant tissue levels have been developed using either yield data from replicated field trials or visual symptoms in immature, often potted, trees. Most of the critical levels that appear in table 16.3 were developed using visual symptoms on young trees due to the difficulty in establishing large field trials and harvesting and drying crops from individual trees. Yield reductions often occur before the appearance of visual symptoms, so it should be stated and recognized that leaf

Table 16.3. Critical nutrient levels (on a dry-weight basis) for nonbearing spur leaves of French prune sampled in July

Nutrient	Excessive level	Adequate level	Deficient level
nitrogen	> 2.8%	2.3–2.8%	< 2.2%
phosphorus	—	0.1–0.3%	—
potassium	> 2.0%	1.3–2.0%	< 1.0%
calcium	—	> 1.0%	—
magnesium	—	> 0.25%	—
sodium	> 0.2%	—	—
sulfur	—	> 0.15%*	< 100 ppm SO_4^{2-}–S*
chlorine	> 0.3%	—	—
boron	> 100 ppm	30–80 ppm	< 25 ppm
manganese	—	> 20 ppm	—
copper	—	> 4 ppm	—
zinc	—	—	< 18 ppm

Sources: Uriu 1981 for all values except sulfur. Sulfur source: Ul Haq and Carlson 1993.
Note: *Use fully expanded shoot leaves, not spur leaves.

critical levels are conservative and represent the minimum concentration necessary for optimal growth. At the same time, excessive plant nutrient availability can and often does have a toxic effect and reduces crop load. Effective and efficient prune orchard management seeks to achieve balanced nutrient levels identified in table 16.3 for maximum economic gains and minimum negative environmental impact.

Seasonal changes of mineral levels

Nutrient levels in prune leaves change over time (fig. 16.2). This change can be helpful when interpreting leaf analysis results. Seasonal trends in leaf nutrient concentrations follow three general patterns dictated by the mobility of the nutrient within the plant and whether the nutrient is stored over winter in the plant to any great degree.

Nitrogen, phosphorus, zinc, and potassium are stored from one year to the next in woody tissues of prune trees. At bud break, these reserves drive growth at a time when soils are often cool and root activity and nutrient absorption are limited. The concentration of these nutrients in leaves is usually highest in the spring and decreases through the spring as leaves grow and their dry weight increases. The concentrations generally reach a plateau in summer and decrease in the fall as nutrients move out of the leaves and into woody tissue to be stored for the next year.

Calcium, on the other hand, is not stored over winter; is readily absorbed from the soil and is not mobile in the tree. Thus, leaf tissue becomes a site of tree calcium accumulation as the season progresses (see fig. 16.2).

Leaf concentrations of other nutrients, including manganese, boron, chloride, and sodium, increase slightly with time, but not as readily as calcium, and they do not drop as do concentrations of nitrogen, phosphorus, zinc, and potassium. The exception is when toxic levels of chloride and sodium accumulate in leaves; leaf concentrations of these nutrients continue to increase through the season.

Leaf nutrient concentrations of all essential elements are generally stable during midsummer, and that is the time when routine sampling and analysis are recommended.

Leaf sampling whenever leaves are present can help diagnose local tree health problems. Comparison of lab analysis results from apparently healthy leaves with those from leaves showing symptoms can point to specific conditions. For example, a general reduction

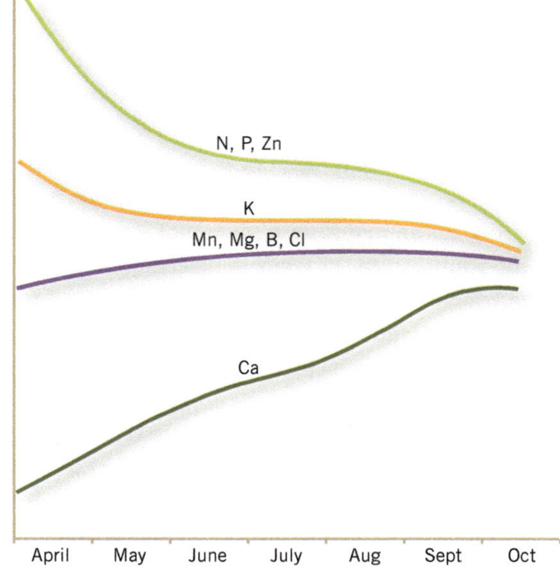

Figure 16.2 Generalized shapes of nutrient concentration curves of prune leaves during the growing season. Trends, not actual values, are shown. *Source:* Uriu 1981.

in all nutrient concentrations in leaf samples compared with levels in healthy leaves often indicates poor root health. When sampling apparently healthy and unhealthy leaves for comparison, sample leaves from similar locations around the tree. This will help avoid adding additional unwanted and confounding differences between the samples.

Because the critical nutrient levels for prune trees in table 16.3 were developed using nonbearing spur leaves, only these specific leaves should be sampled for leaf analysis. Once the area of the orchard to be sampled has been determined, sample one or two nonbearing spur leaves per tree from the ground at different heights. A total of 50 to 75 leaves are needed per sample. Once submitted to a laboratory, samples are dried at 65°C, ground, and analyzed for mineral nutrient content. Results are presented on a percent dry weight (macronutrients) or parts per million (micronutrients) basis. Micronutrient concentrations from leaves treated with fertilizers or pesticides containing zinc, copper, iron, or manganese are not useful for evaluating the orchard nutrient status for those nutrients. Deposits of these micronutrients are very difficult to remove from leaves before analysis and can affect the results.

Visual Evaluation of Nutrient Status

When done by an experienced individual, visual diagnosis of nutrient deficiency symptoms is the quickest way to determine orchard nutrient status. However, the presence of visual symptoms may mean that crop yield or tree health has already been reduced or harmed. In addition, multiple deficiencies in the same tree can confound accurate visual analysis, as can adverse soil physical conditions (too wet, too dry, poor pH, etc.) and leaf damage due to wind or pesticide application.

In general, deficiencies of immobile nutrients (iron, sulfur, manganese, and zinc) appear first in the youngest leaves, while deficiencies of mobile nutrients (nitrogen, potassium, phosphorus, and magnesium) appear first in the older leaves. Descriptions of essential nutrient deficiency symptoms appear under the appropriate nutrient in the following pages. Photographs of deficiency symptoms in prunes can be found in figures 16.3 through 16.11.

Finally, a general visual evaluation of orchard growth is very helpful in assessing orchard nutrient status and interpreting laboratory results. For example, low tree vigor combined with barely sufficient leaf nitrogen levels may indicate that the orchard could benefit from additional nitrogen. Leaf phosphorus levels are often inversely related to shoot vigor due to growth dilution of plant-available phosphorus. Low leaf zinc levels with no zinc deficiency symptoms and high tree vigor could indicate that the current zinc fertilizer strategies are sufficient. The best place to assess laboratory results is standing in the orchard.

Orchard Nutrient Management Strategy

Application of plant tissue analysis results requires an understanding of the practices used to sample and manage orchards. Research is usually conducted by treating and sampling leaves from individual trees. Orchards contain hundreds to thousands of trees and are often managed in large areas; commercial leaf samples often consist of one or two leaves from more than 50 trees from across entire blocks. In general, growers should target leaf nutrient levels to fall in the middle of the adequate range (see table 16.3) to maximize orchard health and crop production. The goal should be to maintain the nutrient status of every tree in a block above the deficient and below the excessive levels.

A more detailed orchard nutrient management strategy involves establishing management zones within orchard blocks. Management zones provide growers an option to optimize tree health and production while minimizing fertilizer costs in a block. These zones may be particularly attractive to growers farming across variable soil types. Weaker trees (due to poor soil and overall growing conditions) have a lower production potential and may not need as much fertilizer as vigorously growing, heavily cropped trees elsewhere in the same block. Using aerial or satellite images or extensive ground surveying, growers can determine management zone boundaries within a block. Once these boundaries are established, growers can develop cost-effective fertilizer strategies that may include nutrient sampling in blocks and differential fertilizer rates between blocks. Differential fertilizer management between zones within a block may be as simple as different tractor speeds when applying fertilizer in different zones. Zones can be identified by simple means such as different tree trunk paint colors, as is done in disk and berm irrigation, to identify different zones.

Until site-specific practices are developed that permit fertilization of each tree, growers may wish to consider orchard management practices that attempt to address natural tree size variability in the orchard by targeting a slightly higher orchard leaf nutrient level and establishing management zones.

Descriptions and Treatments for Nutrient Deficiencies in Prunes

Potassium

Potassium (K) is the essential nutrient required in the largest amount by a mature, producing prune orchard (see table 16.2), and it can be the nutrient most limiting to the health of a mature prune orchard in California. It is a mobile element required for many enzymatic reactions in plants, sugar transport, effective stomatal function, and the salt balance of cells. Potassium is absorbed from the soil solution as the cation K^+. Inadequate potassium nutrition in prune trees reduces fruit size, quality, and yield.

Deficiency symptoms

Potassium deficiency symptoms typically appear in mid- to late summer, but they can appear in spring under severe deficiency situations. Leaves first appear pale and may develop a tan "buckskin" color with some scorching of leaf tissue. Entire leaves may become scorched suddenly after very hot temperatures or heavy cropping. Leaf drop of scorched leaves may follow. Severe bark sunburn can follow defoliation, and *Cytospora* spp. infection and scaffold death may subsequently occur, causing significant reduction in yield for several years and threatening the health of the tree. Cytospora canker growth is more rapid in potassium-deficient prune trees than in potassium-sufficient trees.

Prunes contain significant amounts of potassium, roughly 1.0% potassium on a dry-weight basis, regardless of tree potassium status or health. The fruit's potassium requirement in heavily cropped trees (see table 16.2) can result in remobilization of potassium from leaves to fruit, with subsequent development of severe leaf potassium deficiency symptoms (fig. 16.3), followed by significant canopy defoliation (fig. 16.4). Potassium is very rarely a limiting nutrient in nonbearing or young orchards, as orchard potassium use is largely driven by crop load (see table 16.2).

Treatment

In most California prune orchards, there is little relationship between soil potassium levels and the development of potassium deficiency in prune trees. Most soil potassium is held on the colloidal surfaces (clay and organic matter) in equilibrium with soil solution potassium, and it is released into the soil solution to maintain that equilibrium as roots absorb potassium.

Potassium is not highly mobile in soils, and roots continually "mine" the soil for new sources of potassium. Root growth is limited in heavily cropped years, and this may contribute to reduced potassium availability to plants. In heavily cropped years, tree potassium demand can exceed the rate of release of soil potassium from the cation exchange phase, and trees can become potassium deficient. High-magnesium soils, while not directly competing with potassium for root uptake, may degrade soil structure and reduce water and nutrient movement and root growth, contributing to reduced potassium nutrition.

Figure 16.3 Potassium deficiency symptoms in prune leaves. A normal leaf appears at far right. *Photo:* D. Edwards.

Figure 16.4 A severely potassium-deficient prune tree in midsummer showing extensive leaf scorching in the mid to upper canopy. *Photo:* D. Edwards.

Managing orchard potassium nutrition is best done through an integrated, whole-orchard approach. Thinning heavy crop loads can help reduce excessive tree potassium demand. Adequate irrigation is also important to maximize root growth and improve tree potassium status. Finally, potassium fertilizer can be applied to the soil or leaves.

Two general fertilization practices have been shown to be the most effective way of avoiding potassium deficiency in mature California prune orchards: localized soil applications (banding or fertigation) and foliar sprays. Regular leaf sampling is important when evaluating the potassium fertilizer program in a prune orchard.

Localized soil application of potassium fertilizers can be economically beneficial to prune growers (W. H. Olson et al. 1987). Shanking or banding potassium fertilizer into or onto the soil or applying potassium fertilizer through a microirrigation system can improve leaf potassium levels, eliminate or significantly reduce deficiency symptoms in the canopy, and improve grower returns. Soil texture influences the amount of potassium fertilizer needed to supply tree needs, as heavier clay soils contain more colloidal surfaces that bind potassium and limit plant potassium availability. An annual maintenance program of 400 to 500 pounds of potassium sulfate or potassium chloride per acre (more for heavier soils, less for lighter, sandier-textured soils) is sufficient to avoid potassium deficiency under most orchard conditions. Less fertilizer material (e.g., 300 pounds of potassium sulfate per acre) may be sufficient if potassium is delivered through fertigation (injecting fertilizer with irrigation water). The historical recommendation to improve potassium status of deficient orchards on heavy soil (clay) for a number of years was a very high rate of potassium sulfate (1,500 to 2,000 pounds per acre) banded in or on the soil (Aldrich et al. 1978). However, current prices of potassium fertilizer make the cost of this practice virtually prohibitive.

Repeated fertilizer applications on the same location on the orchard floor help saturate the potassium binding sites of the soil and increase plant potassium availability. In repeated studies across different tree crops, broadcasting potassium fertilizer under flood or solid-set sprinkler irrigation was an inefficient potassium fertilization practice compared with localized potassium fertilizer application. Gypsum can be applied after fertilization at the same location to move potassium fertilizer deeper into the soil.

Several potassium fertilizers are available for use by California prune growers. Two have been extensively tested in UC research with prunes. Either potassium sulfate (sulfate of potash, 0-0-52) or potassium chloride (muriate of potash, 0-0-60) is an effective potassium source when applied to the soil. Excess chloride can be toxic to prune trees, so using potassium chloride as a potassium source in California must be approached with caution. Potassium sulfate is a safe and effective potassium fertilizer material. Most potassium chloride application occurs in the early dormant season, when root uptake has ceased and winter rains can flush the chloride out of the root zone. If only limited winter rainfall occurs by late January, growers must apply sufficient irrigation water to push the chloride below the root zone and avoid potential tree damage. Additional potassium fertilizers available for use in prunes include potassium thiosulfate, potassium carbonate, potassium nitrate, and potassium phosphates.

Regular foliar sprays of potassium nitrate (KNO_3) can maintain potassium nutrition of prune trees. Apply multiple sprays of potassium nitrate at 2- to 3-week intervals beginning in late April to early May before deficiency symptoms appear. A multiple-application foliar program can be as effective as a soil-applied maintenance potassium fertilizer application. Effective rates include 7 pounds of potassium nitrate per 100 gallons sprayed at full dilute (400 gallons per acre) and as much as 30 pounds of potassium nitrate in 100 gallons at a semi-concentrated volume. Experienced managers caution against using rates of potassium nitrate greater than 25 pounds per 100 gallons during hot weather to avoid leaf damage. A total annual rate of 100 pounds per acre of potassium nitrate, applied as multiple sprays, is required to avoid potassium deficiency in heavily cropping trees.

Nitrogen

Nitrogen (N) is the mineral nutrient often most limiting plant growth in the natural environment and in many agricultural systems as well. It is an essential component of the proteins and other molecules that are the building blocks of every plant cell. The annual nitrogen requirement for a heavily cropped prune orchard, in pounds of nutrient removed with the crop, is second only to potassium (see table 16.2). Nitrogen deficiency reduces yield by reducing flower numbers, not percentage of fruit set.

Nitrogen is highly mobile in most well-drained

agricultural soils. It is also essential to the health of many organisms that compete with plants for available nitrogen. Soil nitrogen absorbed by microorganisms or noncrop plants (weeds) is not readily available for prune tree uptake. Soil nitrate is readily leached from the root zone by excess rainfall or irrigation water. Nitrate nitrogen loss from soils can also occur through denitrification, a process that is the result of bacterial activity favored under conditions of warm, high-moisture soils that contain plenty of soil organic matter. Thus, plant-available soil nitrogen availability can change rapidly, and at least annual assessment of orchard nitrogen status (leaf analysis, etc.) is vital to orchard health and productivity. Nitrogen inputs must be carefully managed to ensure good return on this costly investment in orchard health and productivity.

Deficiency symptoms

Symptoms of nitrogen deficiency include reduced shoot growth and smaller, slightly pale green leaves (fig. 16.5). However, other factors can cause reduced shoot growth, and objective evaluation of leaf nitrogen status based on leaf color alone is difficult. Severe nitrogen deficiency significantly reduces flower numbers and fruit yield and can cause scorching or burning of leaves, a symptom that slightly resembles potassium deficiency. Midsummer leaf analysis is currently a key part of orchard nitrogen status assessment (table 16.4).

Figure 16.5 Nitrogen-deficient compared with nitrogen-sufficient prune leaves. The deficient leaves are in the top row, the sufficient leaves are below. *Source:* Uriu 1981.

Table 16.4. Prune tree nitrogen content (lb) in October and midsummer leaf nitrogen levels (% N on dry-weight basis) for trees receiving no fertilizer nitrogen or 1 lb actual nitrogen per tree annually beginning in the second leaf and continuing for 4 years

Tree age	Unfertilized tree N content (lb N)	Unfertilized tree midsummer leaf level (% N)	Fertilized tree N content (lb N)	Fertilized tree midsummer leaf level (% N)
1st leaf	0.02		0.02	
2nd leaf	0.08	2.40 ab	0.11	2.61 a
3rd leaf	0.25	2.29 a	0.35	2.41 a
4th leaf	0.29	2.27 c	0.54	2.66 a
5th leaf	0.57	1.96 c	1.04	2.45 a
6th leaf	—	2.19 b	—	2.53 a

Source: Southwick et al. 1996.

Note: Trees were planted on a spacing of 17 feet by 14 feet. No fertilizer was applied in the year the orchard was planted. Leaf nitrogen levels followed by the same letter are not statistically different from the other value in the same row, based on mean separation analysis by Duncan's MRT, 5% level.

Treatment

Sufficient orchard nitrogen status is vital to orchard health and quality fruit production. Excess tree nitrogen can cause excessive tree vigor and increase the risk of brown rot infection. In addition, excess nitrogen fertilization promotes shoot growth that can shade the lower canopy, reducing fruit bud initiation and fruit sugar levels in that part of the tree. Tree nitrogen deficiency increases the risk of bacterial canker infection. Maintaining balanced nitrogen nutrition in prune production is important to avoiding economically damaging diseases in the orchard.

Prune trees absorb nitrogen from the soil solution primarily in the form of nitrate (NO_3^-) or ammonium (NH_4^+), with nitrate the most important source under well-drained soil conditions. Nitrate is an anion that moves readily with soil water. Ammonium is a cation that is fairly immobile in soils, although it can be transformed into nitrate in aerated soils by certain soil bacteria. Tree nitrogen uptake capacity changes during the growing season, driven largely by shoot growth, fruit growth, and nitrogen storage needs. Research using young, potted, nonbearing prune trees clearly shows distinct differences in the prune tree's capacity to absorb nitrate throughout the year (table 16.5). In that study, prune trees absorbed relatively large amounts of nitrate when shoots were growing most rapidly, limited amounts in the fall during leaf drop, and virtually none during the dormant season through bud swell. Nitrogen absorption through the growing season of mature,

heavily cropped prune trees should be roughly similar to that in the study shown in table 16.5, with fruit nitrogen accumulation increasing tree nitrogen demand in the late spring and summer. Research in peach, a crop similar to prune, shows that mature tree crop nitrogen use closely follows shoot and fruit growth (Muñoz et al. 1993). Tree nitrogen demand after harvest is significantly reduced compared with that during fruit and shoot growth, and limited nitrogen fertilization is recommended in that period. Nitrogen fertilizer applied in late fall through early spring is at a much higher risk of loss from the root zone by leaching or denitrification and is not recommended. To be effective, nitrogen fertilization intended to increase flower nitrogen levels must be absorbed by the tree prior to natural leaf drop the previous fall.

Most soil nitrogen is unavailable to plants in the short term. Soil nitrate and ammonium constitute only a fraction (2 to 5%) of the total nitrogen in soils. The vast majority of soil nitrogen is found in various organic nitrogen compounds that must be decomposed by microbes and the nitrogen transformed though several steps before a plant-usable form is available. Increasing soil organic matter, especially by using materials such as legume (clover, alfalfa, etc.) residues, manures, and composts, can improve plant-available soil nitrogen levels. A sustained increase in plant-available nitrogen from organic inputs such as these requires regular annual inputs.

Deciduous fruit trees store nitrogen in woody tissues over winter and use this stored nitrogen for flower and shoot growth the following spring. This process allows trees to begin the growing season without relying completely on current root nutrient uptake when root activity can be limited by cool, wet soil. Storage nitrogen levels do not appear to limit root nitrogen uptake in the spring, although as much as 20 pounds of stored nitrogen per acre may be available for tree growth in mature prune orchards. This recycled nitrogen is valuable in prune orchard management and is equivalent to 50 to 100 pounds of fertilizer nitrogen material, assuming 20 to 40% fertilizer use efficiency (the amount of fertilizer absorbed by the tree divided by the amount of fertilizer applied).

Annual prune tree nitrogen requirements are largely dictated by tree age and crop load, assuming a healthy, uniform orchard. Fruit nitrogen concentration at harvest in healthy, productive orchards is in the range of 0.5 to 0.6% nitrogen on a dry-weight basis. Immature trees require less nitrogen than mature, cropping trees due to smaller size and reduced crop (compare the values in table 16.2 with those in table 16.4). Excessive nitrogen fertilization early in the development of the orchard wastes money, risks environmental contamination, and can promote excessive vigor that can delay orchard productivity. General recommendations for annual prune tree nitrogen fertilizer needs are shown in table 16.6; they illustrate the general increase in tree nitrogen demand and subsequent fertilizer inputs as trees grow and cropping begins. Individual orchards should be managed using visual assessment of tree vigor, annual leaf analysis, and crop load (where appropriate) to determine actual annual orchard needs and nitrogen fertilizer application rates. Any conditions that limit tree growth or health reduce orchard nitrogen requirements.

Developing an orchard nitrogen fertility program should begin with estimating plant-available nonfertilizer nitrogen sources for the current growing season. These include tree storage nitrogen, soil nitrogen mineralized from soil organic matter, and nitrate in irrigation water. Tree storage nitrogen is difficult to estimate due to leaf losses from pests, harvest operations, and postharvest drought. A

Table 16.5. Nitrate uptake capacity of nonbearing French prune/Mariana 2624 trees following 10-day exposures to tagged nitrogen at different phonological periods

Tagged N application periods	Growth stage during application	Nitrate uptake capacity (%)*
Jan 16–26	dormant	4 a
Mar 5–15	bud swell	4 a
Apr 2–12	rapid shoot growth	30 c
May 14–24	shoot growth cessation	39 c
July 9–19	shoot growth cessation	32 c
Aug 6–16	shoot growth cessation	35 c
Sep 10–20	shoot growth cessation	32 c
Oct 22–Nov 1	mid–leaf fall	16 b
Dec 3–13	dormant	3 b

Source: Weinbaum et al. 1978.
Note: *Nitrate uptake capacity is the amount of nitrate absorbed by the trees divided by the amount nitrate applied. Mean separation within the columns by Duncan's multiple test range, 1% level.

Table 16.6. Suggested general nitrogen fertilization schedule for highly productive French prune orchard with 183 trees per acre (spacing of 14 feet by 17 feet)

Year	Nitrogen (lb/ac)	UN-32 (gal/ac)
1	5–10	1.4–2.8
2	25	7
3	30	8.5
4	40	14.1
5	75	21.2
6	100	28.2
7	100	28.2
8+	150	42.3

Source: Based on results from Southwick et al. 1996.
Note: Expected yield in years 8+ is 4 dry tons per acre.

rough estimate is 0 to 20 pounds, depending on summer leaf analysis results and the degree of defoliation in the fall. Soil sample analysis can be used to estimate soil nitrogen that is potentially mineralized during the growing season and available for tree uptake (table 16.7). The estimate in table 16.7 is produced from experience in the lab when moist soils are incubated for a 2-month period at 77°F. This mineralized nitrogen is potentially available for uptake by prune roots. These data represent an estimate of the maximum soil nitrogen available for plant growth under optimal field conditions from a well-mixed soil sample taken from a depth of 0 to 12 inches. These conditions include constant moisture and temperature and no weed competition. Nitrate in well water can also contribute to plant-available nitrogen (table 16.8). Both mineralized soil nitrogen and irrigation water nitrate are potentially available for tree uptake, but they can also be absorbed by weeds, denitrified, leached, or immobilized by soil microorganisms. However, tree storage nitrogen, nitrogen from soil organic matter decomposition, and nitrate-nitrogen in irrigation water can constitute a large portion of an annual prune orchard nitrogen budget in some years.

While "natural" sources of nitrogen can deliver significant amounts of nitrogen to prune trees, research has shown that additional nitrogen inputs (fertilizers) are needed to maximize fruit production and maintain orchard health (Olson et al. 1980; Southwick et al. 1996). Field trials with mature, well-fertilized prune trees in numerous locations in northern California showed that nitrogen deficient developed within 3 years after nitrogen fertilization was stopped (B. L. Olson et al. 1987).

Deciduous fruit trees commonly absorb less than 50% of fertilizer nitrogen applied under the best conditions of high tree nitrogen demand, adequate irrigation, warm soils, and good orchard health. To maximize the benefit from fertilizers and minimize losses to the environment, careful fertilizer nitrogen application is required. Soil pH should be maintained in the neutral range (pH 6.0 to 7.5) to help maintain good root health and absorption activity. All nitrogen fertilizers should be incorporated into the soil by cultivation or irrigation water as soon as possible after application. Unincorporated ammonium or urea materials on the soil surface are at risk of loss through volatilization of ammonia, especially on soils with pH greater than 7.0. Split nitrogen fertilizer applications should be considered. A good general rule is two-thirds of the nitrogen applied in mid- to late summer and one third in early spring in orchards where tree vigor is adequate or high. In situations where tree vigor is low, an increase in the percentage of fertilizer nitrogen applied in the spring (March and April) could be made, with the remainder applied during June, July, and August or shortly after prune harvest. On sandy soils, more-frequent applications of smaller amounts of fertilizer nitrogen, following the same general pattern, are recommended. Adding organic matter to sandy soil can reduce soil nitrogen losses by increasing the soil water-holding capacity and increasing the ability of the soil to hold nutrients. However, addition of organic matter can lead to increased soil nitrogen losses through denitrification if the soil is initially well aerated and then becomes waterlogged (anaerobic).

Table 16.7. Estimated amount of soil nitrogen (N) mineralized from 1 acre-foot of soil containing a range of different concentrations of soil organic matter

Soil organic matter (% dry wt)	N mineralized (lb)
0.5	28
1.0	56
1.5	84
2.0	132

Note: It is assumed that 1 acre-foot of soil weighs 4,000,000 pounds; that field soil has a uniform soil organic matter content; that 7% of soil organic matter is nitrogen; and that 2% of soil organic matter nitrogen is mineralized in a lab incubation at 77°F and constant soil moisture lasting 2 months.

Table 16.8. Amount of nitrogen (N) applied in irrigation water as a function of nitrate (NO_3-N) concentration and the amount of irrigation water applied*

Concentration		Amount of N applied (N/ac)			
ppm NO_3-N	ppm NO_3^-	2.5 ac-ft	3.0 ac-ft	3.5 ac-ft	4.0 ac-ft
2.26	10	15.4	18.4	21.5	24.6
4.52	20	30.8	36.8	43.0	49.2
6.78	30	46.1	55.2	64.5	73.8
9.04	40	61.5	73.6	86.0	98.4
11.30	50	76.9	92.0	107.5	123.0
13.56	60	92.2	110.4	129.0	147.6
15.82	70	107.6	128.8	15.05	172.2

Note: *Laboratories report results of water analysis as either NO_3-N(ppm N) or ppm NO_3^-. Use the following conversion factors to calculate the amount of N applied annually in irrigation water for N concentrations or levels of applied irrigation ater other than those listed above.
1 ppm NO_3-N in the water = 2.72 lb N/ac-ft of water applied.
1 ppm nitrate (NO_3^-) = 0.614 lb N/ac-ft of water applied.
Since the atomic weight of the N atom is 22.59% of the atomic weight of NO_3^-, then ppm NO_3^-, 0.2258 = ppm N.

Table 16.9. Components of various nitrogen fertilizers and their characteristics

Fertilizer	Formulation	Nitrogen %	Equivalent acidity or basicity (in = lb $CaCO_3$) Acid	Base	Leaching risk*	Volatilization potential	Comments
ammonium nitrate	NH_4NO_3	33.5–34.0	100	—	M	L, M‡	half immediately available, half delayed
ammonium sulfate	$(NH_4)_2SO_4$	21.0	280	—	L	L‡	acidic source of sulfur
calcium-ammonium nitrate solution (CAN-17)	$Ca(NO_3)_2 \bullet NH_4NO_3$	17	9	—	M	L	—
calcium nitrate	$Ca(NO_3)_2$	15.5	—	20	H	L	immediately available, source of calcium
urea	$CO(NH_2)_2$	45.0–46.0	80	—	M	M‡	—
UAN-32†	$NH_4NO_3 \bullet CO(NH_2)_2$	32.0	57	—	M	M‡	some immediately available, some delayed

Source: California Plant Health Association 2002; Jones et al. 2007; Hanson et al. 2006.

Notes:

*L = low, M = medium, H = high. These terms are relative. All ammonium forms will leach after being converted to the nitrate form. This will take place in about 2 to 4 weeks in most soils. Leaching can then be severe on sandy soils and moderate on silt loams and clays.

†UAN-32 is often injected through low-volume irrigation systems.

‡If not incorporated or banded below the soil surface, volatilization can be high on warm soil with a pH over 7.0.

Common nitrogen fertilizer materials include ammonium sulfate, urea, UAN-32, calcium ammonium nitrate (CAN-17), calcium nitrate, and potassium nitrate. Materials vary in their nitrogen content, potential to acidify the soil, leaching risk, and risk for volatilization (table 16.9). Urea has no electrical charge immediately after injection and can move with irrigation water during fertigation. Care must be taken to ensure that neither urea nor nitrate leach out of the root zone during fertigation. Urea-containing liquid fertilizer materials such as UAN-32 should be injected during the middle of the irrigation set, not too early or too late (Hansen et al. 2006). Nitrogen fertilizers containing calcium can increase water infiltration into slowly permeable soils (Wildman et al. 1989).

Organic matter inputs can provide significant nitrogen to prune trees. Cover crop residues, compost, manure, and organic fertilizer materials represent organic nitrogen sources. Generally, organic nitrogen sources contain lower nitrogen concentrations than do synthetic nitrogen fertilizers, and only a small portion (< 15% total N) of the total nitrogen content of the material is expected to be available for plant uptake in the year applied (Hartz et al. 2000). Consequently, organic nitrogen sources must be used at much higher rates than conventional nitrogen fertilizers to deliver similar amounts of plant-available nitrogen. Like synthetic fertilizers, organic nitrogen sources must be incorporated into the soil to maximize plant-available nitrogen. Also, as with synthetic nitrogen sources, nitrate mineralized from organic nitrogen sources can be lost from the root zone through leaching or denitrification. The carbon to nitrogen ratio (C:N) of organic nitrogen sources is inversely related to plant-available nitrogen. In general, among soil-incorporated organic nitrogen sources, the order of highest to lowest mineralization of plant-available nitrogen is manure, composted manure, and plant residue manure. Growers should be aware that organic nitrogen sources, especially manures, may contain significant amounts of salt.

With foliar nitrogen fertilization, the percentage of fertilizer nitrogen absorbed into the tree relative to the amount of fertilizer nitrogen applied is greater than that with soil-applied nitrogen fertilization. However, the risk of leaf burn limits the amount of nitrogen that can be safely applied in any single application between bloom and harvest. Multiple applications are required to deliver significant amounts of nitrogen. Thus, supplying all tree nitrogen needs via foliar fertilization is more expensive than by soil application. Commonly applied foliar nitrogen fertilizers include high-quality urea (low biuret content) and potassium nitrate.

Zinc

Zinc (Zn) is the micronutrient most commonly deficient in California prune orchards. Required in very small amounts by prune trees (see table 16.2), zinc's primary function is to catalyze enzyme reactions. For example, it helps catalyze the production of auxin, a plant growth regulator essential for normal plant growth and development, including leaf and shoot growth.

Deficiency Symptoms

Zinc deficiency symptoms include delayed opening of flower and leaf buds in the spring and small, yellow leaves on shortened stems, often referred to as "little leaf" (fig. 16.6). Symptoms are most obvious in the spring. Terminal dieback can occur in severely zinc-deficient orchards. Soil zinc availability is often low in California prune orchards, as its availability is limited in soils with pH over 6.5.

Treatment

While reducing soil pH may increase soil zinc availability, zinc needs in orchards are most often met by use of foliar fertilizers. Cost-effective zinc foliar fertilization should be approached with care. Zinc absorption by leaves is a passive process positively related to the concentration of the zinc solution on the leaf: the higher the concentration of zinc in the spray solution on the leaf, the more zinc enters the leaf. Underapplication (low zinc concentration on the leaf) leads to reduced zinc uptake by the leaf and possibly an ineffective treatment; overapplication (high zinc concentration on the leaf) can cause leaf damage. Foliar zinc applications may be required every year or every other year, depending on the site. Foliar zinc deficiency symptoms, but not necessarily leaf zinc concentrations, can be corrected by foliar zinc application.

Spring, fall, or dormant period are effective timings for correcting zinc deficiency symptoms by foliar fertilization. Different strategies should be used for these different timings.

Spring applications are usually more cost effective than fall or dormant sprays, as this timing uses less zinc material and can often be tank-mixed with a pesticide or other nutrients. Effective materials for spring application include basic zinc sulfate (52% zinc), zinc nitrate, zinc oxide, zinc EDTA, and numerous other zinc chelate products. Generally, rates of basic zinc sulfate and zinc oxide range from 3 to 5 pounds per 100 gallons of water for spring applications. Combining zinc materials with early-season potassium nitrate application is an effective program, but lower rates of potassium nitrate (10 pounds of potassium nitrate in 100 gallons per acre) are recommended when tank-mixing with zinc. Rewetting of dried spray deposits by rainfall or dew can cause further absorption but may also lead to "shot-holing" of the leaves from too much zinc absorption. Zinc should not be applied in the spring if rain is forecast within 48 hours. The use of spray adjuvants may increase uptake but may cause marginal leaf burn if the spray volume is not reduced and the spray material spreads too well and accumulates on leaf edges.

Fall and dormant applications are usually limited to high rates of zinc sulfate historically applied alone. Rates per 100 gallons of water range from 10 to 15 pounds of zinc sulfate, or a maximum of 20 to 25 pounds per acre, and it should be applied at the beginning of normal leaf drop or during the dormant season. Oil should not be applied within 30 days of zinc sulfate applications. Fall zinc sulfate application can be tank-mixed with a pesticide for aphid control, although oil should be eliminated from the pesticide application. A fall zinc sulfate application, applied as natural leaf drop begins at the rates given above, can accelerate defoliation but will not harm buds the following year. Fall-applied zinc sulfate may not defoliate trees that are severely water stressed, as can occur in some years with minimal fall rains.

Figure 16.6 Zinc-deficient compared with zinc-sufficient prune shoot. *Photo:* D. Edwards.

Figure 16.7 Manganese-deficient prune leaves. *Photo:* D. Edwards.

Manganese

Manganese (Mn) is another micronutrient (see table 16.1) essential to plants due to its role in certain enzymes and membrane systems as well as in photosynthesis. Plant manganese availability is limited by high soil pH and is increased to toxic levels by low soil pH. Elevated leaf manganese levels can be an indication of low soil pH.

Symptoms of manganese deficiency are pale leaves with interveinal chlorosis (fig. 16.7). Yield and fruit size are not affected until more than 50% of the leaves in the canopy show symptoms.

Soil applications of acidifying fertilizer, such as ammonium sulfate, potassium or ammonium thiosulfate, or elemental sulfur (5 to 10 pounds per tree) can correct manganese deficiency in prune trees. This is accomplished by reducing soil pH and increasing soil manganese availability.

Since amending soil to lower soil pH is time consuming and generally detrimental to plant nutrition (see fig. 16.1), foliar fertilization is often the preferred approach to correcting manganese deficiency in prune. Spring foliar applications of manganese sulfate (2 to 3 pounds per acre) or MnEDTA (1 to 2 pounds per 100 gallons of water) can correct manganese deficiency. A manganese spray should be applied early in the spring before leaves reach full size.

Iron

Iron (Fe) is vital to chemical processes during photosynthesis and respiration. Iron deficiency in prunes is due to limited plant availability of soil iron, not limited soil iron levels. Iron occurs in soils in two forms: ferric iron (Fe^{3+}) and ferrous iron (Fe^{2+}). Plants absorb ferrous iron, but they use ferric form. Under well-aerated soil conditions, soil iron is primarily in the ferric form. Elevated soil pH, bicarbonate, and carbonates have been linked to iron deficiency in perennial crops.

Deficiency Symptoms

Iron is very immobile in plants, and deficiency symptoms appear first in young leaves. Iron deficiency symptoms are characterized by severe interveinal chlorosis, with only the veins themselves remaining green while all the remaining leaf tissue becomes yellow to very pale yellow, almost white (fig. 16.8). Because availability of iron is not usually the cause of iron deficiency symptoms, the condition of iron deficiency is often referred to as induced iron chlorosis.

Symptoms of iron deficiency often appear early in the spring, especially in years with late rains or premature irrigations that saturate the soil when the roots are active. Well water high in bicarbonates can exacerbate the problem. Regardless of the water source, when soils remain saturated during cool spring weather when orchard water use is low, soil bicarbonate levels increase and iron deficiency can occur. Elevated soil bicarbonate levels may be the specific culprit in what is referred to as lime-induced iron chlorosis.

Figure 16.8 Prune leaves showing iron deficiency symptoms, with virtually complete interveinal chlorosis compared with that from manganese deficiency seen in figure 16.7. *Photo:* F. J. A. Niederholzer.

Treatment

In alkaline soils, bicarbonate and carbonate levels are high. Numerous acidifying amendments can be used to lower soil pH and increase soil iron availability under these conditions. These include elemental (yellow, or popcorn) sulfur, acidifying nitrogen fertilizers, and sulfuric acid–urea fertilizers (e.g., N-pHuric or Nitro-sul). Sulfuric acid provides a quick fix but is not a long-term solution. Banding on the soil surface or locally incorporating (shanking) acidifying materials is more cost effective than broadcasting and will often provide faster results, as more material can be applied to a smaller soil surface. Only a small amount of soil with proper iron availability will provide ample iron for tree needs.

Foliar application of chelated iron can provide some short-term benefit, but this is expensive, can cause leaf phytotoxicity if applied improperly, and does not address the fundamental problem of high soil pH or saturated soil conditions that often limit iron availability.

Calcium

Calcium (Ca) is required for normal cell function in the meristems (growing tips) of stems and is also important in cell wall structure. Calcium deficiency in prune trees has not been reported in California. Calcium, often found in significant amounts in California well water, is highly immobile in plants and accumulates in the leaves, not the fruit (see table 16.2). Calcium sprays, often used in apple production, have not been shown to delay prune harvest or improve prune fruit quality.

Magnesium

Magnesium (Mg), a mobile element within plants, is a key part of the chlorophyll molecule and

Figure 16.9 Magnesium deficiency symptoms in young prune trees. *Photo:* F. J. A. Niederholzer.

enzyme systems involved in photosynthesis and respiration. Magnesium deficiency symptoms (fig. 16.9) are seen in trees growing on low-pH soils, or, more commonly, in very vigorous young trees. The characteristic V-shape of the chlorotic tissue is found on the lower (basal) leaves of vigorous shoots of trees with high potassium levels. High levels of soil potassium can interfere with root absorption of magnesium, and leaf magnesium needed for new growth is remobilized from old leaves on rapidly growing shoots. Magnesium-deficient leaves on these trees can drop early, but the tree health is not affected; these symptoms are uncommon on mature trees.

Boron

Boron (B) is a micronutrient that is essential to prune tree health at low levels but toxic to trees at high levels. It is used in all plants for the formation of cell walls. More important for orchard production, at low to moderate levels of plant-available boron, increased flower boron levels can improve fruit set and yield in many tree crops, including Italian prune. While immobile in many crop plants, boron is very mobile in prunes (and other stone fruits and almonds) and can readily be moved from leaves to fruit within the tree.

Boron toxicity in prunes is usually due to elevated boron in irrigation water. Boron toxicity symptoms include shoot dieback, profuse gumming, and lack of tree vigor.

Severe boron deficiency, although rare in prunes, has been reported in coastal regions and along the foothills of the Sierra Nevada. Symptoms appear as scorched terminal and subterminal leaves followed by defoliation and dieback of the affected shoots. Under these conditions, the flower number per tree decreases and yield is significantly reduced. Elevated boron levels in almond flowers can increase crop set and yield even when midsummer leaf levels do not indicate deficiency. However, boron nutrition research in prunes has been limited, and improvement in crop set and yield after boron fertilization have not been reported in French prunes in California.

Where present, boron deficiencies can be corrected with soil or foliar boron applications. Care must be taken to avoid excess applications. Review leaf analysis results prior to any fertilization decision. Agricultural borax broadcast at the rate of 50 pounds of material per acre will correct boron deficiencies for 3 to 5 years. Do not apply boron in concentrated bands. Foliar applications of 1 to 2 pounds per acre of a soluble boron source (Solubor, etc.) when foliar symptoms appear will benefit the current season's growth. Applying higher rates per acre of a soluble boron source at bloom may reduce set and yield.

Sulfur

Sulfur (S) is a component of the essential amino acids cysteine, cystine, and methionine, as well as certain vitamins and enzymes. Approximately 90% of sulfur in plants is found in amino acids. Sulfur is absorbed from the soil as the anion sulfate (SO_4^{2-}) and is fairly immobile in plants. Consequently, sulfur deficiency symptoms appear first in the newest (youngest) leaves and are similar in appearance to iron deficiency, with slight similarity to nitrogen deficiency. However, unlike iron-deficient leaves, extremely sulfur-deficient leaves show necrotic spots (fig. 16.10). Sulfur deficiencies that affect production in California prune orchards have not been reported to date.

To sample prune leaves for lab analysis to assess orchard sulfur status, select fully expanded mature leaves. Sulfate sulfur, not total sulfur, is the preferred form to measure. The critical concentration of sulfate sulfur, below which the tree is determined to be sulfur deficient, is 100 ppm. Total leaf sulfur should be above 0.15% (1,500 ppm) to allow adequate sulfur status.

Figure 16.10 Extreme sulfur deficiency symptoms in prune. *Photo:* D. Edwards.

Figure 16.11 Copper deficiency symptoms in prune. *Photo:* D. Edwards.

Several fertilizers and soil amendments contain sulfur. These include ammonium sulfate, potassium sulfate, ammonium and potassium thiosulfates, and gypsum. Elemental sulfur, which is often applied to reduce soil pH, also is a good source of sulfur. Well water can be a source of sulfate sulfur.

Copper

Copper (Cu) is needed by prune trees in very small amounts (see table 16.2), but it is an essential catalyst in several key enzyme reactions in plants. Plants absorb copper as the cation Cu^{2+}, which is less available to plants at soil pH values above 6.5.

Copper deficiency in prune is rare. Deficiency symptoms are similar to those of boron, with death of terminal buds and the subsequent growth of several buds below the terminal bud, producing a "bushy" look (fig. 16.11). Gumming may also occur along the shoot. Leaf analysis may be needed to differentiate between copper and boron deficiencies.

Phosphorus

Phosphorus (P) is a macronutrient that is not believed to limit prune production in California. It is found in numerous vital compounds in plants, including nucleic acids, phospholipids, coenzymes, and adenosine triphosphate (ATP). Prune trees remobilize phosphorus and store it over winter for use the following year. Mycorrhizal fungi can improve soil phosphorus availability.

Phosphorus is absorbed by plants from the soil solution as monovalent [$H_2PO_4^-$] and/or divalent [HPO_4^{2-}] phosphate. However, unlike other anions, these molecules are very strongly held in the soil and are not readily leached. Soils high in available iron, aluminum, or calcium can form very insoluble compounds with phosphorus and reduce plant-available soil phosphorus. These cations are not commonly found in high levels in neutral-pH soils in prune-growing regions of California.

Chlorine

Chlorine (Cl) is an element that, like boron, is both essential and potentially toxic to prune trees. It is present in soils and absorbed by trees as the highly mobile anion chloride (Cl^-). There have been no reports of chloride deficiency in California prune orchards, but chloride toxicity can be significant in soils with poor drainage, high-chloride irrigation water, or both. Insufficient leaching of a fall potassium chloride fertilizer application can also cause chloride toxicity symptoms the following season.

Chloride toxicity symptoms appear as leaf tip burn, with the necrotic tissue covering as much as half of the leaf under severe cases and having a scalded appearance.

Chloride accumulates in woody tissue, shoots, and leaves during the growing season. Over several years, high levels of plant-available soil chloride can increase tree chloride accumulation and tree damage. However, chloride is not remobilized from leaves back into the woody tissue of the tree prior to leaf drop. Consequently, tree chloride levels can decrease in the years after chloride in the soil is reduced (e.g., if high-quality irrigation water with no to low chloride is used) as the chloride exits the tree in fallen leaves. Thus, leaf drop can act as a detoxifying process in prunes and can help orchards recover from certain levels of chloride exposure, as long as tree chloride uptake is decreased. Flushing chloride from the root zone with clean water (rainwater or low-salt irrigation water) is key to orchard recovery after exposure to high levels of chloride.

Sodium

Sodium (Na) is toxic to plants and has no benefit to prune production. Sodium toxicity symptoms are similar to those of chloride, but sodium-affected leaves show a striated pattern perpendicular to the midrib. Leaf analysis is necessary to differentiate sodium from chloride toxicity. Sodium-affected soils are much more difficult to reclaim than chloride-affected soils, as sodium is more closely held to soil particles than is chloride.

Prune trees naturally exclude sodium from their roots. However, sodium toxicity symptoms may appear in prune leaves in trees defoliated due to potassium deficiency, spider mite damage, or prune rust infection and no longer have carbohydrate resources needed to maintain healthy, functioning roots.

References

Aldrich, T. M., R. M. Carlson, P. B. Catlin, L. B. Fitch, D. W. Henderson, D. E. Ramos, R. S. Rauschkolb, H. Schulbach, G. S. Sibbett, and K. Uriu. 1978. Irrigation, fertilization, and soil management of prune orchards. Berkeley: University of California Division of Agricultural Sciences Leaflet 21016.

California Plant Health Association. 2002. Western fertilizer handbook. 9th ed. Danville, IL: Interstate Publishers.

Hansen, B., N. O'Connell, J. Hopmans, J. Simunek, and R. Beede. 2006. Fertigation with microirrigation. Oakland: University of California Agricultural and Natural Resources Publication 21620.

Hartz, T. K., J. P. Mitchell, and C. Giannini. 2000. Nitrogen and carbon mineralization dynamics of manures and composts. HortScience 35(2): 209–212.

Jones, C. A., R. T. Koening, J. W. Ellsworth, B. D. Brown, and G. D. Jackson. 2007. Management of urea fertilizer to minimize volatilization. Montana State University Extension Service website, http://msuextension.org/publications/AgandNaturalResources/EB0173.pdf.

Marschner, H. 1986. Mineral nutrition of higher plants. New York: Academic Press.

Muñoz, N., J. Guerri, F. Legaz, and E. Primo-Millo. 1993. Seasonal uptake of 15N-nitrate and distribution of absorbed nitrogen in peach trees. Plant and Soil 150:263–269.

Olson, B., L. Fitch, D. Ramos, J. Yeager, K. Uriu, J. Pearson, and R. Snyder. 1980. Efficient nitrogen application timing in prune production. California Dried Plum Board Research Report. UCCE website, http://ucce.ucdavis.edu/files/repositoryfiles/19805.PDF-78655.pdf.

Olson, W. H., K. Uriu, R. M. Carlson, W. H. Krueger, and J. Pearson. 1987. Correcting potassium deficiency in prune trees is profitable. California Agriculture 41(5): 20–21.

Olson, B., K. Uriu, S. Sibbett, B. Krueger, and C. Weakley. 1987. Verification of field nitrogen timing trial results, 1987. California Dried Plum Research Report. UCCE website, http://ucce.ucdavis.edu/files/repositoryfiles/1987-14.pdf-78780.pdf.

Southwick, S., M. E. Rupert, J. Yeager, K. Weis, T. DeJong, K. Shackel, and A. Bonin. 1996. Nitrogen fertigation of young prune trees and effects on horticultural performance: 1996 final report. California Dried Plum Research Report. UCCE website, http://ucce.ucdavis.edu/files/repositoryfiles/1996-47.pdf-78994.pdf.

Ul Haq, I., and R. M. Carlson. 1993. Sulfur diagnostic criteria for French prune trees. Journal of Plant Nutrition 16(5): 911–931.

Uriu, K. 1981. Soil and plant analysis and symptomology for diagnosis of mineral deficiencies and toxicities, In D. E. Ramos, ed., Prune orchard management. Berkeley: University of California Division of Agricultural Sciences. 89–95.

Weinbaum, S. A., M. L. Merwin, and T. T. Muroaka. 1978. Seasonal variation in nitrate uptake efficiency and distribution of absorbed N in non-bearing prune trees. Journal of the American Society for Horticultural Science 103:516–519.

Weinbaum, S. A., F. J. A. Niederholzer, S. Ponchner, R. C. Rosecrance, R. M. Carlson, A. C. Whittlesey, and T. T. Muroaka. 1994. Effects of fruit growth and crop nutrient demand on tree capacity for nutrient uptake in "French" prune. Journal of the American Society for Horticultural Science 119(5): 925–930.

Wildman, W. E., W. K. Krueger, and R. E. Pelton. 1989. Calcium amendments for water penetration in flooding systems. California Agriculture 43(3): 14–15.

Part 4 Prune Pest Management

17 Insect and Mite Pests

• Carolyn Pickel and William H. Olson

This chapter describes insect and mite pests encountered in French prune *(Prunus domestica)* production in California. For more comprehensive information on sampling tools and treatment thresholds, see *Integrated Pest Management for Stone Fruits* (University of California Agriculture and Natural Resources Publication 3389). For more information on pesticide recommendations and the year-round plan, ,see the UC IPM Pest Management Guidelines for Prunes at the UC IPM website, http://www.ipm.ucdavis.edu.

Indiscriminate use of pesticides is costly and leads to pest resistance and outbreaks of secondary pests. An integrated pest management program combines

- the use of beneficial insects and mites, pesticides least harmful to these beneficials
- timing pesticides to be most effective on the pest and the least harmful to the environment especially as pesticide runoff into water
- some tolerance of insect injury
- cultural techniques that reduce pest populations, eliminate pest outbreaks, and reduce pesticide applications

However, such a regime requires correct pest identification, knowledge of the pest life cycle and the life stages susceptible to control procedures, and thorough pest monitoring in the orchard.

Insect and mite pests that attack prunes are divided into major pests that can cause serious economic damage if not controlled, occasional pests that occur sporadically and cause economic damage, and minor pests that are usually found in low numbers and are typically controlled by biological means but in some years may cause economic damage. French prunes that are dried can typically tolerate some pest pressure. However, the fresh market for prunes has a very low tolerance for insect parts or fruit damage.

Pest control strategies for prunes are based on sound pest-sampling techniques. These techniques include pheromone traps for peach twig borer, San Jose scale, and obliquebanded leafroller, and visual searches for insect pests on fruit and leaves.

Major pests listed below cause economic damage and require annual management. Occasional pests require monitoring but do not occur annually, and minor pests are listed primarily for identification.

Major Pests

Mealy Plum Aphid *Hyalopterus pruni*

Description

Wingless adult aphids on leaves are pale green or whitish green with three dark green longitudinal stripes on their backs. The bodies are covered with a white mealy wax. Eggs are black with a mealy coating. The winged form has a dark thorax and transverse bands on the abdomen. Mealy plum aphid builds up in large numbers on leaves and can completely cover the underside of prune leaves. This causes cupped leaves that become coated with honeydew and sooty mold (fig. 17.1). High populations can devitalize the tree, retard growth, reduce the sugar content of the fruit, and cause fruit cracking.

Seasonal development

Overwintering eggs found at the base of buds hatch during bloom, becoming "stem mothers" that produce live young that develop into wingless adults and create population outbreaks. Winged adult aphids appear from May to June, and as warm weather approaches they migrate from mature, nonvigorous orchards to reed grass or cattails. In vigorously growing young orchards, the aphids may stay all summer; the winged adults will migrate to other vigorous

Figure 17.1 Honeydew and sooty mold damage from mealy plum aphids. Waxy aphids can be seen in lower right corner. *Photo:* J. K. Clark.

prune trees only to start new colonies. In the fall, the winged adult females are the first to return to prune trees, where they produce live young. Wingless adult females mate with winged male aphids, which return to the orchard about 2 weeks later than females. These mated females lay the overwintering eggs.

Leaf Curl Plum Aphid
Brachycaudus helichrysi

Description

The leaf curl plum aphid is found inside curled leaves (fig. 17.2). The body is shiny yellow-green (fig. 17.3); the eggs are shiny black. Leaves infested with this aphid curl tightly perpendicular to the midvein, and the aphids feed inside the curled leaves. Often only one limb or a portion of a limb is infested. Prunes on infested limbs may become distorted and drop. The remaining prunes are unaffected and may be a larger size. Honeydew is secreted but not as much honeydew as is created by mealy plum aphid.

Seasonal development

This aphid overwinters in the egg stage near the base of buds. Young aphids begin to hatch at bud break and seek new foliage growth on which to feed. Within the curled leaves, the aphids produce multiple generations. In the absence of natural enemies or chemical treatments, large colonies rapidly develop on the leaves. As curled leaves become an unacceptable food source, the aphids move out to the shoot tips to feed on the new growth. In May, winged adults migrate to summer hosts, plants of the safflower and fiddleneck families. Leaf curl plum aphid is present on prunes in the spring and fall. In the fall, winged females that developed on the summer host return to the prune trees, where they produce more wingless females. Winged males are produced slightly later on the summer host and return to the prune trees in time to mate with the wingless females. Mated females lay overwintering eggs in late October on the bark near buds.

Figure 17.2 The leaf curl plum aphid is found inside curled leaves. Colonies of leaf curl aphid cause leaves to curl tightly. *Photo:* J. K. Clark.

Figure 17.3 Leaf curl plum aphids are yellowish green and shiny. *Photo:* J. K. Clark.

Management

The dormant period has traditionally been the time to apply sprays for management of aphids in prunes. However, the concern about dormant-applied insecticide movement out of orchards by drift and surface runoff because of winter rains has forced growers to look at alternative management practices. One very successful aphid control technique is to apply the dormant spray in November before the soil becomes saturated with rainwater. Once the soil is saturated, there is more likely to be runoff, taking pesticides with it. Growers can monitor dormant spurs to determine whether aphid eggs are present or review orchard history to determine whether aphids have been a problem. Very low rates of specific insecticides have been highly effective for aphid management in prunes. Beneficial insects often control leaf curl plum aphid before populations become damaging. Low populations in the spring do not cause economic damage. Review the current UC IPM Pest Management Guidelines for Prune (available at the UC IPM website, http://www.ipm.ucdavis.edu/) for sampling methods, thresholds, treatment timing, and insecticide rates.

Because these pests cannot complete their life cycle without passing part of the year on their alternate host, removal of the secondary host would appear to be an important means of control. However, this pest can migrate 20 to 30 miles, making removing the alternate host impractical.

San Jose Scale
Quadraspidiotus perniciosus

Description

This tiny major pest of fruit trees is usually not noticed until it builds up to large numbers on limbs. It can also infest shoots, leaves, and fruit. Adults and nymphs suck plant juices and cause considerable damage if left unmanaged. They have been known to seriously weaken branches and main scaffold limbs, causing permanent injury to mature trees. Immature scales, called crawlers, settling on fruit may cause fruit spotting.

Limbs supporting large populations of San Jose scale often ooze gum and exhibit rough bark and dieback. Dead leaves adhering to fruit spurs during the dormant season indicate the presence of scales.

The shell-like covering over the pest is ash-gray, almost circular in outline, with a nipple nearly centrally located. The adult's body is lemon yellow beneath the shell covering (fig. 17.4).

Figure 17.4 Adult female San Jose scale with cover removed (right) and a pupae of the parasite *Aphytis* sp. above the remains of a parasitized scale (left). Mature scale with nipple in center (upper right). *Photo:* J. K. Clark.

Figure 17.5 Tiny yellow crawlers that hatch from San Jose scale eggs. *Photo:* J. K. Clark.

Seasonal development

Partially mature nymphs overwinter on limbs and trunks. In the spring, the nymphs develop into winged male and sessile female scale insects that mate. In May, the females produce live young. The young crawlers (fig. 17.5) settle on shoots, feed, and become adults or overwinter as partially grown scales. In California, there are four overlapping generations per year. Crawlers first appear in late April and May, followed by continuous overlapping emergence from late June through December.

Management

San Jose scale populations can be monitored by sampling dormant spurs for the presence of scales and using pheromone traps in the spring. The dormant spur sample detects the presence of scales and helps determine dormant treatment needs and rates of oil and insecticides. The pheromone traps catch male San Jose scales and their parasites and

can help growers determine whether biological control is working in their orchard. In the Sacramento Valley, summer generations of San Jose scale rarely have to be treated. However, in the San Joaquin Valley, the populations sometimes require summer applications; these can be determined using degree-day models or double sticky tape to monitor for crawlers. These techniques and treatment thresholds are described in the UC IPM Pest Management Guidelines for Prune.

Thorough coverage with insecticide sprays in the dormant period, or during the crawler hatch in early May, has given excellent control. Treatments for San Jose scale are rarely needed every dormant season.

Occasional Pests

Spider Mites

Description

Although four species of spider mites are found on prunes, the brown mite *(Bryobia rubrioculus)* and the European red mite *(Panonychus ulmi)* rarely cause economic damage and can be beneficial. The twospotted mite *(Tetranychus urtcae)* and the Pacific mite *(Tetranychus pacificus)* are the most damaging to prunes.

The brown mite can be recognized by its flattened body and long front legs. The adult is blue-green; the nymphs are red at first. Eggs, often laid in clusters, are oval and red. The European red mite is bright red and has a round body with prominent large white spots at the base of hairs on its back. The red, oval eggs are often laid singularly and have a "stipe" (spike) on one part of the egg. The stipe is often broken off, making these eggs difficult to distinguish from brown mite eggs. Neither the brown mite nor the European red mite spins a web. The twospotted mite and Pacific mite produce abundant webbing on both sides of the leaves (fig. 17.6). These two mites are not easily distinguished. Both have two black spots on their yellow-green bodies (fig. 17.7). In the fall, they turn orange-red. The white, oval eggs of both mite species are laid singularly. The twospotted mite is most common in the Sacramento and San Joaquin Valleys.

Spider mites feed by sucking the contents out of leaf cells. This leaf damage reduces tree vitality and can adversely affect prune size. Defoliation caused by twospotted and Pacific spider mite damage often allows the tree and prunes to become sunburned. Twospotted mite and Pacific mite do not cause economic damage to prunes every year.

Figure 17.6 Twospotted web-spinning mites can produce copious amounts of webbing with high populations. *Photo:* J. K. Clark.

Figure 17.7 Viewed through a hand lens, twospotted spider mites are pale green or yellowish white with dark blotches on the sides and two red eyespots at the anterior end. *Photo:* J. K. Clark.

Seasonal development

The brown mite and European red mite overwinter as eggs on twigs and branches. Then eggs hatch in spring, and the young move out to the leaves. During summer, there are numerous overlapping generations, with eggs being laid on twigs and the upper and lower surfaces of leaves.

The twospotted mite and Pacific mite overwinter as adult females under tree bark on the lower trunk and in the soil around the trunks. When weeds dry up in spring, the mites move to trees and feed on lower leaves first. There are many overlapping generations each summer, with eggs being laid on the undersides of leaves. Males generally do not survive the winter except in the San Joaquin Valley, where they feed and may reproduce with females on weeds during the winter.

Management

The European red mite has long been considered a pest because it is easily found and feeds on prune leaves. However, there has been little documented evidence that this mite causes economic damage in prunes. The European red mite serves as a food source for beneficial insects including predatory mites in the spring. This allows the predator population to increase, making them better able to control the twospotted mite in the summer. Both the European red mite and the brown mite die out naturally when the weather gets hot. The brown mite hides in spurs on hot days and is difficult to monitor in the middle of hot days. Dormant sprays that include oil provide effective control of the overwintering eggs of brown mite and European red mite. Dormant treatments do not control twospotted or Pacific mite.

Beneficial mite predators are important in keeping mite populations at low levels. Miticides or oil may be necessary in some orchards in the summer, but only when twospotted and Pacific mite populations begin damaging foliage. Successful mite management requires good cultural practices to maintain healthy tress and regular monitoring for pest mites and mite predators. Some insecticides such as pyrethroids reduce beneficials and increase mite populations. Choosing insecticides that are less disruptive to beneficials reduces the need for more expensive miticides. The "timed search" for spider mites was developed for prune orchards and can give growers information on the status of spider mites in the orchard. This monitoring tool can often allow the use of low rates of miticides before mite populations get to damaging levels. Monitoring should be done in prunes from June to mid-July. See the UC IPM Pest Management Guidelines for Prune for a description of this monitoring technique and treatment thresholds.

Peach Twig Borer *Anarsia lineatella*

Description

This pest damages in two ways: larvae can burrow down tender twigs and kill shoots, and they can also feed on prunes, causing them to become off-grade or drop before harvest (fig. 17.8).

Larval mining of shoots is evidenced by presence of wilted shoots (flagging) on vigorous terminal growth (fig. 17.9). Shoot strikes cause economic damage mostly on young trees, particularly where damage interferes with tree growth and training. Fruit feeding is seen as roughened holes in the prune and by presence of gum, insect frass, or both.

The adult moth is small, gray, and inconspicuous and can be distinguished from other moths by the prominent snout on its head. Larvae have black heads and brownish bodies with white portions between each body segment, giving the appearance of stripes (fig. 17.10).

Figure 17.8 Fruit damage caused by peach twig borer. *Photo:* J. K. Clark.

Figure 17.9 Shoot flagging caused by peach twig borer. *Photo:* J. K. Clark.

Figure 17.10 Peach twig borer larva on prune leaf is easy to identify because of the distinctive ringed appearance. *Photo:* J. K. Clark.

Seasonal development

This insect overwinters as young larvae in hibernacula (chimney like structures) in the crotches of limbs (fig. 17.11). In the spring, larvae move out and infest buds and young shoots, pupate, and turn into adults in May. Adults deposit small white eggs on leaves, twigs, and young fruit that give rise to later generations that infest other shoots and fruit. There are three or four generations per year.

Management

The dormant and delayed dormant periods have been the preferred time to apply sprays for this pest, but the concern about dormant-applied insecticides leaving orchards with winter rains has forced orchard growers to look at alternative management practices. Recent research has shown that peach twig borer does not always cause economic damage in dried prunes but can cause damage in prunes grown for the fresh market.

Several alternatives have been developed for control when populations warrant it. These include *Bacillus thuringiensis* bloom sprays, a dormant spray every other year (this works where there is no aphid problem), and fruit monitoring conducted at 400 degree-days from first moth catch in pheromone traps to determine whether peach twig borer is causing economic damage. Sprays applied at petal fall or in late May are effective against emerging larvae and are typically needed only on trees under 3 years of age and for prune crops being used for the fresh market. For more details, see the UC IPM Pest Management Guidelines for Prune.

Codling Moth
Cydia pomonella

Description

Although not a primary cause of damage to prunes, fruit feeding by codling moth larvae has caused a high percentage of unmarketable prunes in some orchards (fig. 17.12). Damage by insects that have tunneled all the way to the prune pit is characteristic of codling moth larval feeding. Larvae have white or pinkish bodies with brown to black heads and can be confused with oriental fruit moth, which can also be found in prunes. Codling moth larvae can be distinguished from oriental fruit moth by looking for the anal comb after the third-instar larvae. Adults have gray wings with a copper spot on each wing tip.

Seasonal development

After overwintering as mature larvae in silken cocoons under loose bark on the tree, the insects emerge as adult moths during April and May. Adults mate and lay eggs that produce larvae that feed on small fruit. A second and often third generation appears in July and August.

Management

Cleanup of uncared-for trees, particularly apple, pear, and walnut, helps reduce damage caused by codling moths. Insecticide sprays timed with the use of pheromone traps and degree-days improves control.

Obliquebanded Leafroller
Choristoneura rosaceana

Description

Rolled leaves webbed together to form protective cases or leaves webbed to prunes indicate the presence of tortricid leafrolling caterpillars. The obliquebanded leafroller is the predominant

Figure 17.11 Peach twig borer overwintering in a hibernaculum that forms a chimneylike pile of frass. *Photo:* J. K. Clark.

Figure 17.12 Codling moth damage on a prune. *Photo:* J. K. Clark.

tortricid leafrolling caterpillar found in prune orchards. When disturbed, obliquebanded leafrolling larvae wiggle backward and drop toward the ground on a silken thread (fig. 17.13). Obliquebanded leafroller larvae are pale to dark green, usually with black heads but sometimes with lighter heads. Adults are light brown moths with a darker oblique band on the wings. Moths are medium sized and bell shaped (fig. 17.14).

Larvae feed on leaves and buds (fig. 17.15) and tunnel throughout the surface of the prune, causing severe surface scarring, which is more serious on fresh market prunes than on dried prunes.

Figure 17.13 All leafroller larvae have the habit of dropping on a silken thread when disturbed. *Photo:* J. K. Clark.

Seasonal development

Obliquebanded leafroller overwinters as third-instar larvae under loose scales or pieces of bark. The overwintered larvae pupate in the spring, and the first generation of adults emerges in late March or April. There are two generations of this moth per year in the Sacramento Valley. In the San Joaquin Valley, it can have three generations.

Management

Spray programs for other insects generally help reduce populations of leafrollers. As growers reduce their insecticide programs, obliquebanded leafroller populations have begun to increase. Pheromone traps are available to monitor adult emergence, but treatments should not be based on pheromone trap catches because these moths are widely distributed, they are readily attracted to pheromone traps, and there is no correlation between trap catch and the incidence of damage in the orchard. Reduced-load pheromone traps are now available and can be used as an indication of when to monitor prunes for the presence of damage. Details on this monitoring technique and treatment guidelines are available in the UC IPM Pest Management Guidelines for Prune.

Figure 17.14 Obliquebanded leafroller adult. *Photo:* J. K. Clark.

Oriental Fruit Moth
Grapholita molesta

Description

Occasionally, oriental fruit moth larvae can cause damage by feeding on shoots and prunes. Shoot damage, like that caused by peach twig borer, can be severe on young trees. Larvae usually enter the prune near the stem end or where two prunes touch. The larvae bore into the prune and feed around the pit. When mature, they leave the prune to pupate. Feeding wounds can cause prunes to be off-grade and serve as infection sites for brown rot. The adults are dark gray with alternating black and

Figure 17.15 Prune leaf attached to fruit by a leafroller. *Photo:* J. K. Clark.

gray markings that form a chevron pattern on the wings, which are folded tentlike when the moth is at rest. Moths are smaller than adult codling moth or peach twig borer. Young larvae are white with black heads. They turn pink as they get larger and resemble codling moth larvae, but they can be distinguished by the presence of an anal comb on the last abdominal segment on third-instar and older larvae. Eggs are white or cream colored, disc shaped, and laid singly on the underside of leaves of vigorous shoots or, in later generations, on the prunes.

Seasonal development

Oriental fruit moths overwinter as prepupae inside cocoons in protective places on the tree or on the ground around the base of the tree. After spring pupation, moths of the first generation emerge in March. Eggs are laid and larvae hatch. The larvae feed on shoot tips, causing shoot strikes, or flagging. They often pupate in cracks in the bark and produce the second-generation adults. Larvae from these adults usually appear during the first 2 weeks of May. These larvae also primarily feed on shoot tips, while later-generation larvae feed more on prunes. There are usually five generations per year.

Management

Oriental fruit moth has rarely been identified as an economic pest in prunes. It can be monitored by looking for shoot strikes and fruit damage and by using pheromone traps to monitor adult emergence. Although dormant pesticide treatments are not effective, growing-season pesticide sprays are effective. These techniques are described in *Integrated Pest Management for Stone Fruit* and in the UC IPM Pest Management Guidelines for Prune.

Minor Pests

Other Scale Insects

European fruit lecanium scale *(Parthenolecanium corni)* is a round, brown, hemispherical-shaped scale insect that is larger than other scales. It is not a serious pest in most prune orchards. It is often present where growers have not applied a dormant oil application, and it is frequently parasitized in the summer. The crawlers are flattened yellowish or brown. Italian pear scale *(Epidiaspis leperii)* is circular, light gray, and shiny with a brown nipple slightly off center (fig. 17.16). The body is similar to San Jose scale, except under the scale covering the insect body is dark reddish purple. Italian pear

Figure 17.16 Adult Italian pear scale with cover removed. *Photo:* J. K. Clark.

Figure 17.17 Fruittree leafroller adult. *Photo:* J. K. Clark.

scale is often found under moss and lichens on old prune trees; it usually does not cause economic damage, but if needed, it can be controlled with fixed copper materials that reduce the moss and lichens.

Other Leafrolling Caterpillars

In addition to the obliquebanded leafroller, three other species of tortricid moths may be found attacking prunes. They are the fruittree leafroller *(Archips argyrospila)*, the omnivorous leafroller *(Platynota stultana)*, and the orange tortrix *(Argyrotaenia citrana)*.

Fruittree leafroller larvae are dark green and have black heads. The adult fruittree leafroller wing is mottled with gold and white flecks (fig. 17.17). The fruittree leafroller overwinters in egg masses on twigs. Larvae emerge in the spring and feed on leaves and fruit. Adults emerge in June or July and deposit overwintering eggs. There is one generation per year.

Omnivorous leafroller larvae have tan heads and creamy white to tan or orange bodies with light stripes along each side. Mature larvae have raised whitish spots on the upper surface of each abdominal segment. Adults are tan and have light spots on the forewings (fig. 17.18). They overwinter as larvae in a nest of webbed leaves or in mummy fruit on the tree or ground. The first moths generally emerge in March; they eggs laid by female moths hatch to larvae that feed on leaves and prunes. There are four or five generations per year.

The orange tortrix is found mainly in coastal areas. Its larvae are straw colored to light green and have yellow to brown heads. Adults are fawn or gray with darker mottling on the forewings. Three generations of this moth occur per year in coastal areas. Although many overwinter as larvae on an alternate host, all development stages are present throughout the year and the three generations overlap.

Like most tortricid moths, the adults of these three species appear bell shaped when at rest; their eggs are laid in overlapping rows resembling fish scales (fig. 17.19). When disturbed, the larvae wiggle backward and drop toward the ground on a silken thread.

Leaf-Eating Caterpillars

Redhumped caterpillar *(Schizura concinna)*, fall cankerworm *(Alsophila pometaria)*, tent caterpillar (*Malacosoma* spp.), and fall webworm *(Hyphantria cunea)* are the most common moth caterpillars that feed on prune leaves. Of these, the redhumped caterpillar is the most common.

Damage caused by leaf-eating caterpillars may be serious on individual trees, but it is usually randomly scattered throughout the orchard. On small trees, infested twigs may be cut out and destroyed. Insecticide sprays applied for other pests often keep these leaf-eating caterpillars in check. If insecticide treatments are required, localized treatments on individual trees and branches are generally all that is necessary.

Redhumped caterpillars skeletonize leaves, leaving behind only leaf veins. They form no webbing on the leaves. Redhumped caterpillar is easily recognized because of its striking appearance—the main body is yellow and marked by longitudinal reddish and white stripes, the head is bright red or yellow, and the fourth abdominal segment is red and enlarged. The caterpillars pass the winter as full-grown larvae in cocoons on the ground. In the early summer, moths lay egg masses on the undersides of leaves. The eggs hatch into larvae that begin feeding on the leaves. There are at least three generations per year in northern California. Redhumped caterpillars are often controlled by the parasitic wasp *Hyposoter fugitives* (fig. 17.20). Look for distinctive parasite pupae on leaves before deciding to spray. These caterpillars can easily be controlled by *Bacillus thuringiensis.*

Fall cankerworm larvae also skeletonize leaves. The larvae have three pairs of prolegs near the tip of the body. They are brownish above, with three narrow whitish stripes and one yellow stripe along

Figure 17.18 Adult moth of the omnivorous leafroller. *Photo:* J. K. Clark.

Figure 17.19 Adults and egg masses of orange tortrix. *Photo:* J. K. Clark.

Figure 17.20 The redhumped caterpillar parasite forms a fluffy white mass of pupal cases (left). The young caterpillars on the right are not parasitized. *Photo:* J. K. Clark.

the sides of the body; underparts are light green. Fall cankerworms pass the winter in the egg stage. Caterpillars hatch, feed on leaves in the spring, pupate, and develop into moths in the fall. There is one generation per year.

Tent caterpillars are hairy and dull yellow-brown with a row of blue spots adjacent to orange spots on top of their bodies (fig. 17.21). They build large silken tents over the leaves on which they are feeding and do not eat leaf veins. Tent caterpillars have one generation per year. They overwinter in the egg stage; the eggs give rise to the destructive caterpillars in spring and early summer. Moths emerge in the summer that lay the overwintering eggs.

Larvae of the fall webworm are pale brown or gray. Their bodies are covered with long white hairs arising from black and orange spots. Like the tent caterpillar, they eat leaves (not veins) and also form silken tents. They damage prune trees primarily from July to September, whereas the tent caterpillar is most active from April to June. Fall webworm spends the winter as pupae. Moths emerge in late spring and lay eggs that hatch into caterpillars in late summer. There is one generation per year.

Green Fruitworms

The larvae of several species of noctuid moths, collectively referred to as green fruitworms, occasionally cause economic damage to prunes. Large holes eaten in young prunes are often caused by feeding by the larvae of green fruitworms. The citrus cutworm *(Xylomyges curialis)* is the most common in the San Joaquin Valley, but it also occurs in Sacramento Valley orchards. The species most commonly found in the Sacramento Valley are the humped green fruitworm *(Amphipyra pyramidoides)* and the speckled green fruitworm *(Orthosia hibisci)* (fig. 17.22). Larvae are pale grass-green, often with whitish stripes down each side of the body and a narrow stripe down the middle of the back. Adults of all green fruitworm species are fairly large grayish or brownish moths.

Most species pass the winter as pupae; adults emerge and lay eggs over an extended period beginning in January. The highly reticulated eggs are laid singly or in clusters on twigs. Egg hatch begins at bud break, and larvae feed on leaves, flowers, and prunes. The humped green fruitworm overwinters in the egg stage. Larvae feed on flowers, leaves, and fruit; they mature in June, pupate in the ground, and become adults in the summer or fall and lay eggs on branches and twigs. For all species, there is only one generation per year.

Figure 17.21 Larva of forest tent caterpillar. *Photo:* J. K. Clark.

Green fruitworms are rarely of economic importance. The *Bacillus thuringiensis* program helps provide control if sprayed on young larvae. However, insecticide applications applied at petal fall or during April have been effective when there has been a history of fruitworm damage.

Wood Borers

Larvae of the peachtree borer *(Synanthedon exitiosa)*, Pacific flatheaded borer *(Chrysobothris mali)*, shothole borer *(Scolytus rugulosus)*, and the branch and twig borer *(Melalgus confertus)* are sometimes found boring into the trunks and branches of prune trees. If trees do not die from the larval boring, growth is usually impaired, and trees are weakened and subject to being blown over or having branches broken.

Gum exuding from around the base of the trunk is evidence of peachtree borer and Pacific

Figure 17.22 Speckled green fruitworm larva feeding on leaf. *Photo:* J. K. Clark.

flatheaded borer (fig. 17.23). Larvae of the peachtree borer, found mainly in coastal areas, are white and have brown heads. Adults are clear-winged moths with blue-black bodies and yellow or orange bands across the abdomen. Pacific flatheaded borer larvae are whitish yellow with greatly enlarged and flattened thoraxes (fig. 17.24). Larvae are primarily found in young, weakened trees. Adults are short, broad, flat beetles.

Shothole borers also attack weakened trees. They are tiny brown or black beetles; their white legless grubs mine the tree's sapwood and often reduce it to powder.

Branch and twig borers are not economic pests in prune but are sometimes found attacking prune trees. The adults are slender brown beetles that make round holes at the base of buds (fig. 17.25), fruit spurs, and in the forks of small twigs. They are rarely found in prunes.

The stage that bores into and mines the tree's wood in all cases is the larval stage. The peachtree borer and Pacific flatheaded borer have only one generation per year. The adult peachtree borer may be found from June to September, with larvae present in the tree the rest of the year. Pacific flatheaded borer adults are generally present in May and June; when spring months are warm, they may be seen as early as March or early April. Shothole borer and branch and twig borer have several generations per year.

Insecticides can help control peachtree borer adults if applied when adults emerge. Sex-pheromone traps are available to monitor adult emergence and to time insecticide applications.

Pacific flatheaded borers, shothole borers, and branch and twig borers do not prefer healthy, vigorously growing trees. Sunburn protection and a good irrigation and fertilization program are important components of keeping trees in good health. Prompt destruction of brush piles, which harbor these pests, is also important. Badly diseased or borer-infested trees and branches should be removed from the orchard each winter and destroyed before spring. Insecticides applied when adult beetles are active can be of some value, but no insecticide treatment is currently recommended to control the larvae of these borers.

Figure 17.23 Peachtree borer, location of infestation on tree. *Photo:* J. K. Clark.

Figure 17.24 Larva of Pacific flatheaded borer. *Photo:* J. K. Clark.

Figure 17.25 Sap leaking from shothole borer holes. *Photo:* J. K. Clark.

18 Diseases and Physiological Disorders

• Themis J. Michailides, James E. Adaskaveg, Beth L. Teviotdale, Franz J. A. Niederholzer, Richard P. Buchner, Joseph H. Connell, and William H. Krueger

Prunes, like other stone fruits, are affected by several fungal diseases and physiological disorders. In general, the fungal diseases attacking prune are similar to those attacking other stone fruits, while some of the physiological disorders are unique to prunes (e.g., russeting, slip-skin disorder, wind scab, etc.). Among the diseases, brown rot blossom or shoot blight and fruit rot caused by the brown rot fungi may cause major yield losses. In general, immature prunes are susceptible to latent (invisible) and quiescent (visible) infections by the brown rot fungi. The detection of these infections before harvest can be used to predict the risk of fruit rot at harvest and after harvest. Brown rot, however, as well as the majority of the other diseases discussed in this chapter, can be well managed to prevent losses in fruit yield and quality.

Root and canker diseases are more difficult to control than foliar and fruit diseases, and their management generally depends on availability and selection of the resistant or tolerant rootstock, management of water and soil conditions, as well as other cultural practices. For current efficacy and timing of registered fungicides of foliar diseases of prune, see Adaskaveg et al. 2011.

Diseases

Bacterial Canker

Bacterial canker is a much-studied complex disease that still is poorly understood. All cultivated *Prunus* species are susceptible to this disease, which is caused by the bacterium *Pseudomonas syringae*. In some years, many trees are lost to bacterial canker, and it remains a serious and difficult problem in stone fruit production.

Symptoms

Bacterial canker infections develop in the fall and winter. Symptoms may be found in late winter into spring. Symptoms are generally divided into two categories, the canker phase and the blast phase. Symptoms of the canker phase range from discrete cankers on shoots or limbs to death of the entire tree. Young trees often are killed in a manner called souring out. These trees have a watery exudate emerging from the trunk and scaffolds and a distinctive, pungent, vinegary odor. Older trees usually do not die, but shoots, twigs, and branches or whole scaffolds may collapse. In these cases, gum balls are scattered over the infected area, and the typical smell of bacterial canker is present as well. Internally, there are small reddish orange flecks in the white tissue just under the outer layer of bark. As large areas of the bark expire, the internal tissues turn cinnamon brown. The infection and damage sometimes cross the graft union but do not progress into tissues below ground level. Later in spring, suckers grow from the still-living rootstock. The diseased and dead areas become a discrete canker if less than the complete branch is affected.

The blast phase occurs in late winter or early spring. Buds die, or if the tree has begun to grow, flowers and young leaves turn black; but the necrosis does not extend very far into the attached shoot. Gum may also be present at the infection sites.

Inoculum and infection

Bacteria live on the outer surfaces of many plants, and *P. syringae* is a common epiphytic, facultative parasite of many *Prunus* species. Leaf scars appear to be the primary avenues of entry into the tree, but other natural openings, such as lenticels, also may be involved. Wounds, though sometimes invaded, are seldom important entry sites for the pathogen. Scientists in South Africa recently found that the bacteria sometimes can be systemic, present inside trees without causing symptoms (Roos and Hattingh 1987).

Presumably, rain in the fall season washes bacteria into leaf scars, and from there they move into the tree. As the tree becomes dormant, its defense mechanisms also are dormant, so the bacteria are unrestricted. When the tree resumes growth in spring, infections are walled off or otherwise halted. If this happens in time, before the tree is killed, the tree survives but bears the old canker or dead scaffold for many years. These cankers are annual and do not resume activity in later years.

The physiological condition of the tree seems to be extremely important in disease expression. Generally, healthy, vigorous trees suffer less from bacterial canker than do their weaker counterparts. The disease is often severe in orchards grown on sandy soil, probably because high populations of the ring nematode also flourish there. Still, bacterial canker may devastate trees growing in heavy soils in some areas of the state. Freezing temperatures are not required, but the disease is frequently associated with cold winters or colder regions or portions of an orchard.

Control

No reliable control measures are available for bacterial canker. A combination of several practices, however, may help mitigate the disease. Prune trees on peach rootstock usually are less severely affected by bacterial canker than are those on plum rootstocks, and preplant fumigation is recommended to reduce nematode populations. A high-trunk budding approach to tree propagation helps reduce bacterial canker. Adequate nitrogen, potassium, and phosphorous nutrition and all practices that promote good tree vigor also help protect trees. Although experimental tests failed to document the value of late pruning, many growers routinely delay pruning well into spring to help avoid wet conditions that are favorable for bacterial canker. Similarly, dormant treatment with copper materials, effective in other parts of the world, has not proven beneficial in California prune orchards.

Ceratocystis Canker

Ceratocystis canker, also known as mallet wound canker, was initially reported on prune in 1962. The disease was widespread in California prune and almond orchards until modern harvesting equipment, which causes far less bark damage, was introduced. The disease can cause significant damage over time.

Causal organism

Ceratocystis fimbriata Ell. & Halst (anamorph *Chalara* sp.) forms perithecia with long necks that produce hat-shaped ascospores surrounded by a tapered gelatinous matrix. Fungal isolates from stone fruit are brown to dark olive, and clumps of perithecia develop in concentric rings on malt-extract agar after 6 to 12 days. Optimal growth occurs between 75° and 81°F. The fungus produces three forms of asexual spores: cylindrical and doliform endoconidia and chlamydospores. Hyaline to subhyaline endoconidia are produced in chains of up to 20 septate conidia; doliform endoconidia are at first hyaline but later become light brown and are formed in shorter and wider conidiophores than those producing the cylindrical endoconidia. The conidiophores that form the doliform endoconidia often aggregate, especially around perithecia. The pathogen also produces chlamydospores of variable shapes (oval to subglobose) that have smooth to rough walls. They are pale to olive brown and form singly or in short chains on simple or branched conidiophores. Chlamydospores perhaps travel upward in the transpiration stream and initiate the secondary cankers sometimes found beyond the initial wound canker. The fungus is often difficult to isolate and usually can be cultured only from canker margins. Carrot root disks can be used to selectively isolate the pathogen but do not increase the recovery rate.

Symptoms

Cankers begin as elliptical, water-soaked, darkened areas with distinct external margins. Large, amber gum balls encircle the canker along the margin, and the infected bark becomes water soaked. There is a characteristic reddening of infected tissues, and a brownish black stain permeates the heartwood. Cankers are perennial and expand more rapidly in summer and fall and on small rather than large wood. These infections may eventually kill infected limbs or trunks. Heart rot fungi frequently infect the exposed wood of old trunk cankers. The worst-affected trees are those that are damaged annually because new infections at harvest injuries, year after year, ensure a constant supply of infectious pathogen spores.

Inoculum and infection

The pathogen overwinters in old infections and in association with its insect vectors. The most important vectors are a nitidulid beetle *(Carpophilus freemani)* and a drosophilid fly *(Chymomyza procnemoides)*. Numerous other insects and one mycophagous mite are secondary vectors. Insect frass contains viable spores, and some contaminated larvae produce adults that carry the pathogen. The pathogen's spores germinate and infect injured tissues. Most such injuries were caused by mechanical shakers or harvesters, and some are caused by cultivation equipment. Pruning wounds or broken twigs occasionally are infected, and sunburn injuries may also be sites for infection. Fresh bark wounds attract the insects, where the vectors deposit spores and feed on resultant fungal colonies. Wounds are resistant to infection after 14 days of healing.

Control

Properly adjusted and operated mechanical shakers that do not injure tree bark provide the best protection against Ceratocystis canker. Pruning when rain is least likely may reduce infections. Severely cankered limbs should be removed by cutting at least 6 inches below the canker margin. Surgical removal of infected bark and outer xylem is effective with small to medium cankers and should be performed during winter. Canker removal is often not successful in the first attempt, and surgery must be repeated in following years. Such an eradication procedure, although expensive, is generally effective when properly done. If sunburn is the suspected injury where infection occurred, white paint and careful canopy pruning to shade trunk tissue might be helpful.

Cytospora Canker

Cytospora, or Leucostoma, canker is relatively serious and common on peaches in other parts of the United States, but in California, the disease primarily affects French and Imperial prune trees. The disease has been observed in all prune-growing districts of the state.

Symptoms

Cytospora canker does not kill entire trees but may destroy several branches. In late summer, individual branches die, retaining dead brown leaves. A dark, depressed canker can be found on such branches. The canker is ovoid with U-shaped ends. During the growing season, amber-colored gum may exude around the canker. The causal fungus, *Leucostoma cincta* (anamorph *Cytospora leucostoma*) produces small, olive-gray pycnidia just under the outer surface of the bark, giving the bark a pimpled appearance. When the outermost layer of bark is removed, the pycnidia, each surrounded by a white halo, can be seen. These white dots with dark centers are diagnostic of Cytospora canker.

Inoculum and infection

Conidia are produced in pycnidia in wet environments, and during humid conditions amber cirrhi (thread- or stringlike masses of conidia) are exuded from the pycnidia. Cirrhi contain conidia embedded in a water-soluble matrix and can be dispersed by wind, insects, and animals (e.g., birds). During rainy weather, the conidia are extruded in masses and dispersed by rain. Windblown rain is the most effective means of spread.

Cytospora leucostoma is a weak pathogen of healthy tissue. The fungus invades the tree through damaged tissues. The most common sites for infection are wounds caused by sunburn, insects, or old bacterial cankers. Other injuries sometimes are invaded. Even though the fungus invades a sunburn or bacterial canker wound, a Cytospora canker may not develop: prune trees must be weak or under stress to support canker growth. Cankers are perennial once established, but canker extension is a function of tree vigor. Cankers may extend rapidly in trees under stress and little or not at all in healthy, vigorous trees.

Conditions that predispose prune trees to Cytospora canker include water stress, freeze damage, potassium deficiency, and high soil populations of the pin nematode (i.e., *Criconomella xenoplex*). Trees grown on heavy clay soils also are more likely to have Cytospora canker.

Control

Fungicide treatments have not proven effective for control of Cytospora canker, and none are registered. Orchard habits that reduce sunburn and bacterial canker damage also help protect against Cytospora. Any potassium deficiency should be corrected. Vigorous, full-canopied trees are at least risk to infections by *C. leucostoma*.

Brown Rot Blossom Blight and Rots of Green and Mature Prune Fruit

Brown rot is the most important fungal disease of commercially grown *Prunus* species throughout the world. Depending on the environmental conditions during the growing season, the disease can cause severe crop and economic losses. Mature fruit rot usually causes most of the crop losses, although in years with wet spring conditions, blossom and twig blights can be severe and may reduce yields. Conidia produced on blighted blossoms, twigs, and decayed green prunes are important sources of secondary inoculum for infection of mature prunes.

Causal agents

In North America, brown rot may be caused by either *Monilinia fructicola* or *M. laxa*. Brown rot symptoms of both European plum and French plum include blossom and spur blights, as well as green and mature fruit rots. Of the two fungal species, *M. laxa* usually causes more twig blight than *M. fructicola,* whereas both species can cause preharvest fruit rot. In the past, *M. laxa* was the most common brown rot fungus found on French prunes in California, but recently the populations of *Monilinia* species have changed. Currently, *M. fructicola* is the more prevalent species. Occasionally, both fungi can be present in the same infected flower or fruit.

Symptoms

The symptoms caused by the two species of *Monilinia* are essentially the same. In the spring, shortly after bloom, infected flowers wither or become blighted (fig. 18.1). All blossom parts, including sepals, petals, anthers, and stigma, are susceptible to the brown rot fungi. Usually within a week, as the disease progresses into the spur, terminal leaves wilt, and a canker may form in the subtending shoot. If girdled, the whole shoot dies. Gum may exude from infected shoots, and gray-brown tufts of fungal spore-producing structures (sporodochia) often are found on infected blossoms and blighted twigs. In years with frequent rainfalls in the spring, prune leaves can also be directly infected. Leaf lesions are brown and water soaked, and upon incubation, sporulation of the pathogen may develop on the surface (fig. 18.2). Infected blossom parts (shucks) may cling to enlarging prunes and function as sources of inoculum for the developing green prunes.

Green fruit rot infections begin as water-soaked areas and develop as brown to black necrotic lesions. Immature fruit may also develop latent, or quiescent, infections. These are seen as small (less than 1 mm in diameter) black specks on the fruit surface. *Monilinia* species can be recovered with high incidence from these small specks or from symptomatic prunes incubated under high humidity. A few of these infections may resume growth in the summer and rot the mature fruit. Brown rot infections of mature prunes begin as soft, water-soaked areas. Soon they become dark brown circular lesions that expand over the entire fruit. The affected tissues remain firm and quickly produce sporodochia (spore tufts) with an abundance of gray to buff spores (fig. 18.3). Infected mature prunes shrivel and dry on the tree. Often, several brown-rotted prunes stick together in clumps. Brown rot is also associated with injuries (hail, insect and bird wounds, russet scab, and wind scab) that predispose prunes to the disease.

Figure 18.1 Blossom blight caused by *Monilinia fructicola* (A); shoot and blossom blight caused by *Monilinia fructicola* and/or *M. laxa* (B). *Photos:* T. J. Michailides (A); P. L. Sholberg (B).

Figure 18.2 Leaf infection by the brown rot fungus. *Photo:* T. J. Michailides.

Inoculum sources and disease development

The brown rot fungi overwinter in dead infected flowers, shoot cankers, or mummies on the tree and ground. Mummies consist of both fungal and host tissues that originate from infected green or mature prunes. Most conidia on mummies that are left over from the previous season are not viable. When mummies become wet, however, new viable conidia are produced from hyphae in the mummy that can function as primary inoculum. In addition to conidia, mummies infected with *M. fructicola* that fall to the ground can produce sexual fruiting bodies, known as apothecia, in the spring. A mummy can produce from 1 to 35 apothecia. For any mummy, the more apothecia that are produced, the smaller they are in size. With minor changes in humidity, ascospores (sexual spores) are forcibly discharged from the asci (sacs containing eight ascospores) of the apothecium. The optimal temperature for ascospore discharge and ascospore germination is 59°F. A single apothecium can produce ascospores for about 4 days, but this period is extended when temperatures are below 50°F and is shortened when temperatures are above 59°F. Maturation of the apothecia coincides with bloom and spring lighting conditions, as well as favorable temperatures and wetness conditions. The highest density of spore inoculum generally occurs at full bloom. Invisible clouds of ascospores are carried on wind currents to susceptible tissues. This may be important for initiating the disease in the spring where *M. fructicola* is the pathogen. Apothecia of *M. laxa* have not been observed in California orchards. Therefore, conidia produced on blighted shoots, flower parts, or mummies are considered the only primary inoculum for this species.

Figure 18.3 Brown rot of green fruit (A); brown rot of immature fruit (B); brown rot of green fruit (C). *Photos:* T. J. Michailides.

Infected green prunes may decay early in the season (green fruit rot) or near harvest time (mature fruit rot). Both types of rots can originate from active infections of noninjured and injured prunes or from latent and quiescent infections. Latent infections are not visible, while quiescent infections are visible and occur when environmental conditions are conducive (unusually wet spring). In addition, green prunes on the ground (mainly after mechanical fruit thinning) can also develop brown rot. Sporulation of the brown rot fungus on thinned prunes depends on the water content of thinned prunes and environmental temperature. Spores produced on thinned prunes or prunes rotting on the tree can serve as inoculum for secondary infections of later-maturing prunes. There is evidence that the amount of latent infection may affect the incidence of mature fruit rot. Therefore, measurement of the incidence of latent infections may allow estimation of fruit rot of mature prunes in the field.

Brown rot infection is favored by rainfall, dews, and high humidity. Conidia may germinate in 3 to 4 hours, and a wetness period of several hours is sufficient for penetration of the germination tubes to occur, provided temperatures are also conducive. As the length of the wetness period increases, so does the severity of brown rot. Rot often develops first in clustered fruit rather than in individually hanging fruit because infections occur more readily where fruit touch each other and wetness accumulates. Microscopically, contact areas have a thinner cuticle that predisposes fruit to infection by the brown rot fungi. Infection by both species of *Monilinia* can occur over a wide temperature range (41° to 86°F), with an optimum of 75° to 77°F. Sporulation of *M. laxa* is favored by lower temperatures (50° to 59°F).

Control

Brown rot blossom and twig blights are controlled by application of protective fungicides during bloom. As all flower parts are susceptible, fungicide treatments should be applied beginning at green tip and continuing through white tip and full bloom. Depending on the year and orchard history, one to three applications may be necessary. Delayed bloom applications using systemic fungicides have also been highly effective in reducing brown rot blossom and shoot blight. The most effective materials registered for control of brown rot are thiophanate methyl, iprodione, anilinopyrimidine fungicides (cyprodinil and pyrimethanil) and the DMI (demethylation inhibitors) fungicides such as tebuconazole, fenbuconazole, and propiconazole. Iprodione combined with summer oil showed excellent control of blossom blight and fruit rot. Resistance to benzimidazole or MBC fungicides (e.g., thiophanate methyl and benomyl) is widespread in *M. fructicola* but less common in *M. laxa*. In the United States, the registration of benomyl has been canceled, and its usage is no longer allowed for disease management. Resistance to iprodione or DMI fungicides has not been reported for either fungus. Resistance in populations of *M. fructicola* to the anilinopyrimidines fungicides has been documented in a few orchards in northern California. With several classes of fungicides with unique modes of action registered for prunes in California, no more than a single application of any one class of fungicide per season should be used as a precaution against the development of resistance.

Previously, preharvest treatments were shown to be ineffective. Preharvest applications of several fungicides including thiophanate methyl, iprodione, fenhexamid, and the DMI fungicides, as well as premixtures such as pyraclostrobin and boscalid in mixtures with oil, however, provided pre- and postharvest control of fruit brown rot. Fungicide residues were obtained on prunes treated with oil that reduced the surface tension of the waxy fruit surface. Preharvest fungicide applications made when prunes are near full size but before they touch (before clustering) are also effective in reducing fruit brown rot in the field and after harvest. Apparently, better coverage of the entire fruit surface is important in achieving disease control. Removal and destruction of infected mummies from nectarine and peach orchards significantly reduced brown rot blossom blight in the following growing season. In prune orchards, significant reductions were also shown with prebloom mummy removal, but it was determined not to be economical and not as effective as protective chemical treatments.

Crown Gall

Crown gall is sometimes called the plant cancer. Although this disease is important on other crops, it is not common on plum. On prune, galls can occasionally form on the roots, crown, stem, and even leaves, but they are usually found at the juncture of the main roots and the trunk or scattered along other roots. Galls range in size from microscopic to several inches in diameter. Large gnarled galls are sometimes visible at ground level, but most are growths the size of a golf ball to a softball on the root system; they cannot be detected on the root system until soil is removed.

Symptoms

Galls are composed of disorganized tissues and are soft and punky. Because the development of galls depends on plant growth, they appear and enlarge only during the growing season. Galls initiated in fall and winter appear the following spring; the incubation period in spring is longer than in the summer.

Young gall surfaces are generally smooth and have the color of healthy bark. The internal color is similar to that of normal wood, but annual growth rings are absent. As galls age, they become roughened and may develop internal cavities lined with dead bark and decayed tissue. Sometimes secondary wood-rotting fungi colonize the decayed gall tissue and then rot the internal healthy wood of the tree. These trees may be sufficiently weakened that they fall over in a windstorm.

The severity of damage caused by crown gall depends on several factors, including the number and location of infections, size of galls, age of the host, and establishment of secondary infections. Infections developing within the first 3 years after planting are much more debilitating than those beginning later in the life of the tree. Especially on young trees, clusters of galls may involve the entire crown region belowground. Such severely infected trees are weakened and may cease to grow.

Inoculum, infection, and control

Crown gall is caused by the bacterium *Agrobacterium tumefaciens*, which is soilborne and present in living galls. Because bacteria invade the plant through wounds, the disease is best prevented by avoiding injury to trees at planting and during cultural practices involving machinery. A biocontrol agent for crown gall, *Agrobacterium radiobacter* K-84, gives excellent control when used as a preplant treatment. It is strictly a preventive treatment and does not eliminate established infections, no matter how small. To be effective, nursery trees should be treated when dug, and an additional treatment may be necessary before planting.

Another material, Gallex, can be used to remove existing galls on orchard trees, but this procedure can be difficult, expensive, and is not always successful.

Peach rootstocks are more susceptible than plum, whereas Marianna 2624 is less susceptible than is Myrobalan 29C. Soil fumigation with methyl bromide–chloropicrin for orchard replant only reduces bacterial populations and does not control the disease. Irrigation and fertilization practices have no effect on the disease, but acid soils may be slightly inhibitory. Burning the galls with a blowtorch may cure the trees. The best protection for established trees is to not injure them during any orchard operations.

Phytophthora Crown and Root Rot

Prune trees generally escape serious damage from Phytophthora crown and root rot because many are propagated on the more-tolerant plum rootstocks. However, plum rootstocks are not entirely resistant, and peach rootstocks are subject to infection by these important soilborne organisms that grow in tubes or hyphae and reproduce by spores similar to true fungi.

Symptoms

Infections of the crown during winter and early spring usually girdle the tree, and with the first warm weather, the tree suddenly collapses and dies. Leaves remain attached and have a dry reddish brown color. The tree usually gums along the trunk. The inner bark and wood are deep reddish brown. When *Phytophthora* species attack feeder or small roots, the tree is often able to recover partially from this loss each year but shows the effects of root loss by reduced growth, more open canopy, early senescence, and defoliation. Over several years, the tree slowly declines and finally dies.

Inoculum and infection

Several species of *Phytophthora* attack the roots and trunks of many stone fruit rootstocks. The time of year and environmental conditions, including temperature, soil moisture, available nutrients, and host-pathogen combinations, affect the amount and severity of the disease. In general, when soil is saturated with water for several hours, *Phytophthora* species produce tiny sacklike structures called sporangia. The contents of the sporangia divide into 10 to 30 one-celled units known as zoospores. The zoospores are released from the sporangia and swim short distances through the water present in the soil. Nearby small plant roots attract the zoospores, which congregate on the root surface. There, the zoospores encyst (form a wall and loose their flagella) and soon invade the root to initiate an infection. Once inside the root, the pathogen grows through the tissues and causes root or crown rot.

Phytophthora species have become more widespread as a consequence of irrigated agriculture. The water in many rivers and canals may be now contaminated with these organisms. When this water is used for irrigation, *Phytophthora* species can be introduced into orchards and fields. Water drawn from wells has thus far proven to be free of these pathogens.

Control

Phytophthora crown and root rot is most effectively controlled by water management. Conditions that allow water to stand around tree trunks or that keep soil saturated for 24 hours or more are favorable conditions for establishing root and crown rot. Irrigation regimes should be designed with this in mind. Orchards on sandier, well-drained soils are the easiest to manage from the standpoint of Phytophthora control, but the disease can occur even there. Planting trees on berms and using microsprinklers reduce saturated soils to the shortest time possible while providing the required water amount and keep crowns and trunks dry. Newly planted trees need very little water until they begin to grow and should not be overwatered. Chemical treatments with fungicides such as phosphites, metalaxyl, and mefenoxam are not effective unless coupled with good water management.

Propagation on plum rootstock, especially Myrobalan 29C plum, affords considerable protection against Phytophthora crown and root rot. Lovell peach is more susceptible to some *Phytophthora* species than is Nemaguard; still, Nemaguard is more resistant than Marianna 2624 to other species.

Armillaria Root Rot

Armillaria root rot, also known as oak root rot, mushroom root rot, and shoestring root rot, can be a serious problem in areas where oaks have been removed and the fields planted to prune. The disease can lead to the death of trees. All stone fruit rootstocks are susceptible to Armillaria root rot. The plum rootstock Marianna 2624 has some tolerance and may be useful in some situations. Armillaria root rot is caused by the Basidiomycete *Armillaria mellea*, and is also known as the oak root rot fungus.

Symptoms

Infected roots have white to yellowish fan-shaped mycelial mats between the bark and the wood that become obvious after removing the bark. Dark brown to black rhizomorphs can be seen on the root surface. Foliar symptoms of Armillaria root rot include poor shoot growth, small leaves, premature yellowing, defoliation, and branch dieback. Excavation of the tree crown and major lateral roots is necessary to confirm presence of the causal fungus. Removing exterior bark at or below the soil line usually reveals mycelial fans and rhizomorphs of the pathogen at any time of the year. Sometimes, in autumn, its fruiting structures (honey-colored mushrooms) form at the base of infected trees. The wood decay caused by *Armillaria mellea* is of the white-rot type (soft wood decay) because the fungus removes all cell wall components including cellulose, hemicellulose, and lignin. In advanced stages of decay, the wood is light colored, spongy, and produces a mushroom odor.

Inoculum and infection

Woody debris in soil can harbor viable *A. mellea* inoculum for up to 100 years. Primary spread into an orchard is through root contact with infected roots of the previous crop or native trees. Secondary spread within an orchard between infected and healthy trees is via root-to-root grafting or root-to-rhizomorph contact. This leads to formation of infection centers, or foci. Tree death often starts at these foci and radiates outward, causing localized areas of infection. Rhizomorphs may grow several meters through the soil away from their food source. Basidiospores produced by the mushrooms are not considered to be major sources of inoculum, as their role in the pathogen's disease cycle is limited.

The disease occurs in many soil types but is more common in light, well-drained, sandy soils. Stressed trees, especially by drought, are thought to be more susceptible.

Control

There is no truly effective control for Armillaria root rot once established in an orchard. Dead and dying trees should be removed. Before replanting, all woody roots 1 inch (2.5 cm) or more in diameter should be excavated and the ground fallowed for at least 1 year. Soil fumigation with methyl bromide for orchard replant only when soil is at the correct soil moisture content is somewhat effective, but success depends greatly on soil type and the fumigation depth achieved. Because the fungus can survive in roots away from the fumigation zone or within large root pieces, fumigation rarely completely eradicates the pathogen. Avoid planting prune orchards where forest or oak woodland has recently been cleared or where there is a history of Armillaria root rot. All rootstocks can be attacked by *Armillaria mellea,* but some, such as Marianna 2624, are less affected than others. Maintain tree vigor to help resist infection by *Armillaria* attack. Infested sites can be fumigated, but this procedure often will not prevent recurrence of the disease.

Cultural practices have received relatively little attention in management of Armillaria root rot in prune orchards. Water management may be an important tool; however, there is little data on the reduction of Armillaria root rot by increasing plant vigor. Plastic barriers or trenches dug around an infested area have proven effective in retarding disease spread in Australian stone fruit orchards, but these systems have not been tested on prune trees in California. Almond growers in California have used this technique with some apparent success. Exposing an infected crown and upper root area of a tree infected with *Armillaria mellea* may stop the development of the fungus into the crown area and allow the tree to regrow, but this procedure has not been tested on prune trees.

Before fumigation, remove all infected trees, stumps, and as many roots more than 1 inch in diameter as possible. Symptomless trees adjacent to those showing symptoms are often infected. Thus, close examination, removal of these adjacent trees, and soil fumigation of these areas may be needed. Infected trees, stumps, and roots should be burned at the site or disposed of in areas where floodwaters cannot wash them to agricultural lands. Complete eradication seems to be impossible, and re-treatment may be necessary in localized areas. Eradication is most successful in shallow soils less than 5 feet in depth. Fumigate from late summer to early fall using materials legal for the location.

Plum Pox

Plum pox, also known as sharka, is the most devastating viral disease worldwide of stone fruit, including peaches, apricots, plums, nectarines, almonds, and sweet and sour cherries. Fortunately, this disease has not been detected in California-grown *Prunus* species. The disease significantly limits stone fruit production in areas where it is established. It was first described on plums in Bulgaria in 1915 and has subsequently spread to a large part of Europe, the Mediterranean, the Middle East (Egypt and Syria), India, and Chile. In 1999, plum pox was detected for the first time in North America in a Pennsylvania peach orchard. The disease makes fruit unmarketable and can greatly decrease yields of infected trees. The severity of damage depends on the strain of the virus present and the susceptibility of *Prunus* cultivars.

Symptoms

The virus causes chlorotic rings (plum) or chlorotic blotches (peach) on leaves and similar rings and blotches on the plum fruit. Several cultivars show yellowing line patterns and blotches or necrotic brown lesions. Leaf and fruit distortion has also been recorded. Infected plum fruit in advanced stages are severely deformed and become bumpy. The pits of many infected plums also show dark brown rings. Education of growers, diagnosticians, and nurserymen is considered crucial in early detection of plum pox.

Source of inoculum and infection

Plum pox is caused by the *plum pox potyvirus* (PPV). The pathogen infects various species of the genus *Prunus,* including wild and ornamental species. The disease is spread within and among closely located orchards by up to 14 aphid species including *Myzus persicae, Aphis spiraecola, A. gossypii,* and *A. fabae*. Some of the identified PPV strains, however, are not transmissible by aphids. Aphids spread the virus by carrying it in a nonpersistent manner. The virus can remain on the stylet of the aphids for minutes to perhaps a few hours. Some PPV strains are not transmissible by aphids. Long-distance spread usually occurs by moving infected nursery stock and/or propagation materials.

Control

Although this viral disease has not been recorded in California, stone fruit growers must be on alert, just in case they discover any plum pox or symptoms resembling plum pox. General management guidelines include field surveys for PPV, use of certified clean nursery stock and resistant plants (if available), aphid control, and removal and destruction of infected trees in nurseries and orchards. A genetically engineered PPV-resistant plum (known as C5) has been developed, and its resistance can be transferred through hybridization to other plum trees. Thus, there is a unique source of germplasm available for future plum breeding programs worldwide.

Powdery Mildew

Powdery mildew is very common on prune leaves but less common on the fruit. The fungal biotrophic pathogen develops conspicuous mycelium usually on the underside of the leaves. Numerous chasmothecia (sexual fruiting structures) develop in late summer and fall. Although the economic importance of powdery mildew of prune is not known, the fungus is very common in prune orchards in northern California. At times, prune rust and powdery mildew can coexist on the same leaf.

Causal agent

Based on the descriptions of species of powdery mildew fungi recorded on *Prunus domestica*, the prune powdery mildew fungus in California fits that of *Podosphaera tridactyla* [anamorph: *Oidium passerinii*]. The pathogen has been reported on leaves of *Prunus domestica* in Georgia and other locations. Another species, *Podosphaera clandestina* [anamorph: *Oidium crataegi*], has also been reported on *Prunus domestica* in California, Florida, and Georgia, but it is less common. The sexual reproductive structure of these fungi is known as the chasmothecia. Mature chasmothecia of *Podosphaera tridactyla* are globose to spherical, with an average diameter of 83 μm (range of 72 to 92 μm) (fig. 18.4). Two to four dichotomously branched appendages of unequal length (282 to 330.6 μm) are attached at the upper pole of chasmothecia (fig. 18.5). There is only one spore sac, or ascus, per chasmothecia with eight oval ascospores (sexual

Figure 18.4 Chasmothecia of powdery mildew on the lower surface of prune leaf. *Photo:* T. J. Michailides.

spores) per ascus (see fig. 18.5). Conidiophores are single and 145 μm in length, bearing 4 to 10 conidia in chains; conidia are hyaline, ellipsoidal to barrel shaped, and measure 12.3 by 21.5 μm.

Symptoms

Symptoms of the disease include chlorotic to necrotic irregular leaf lesions. Signs include whitish mycelium that develops mainly on the lower surface of the leaves and on green prune fruit. Chasmothecia are scattered or aggregated and range from reddish brown when immature to black when mature (see fig. 18.4).

Overwintering of prune powdery mildew

Chasmothecia of prune powdery mildew may be splashed by water or washed by rain onto the sustaining shoot and survive in the tree bark during winter. Additional inoculum may survive on leaves of water sprouts growing from the lower trunk until at least January under California winter conditions.

Control

No specific control measures have been developed for powdery mildew of prunes because this is a minor disease.

Prune Brownline

Prune brownline is widespread and serious in some parts of northern California. It has been reported from most of the prune-growing areas of the Sacramento Valley but thus far has not been observed in San Joaquin Valley prune orchards. A disease of Stanley prune in New York, for many years called Stanley constriction disease, also is prune brownline disease.

Symptoms

At first, leaves yellow at the margins and develop interveinal chlorosis. They then droop and curl upward. For several years, the crop is usually heavy, the fruit small and of poor quality, and diseased trees senesce and defoliate early in autumn. Terminal shoots cease to grow and sometimes die back. Overall, the general decline may take several years. Young or vigorous trees usually die more quickly than older or less-vigorous trees.

The diagnostic symptom of brownline is a thin strip of dark brown necrotic tissue in the cambial area at the graft union. Early in the disease, the brown line does not extend the full circumference of the tree. As the disease progresses, the necrosis expands into the bark, causing a girdling that ultimately kills the tree. An overgrowth of the scion often develops at the graft union.

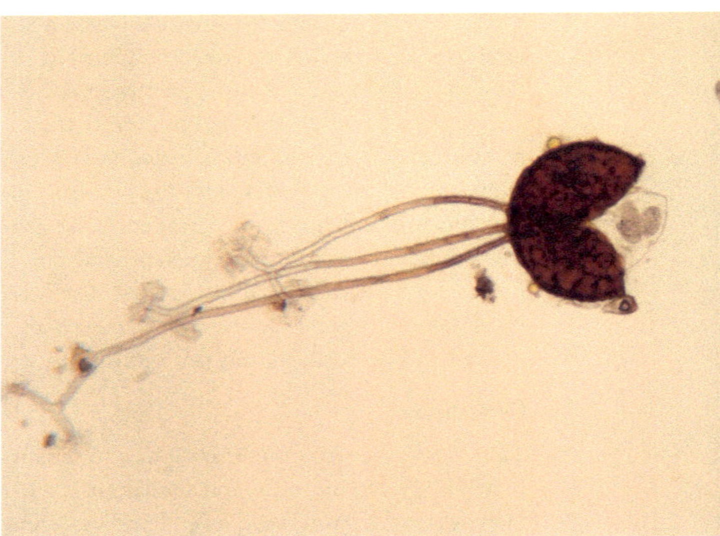

Figure 18.5 Chasmothecium of powdery mildew of prune showing appendages and the single ascus. *Photo:* T. J. Michailides.

Inoculum and infection

Brownline is caused by *tomato ringspot virus.* It is transmitted by the dagger nematodes *Xiphinema americanum*, *X. californicum*, and *X. rivesi.* The disease usually spreads from infected trees into adjacent healthy trees, reflecting the migration of virus-carrying nematodes. The virus survives from year to year in infected hosts (including many weeds) and its nematode vectors. Prune scion wood is free of infection, so the virus is not transmitted by budding or pollen.

The virus moves through peach or Myrobalan 29C plum rootstocks until it reaches the prune scion, which is resistant to the virus. The prune scion responds to the presence of the virus with a hypersensitive reaction that halts the advance of the virus and causes the brown line at the graft union.

Control

The best control of brownline is the use of Marianna 2624 rootstock, which is not susceptible to infection by the virus. In established orchards, infected trees along with adjacent healthy trees should be removed. If Myrobalan 29C plum or any peach rootstock is to be used where brownline is a threat, the suspect areas should be fumigated and kept weed-free for at least 2 years. During this and other processes, care should be taken not to move infested soil into clean areas of the orchard.

Stone Fruit Rust of Prune

Stone fruit rust is caused by the biotrophic fungus *Tranzschelia discolor,* which attacks all species of *Prunus*. It is common in many prune orchards but is usually most serious in the northern regions of California where late spring and summer rains are more likely to occur. There is some host specialization in populations of *T. discolor,* and forms found on prune do not readily attack other *Prunus* species and vice versa.

Symptoms

Rust may develop anytime from late spring to autumn. The first evidence of rust infection is a small yellow angular lesion on the leaf surface, usually visible on upper and lower surfaces (fig. 18.6). Later, when the fungus sporulates, a rusty reddish brown pustule develops on the lower leaf surface within the yellow angular infection (fig. 18.7). Examined with a hand lens, the reddish brown spores (urediniospores) of the fungus can be seen emerging through the cracked epidermis (outer tissue layer) of the leaf. Each pustule can produce thousands of spores. Leaves with multiple infections yellow and then drop. Thus, early defoliation is the major damage to trees from rust disease. Defoliation during harvest does not directly damage the crop but makes separation of fruit and leaves difficult. After harvest, accelerated leaf loss does no harm. The later in the season rust develops, the less harm it does. The prune fruit is not attacked, and rust lesions rarely occur on prune twigs in California.

Inoculum and infection

The prune rust fungus has a complicated life cycle that includes several spore types. The black teliospores develop at the end of the season in the same pustules that produced the reddish brown urediniospores. Teliospores give rise to yet another kind of spore (basidiospores), which can infect only the garden anemone. The infection of the garden anemone produces still another spore type, the aeciospore. Aeciospores in turn infect only *Prunus* species, in this case, prune trees.

In California, the rust fungus apparently does not require passage through the garden anemone to complete its life cycle, and this part of the disease cycle of this rust fungus is not considered important. Instead, the reddish brown urediniospores produced in the pustules are easily wind-disseminated during the growing season. In the fall season, some become lodged in bark crevices and under bud scales. Here, they overwinter, and in the following spring they infect new leaf tissue, and the cycles of infection and urediniospore production are repeated through the spring and summer.

Rust diseases are notorious for their explosive capacity to become epidemic. The urediniospores can germinate and infect within a matter of hours

Figure 18.6 Lesions of rust on the upper surface of prune leaf. *Photo:* T. J. Michailides.

Figure 18.7 Uredinia of prune rust on the lower leaf surface. *Photo:* T. J. Michailides.

when free moisture is present or the relative humidity is 100%. The stone fruit rust fungus develops rapidly over a fairly wide range of moderate temperatures but ceases activity when temperatures exceed 100°F. Years with high levels of rust are characterized by certain rainfall patterns: rainfall in late spring through summer sets the stage for outbreaks, and rust can be exacerbated by morning dews and high humidity.

Control

Application of fungicides for control should be made in late spring or early summer. One to three treatments may be required, depending on environmental conditions and disease history of the orchard. Most importantly, a monitoring method has been developed involving the evaluation of 40 designated trees in an orchard each week in the Sacramento

Valley, and every other week in the San Joaquin Valley, from May until mid-July. When rust is first detected, a treatment is recommended. Because not all orchards that harbor the rust fungus experience economic damage, decisions to treat usually can be based on past experience and the above monitoring schedule. Orchards with a history of early rust and varieties that are especially susceptible should be treated at least once every year, beginning in late May or June. Additional treatments should be added at monthly intervals. Warm rain favors rust, but dew also provides a favorable rust environment. The combination of cover crop, full canopy, and high dew point can cause devastating epidemics. Yield losses have been documented where early preharvest defoliation is severe, but when little or no defoliation occurs before harvest, yields are not compromised.

Sulfur has been an effective material registered for control of prune rust, and it has been used by growers for this purpose for many years. Several registered fungicides including the DMI (e.g., propiconazole, tebuconazole, etc.) and QoI (azoxystrobin, pyraclostrobin, etc.) fungicides protect the healthy leaves from infections and may provide postinfection activity as long as they are applied shortly after infection periods.

Wood Decays

Wood decay is one of the most common problems in older prune orchards that many growers accept as part of the aging process of trees. Wood decay leads to a weakening of trees, a slow decline in productivity, and eventual tree death. Wood decay is caused by numerous fungi in the Basidiomycota. These organisms are commonly called mushrooms, bracket fungi, and conk fungi. On prunes in California, several wood decay fungi have been described that are associated with dieback, broken branches, and overturned trees.

Causal organisms

Wood decay fungi in the Basidiomycota are characterized by the production of sexual spores called basidiospores that develop from basidia. Among the most common and widespread on prune in California are the white rot fungi *Armillaria mellea, Chondrostereum purpureum, Perenniporia medulla-panis Phellinus robustus* and *P. texanus,* and *Trametes versicolor* and the brown rot fungi *Fomitopsis cajanderi* and *Laetiporus sulfureus.* Additional species have been reported from other prune-growing areas.

Wood decay fungi are most reliably identified based on their fruiting bodies (basidiomes) where the basidia are arranged on a smooth surface, on lamellae (i.e., gills), or on the inner lining of tubes (i.e., pores). Basidiomes may be produced during later stages of disease development during favorable environments (e.g., wet conditions). Basidiomes may be annual, biennial, or perennial, depending on the fungal species. In the absence of basidiomes or any other characteristic symptoms, identification of the pathogen is possible only by fungal isolation or by molecular methods.

Symptoms

Symptoms caused by wood decay fungi mainly occur as wood rots on tree roots, trunks, and scaffold branches, as well as cankers and diebacks on aerial parts of the tree. Infection by most wood decay pathogens generally leads to a slow tree decline over several years. Severely affected trees may die. Disease progress with some of the more aggressive pathogens such as *Armillaria mellea* (see the section "Armillaria Root Rot") or *Chondrostereum purpureum* (Pers.) Pouz. (causal agent of silver leaf disease) can be much faster. Structurally weakened trees are subject to limb or trunk breakage and uprooting. Two major types of wood decay can be differentiated, depending on the wood components that are degraded by the pathogen: brown rot and white rot. Fungi causing brown rot degrade only the polysaccharide components of the wood, leaving the wood dry, crumbly, and of brown appearance from the remaining lignin. The majority of wood decay fungi on prune trees cause white rot in which, in addition to polysaccharides, lignin substances are also degraded. This leaves the wood soft or spongy and of a bleached, light-colored appearance.

Inoculum and infection

Wood decay fungi are mostly opportunistic wound pathogens, with some exceptions like *Armillaria mellea* (see above). These fungi can infect a range of other woody host plant genera, and inoculum sources can often be found near orchards. Fungus-host associations range from aggressive pathogens that infect healthy, living trees to fungi that infect weakened trees or that colonize only dead woody tissues. Most of these fungi colonize their hosts by airborne basidiospores that land and germinate on wood-exposing wounds. These spores are produced seasonally during favorable environments and function as primary inoculum. Wood-exposing injuries may be caused by environmental factors such as frost, wind, or sunburn or by biological factors such as damage by animals, including humans. Injuries from pruning wounds or scaffold supports are major entry sites. Some wood decay fungi also infect new hosts by natural tree-to-tree root grafts. Decay in a single tree can be caused by more than one fungal species.

Control

Management of wood decays is mainly based on cultural practices that minimize wood-exposing injuries to trees and reduce favorable environments in orchards for establishing wood decay. For example, irrigation systems should be designed to prevent water from wetting tree trunks and scaffold branches. Pruning should not be done during rainy weather, when spore inoculum is more likely to be present and infections have a better chance to get established. Ideally, several weeks of dry weather should follow pruning activities. Pruning cuts should be made in a way that favors wound healing. Stub cuts, bench cuts, and large wounds should be avoided. Tree paints may not be very effective on pruning wounds because they may crack and then provide a favorable microenvironment for establishment of wood decay. Orchard sanitation, including the removal of dead trees and tree prunings, helps to reduce the amount of fungal inoculum. Systemic fungicides such as myclobutanil or propiconazole, and biological control with species of *Trichoderma (T. harzianum* and *T. viride)* have been successfully used in the management of some diseases caused by wood decay fungi. Host resistance against wood decay fungi is unknown.

Postharvest Diseases

Although the majority of the California prune crop is dried soon after harvest, in the last several years more and more prune growers harvest part of their crop for the fresh market. These prunes can be infected by a number of fungi if not refrigerated or stored properly.

Postharvest Diseases of Fresh Prunes

Prunes are mainly harvested mechanically when they are destined to dry. Prunes for the fresh market, however, are harvested by hand. The method of harvest can affect the incidence of diseases that may develop. For instance, any injury may serve as avenues for infection by spores of *Monilinia, Botrytis, Rhizopus, Cladosporium, Aspergillus, Mucor*, and *Alternaria* species, as well as nonfilamentous yeasts. Mechanically damaged prunes are usually infected first, and decay may expand to surrounding healthy prunes. Losses can be significant when prunes are held in bins for 1 to 2 days or longer without refrigeration. Relatively high water content of the prune is required for these organisms to infect.

Symptoms

Decay by the brown rot fungi (i.e., *Monilinia* spp.) and gray mold (i.e., *Botrytis cinerea*) is firm, whereas decay by the rest of the fungi, including yeasts, is soft and wet, meaning that there is leakage of fruit juice. Species of *Aspergillus, Rhizopus,* and *Mucor* produce active pectinolytic enzymes that degrade cell walls and essentially liquefy the fruit. Decaying prunes quickly lose their skin when being dropped into wash water. Sporulation of the decay causing fungi and yeasts is often apparent.

Inoculum and infection

Many of the fungi that cause postharvest diseases of fresh prunes are soilborne, and some become airborne on their own or with blowing dust from the soil. Dust deposited on prunes can serve as the source for the propagules of these postharvest pathogens. When water drops are present on the fruit surface, spores of some fungi (e.g., *Monilinia* spp.) germinate and infect directly through the epidermis, or fruit skin. Fruit injuries that break the fruit skin facilitate the infection, even without the presence of free water because of the high water content of the fruit itself. Although infections can occur on any part of the fruit, including the stem end scar (if stem is removed) and stylar end, research showed that most of the infections at least by *M. fructicola* are located in the fruit-to-fruit contact surfaces (Michailides and Morgan 1997). These fruit-to-fruit surfaces act as traps in collecting propagules of filamentous and nonfilamentous fungi, have rubbing injuries, and create more favorable conditions for infection. Damage by insects is also common in fruit-to-fruit contact surfaces, which also facilitates fungal infection.

Control

Freshly harvested prunes should be refrigerated immediately after harvest or be dried within 1 day of harvest. Handling of prunes should be minimal. No difference in decay incidence has been observed between washed and nonwashed prunes when prunes are refrigerated at 4°C for 30 days or stored at 20° or 28°C for 5 days. However, the method of harvest has a major effect on decay incidence. For instance, more decay develops on prunes dropped on the ground and then collected than on those harvested on canvas laid on the ground. The least amount of decay develops on prunes harvested mechanically in elevated frames or by hand. Washing prunes with chlorinated water using sodium hypochlorite and refrigerating at 39°F for at least 20 days or storing at 68°F for 4 days reduced decay. No postharvest treatments are registered for prunes.

Postharvest Diseases of Dry Prunes

The fungal species that grow on dried prunes mainly belong to the genera *Aspergillus* (e.g., *A. glaucus, A. chevalieri*, and A. *japonicus*) and *Penicillium*. Yeasts that have been isolated from dried prunes include *Saccharomyces rouxii, S. mellis, Toluropsis magnoliae, T. stellata, Candida krusei, Trichosporon bebrendii, Pichia fermentans*, and *P. membranaefaciens*. Other fungi isolated include *Aspergillus glaucus*, as well as *Monilia* and *Chaetomella* species. Although the dehydration process kills the majority of mycelia and spores of all fungi, spores of *Aspergillus* and *Penicillium* species may recontaminate the prunes. In addition, spores of *Aspergillus chevalieri* are not killed by the dehydration process. This is the reason *A. chevalieri* is the most common mold of dried prunes. For spores of fungi to germinate and grow, prunes must have moisture content greater than 24%. The exact moisture content required for fungal growth depends on the concentration of reducing sugars, ambient temperature, and species of mold present.

Control

Prunes should be dried to a moisture content of less than 24% (preferably 20%) to ensure that molds will not grow on them in storage. Because this is an average value, it is possible that larger prunes in a lot may have a higher moisture content and could be subject to spoilage. Frequent checking of the prunes and turning them periodically will prevent localized moisture accumulation and produce a more uniform distribution of moisture and temperature in the bin. In addition, prevention of insect infestation is important because insect feeding increases moisture content, which allows mold development.

Slip-Skin Maceration Disorder

Prune growers use the term "box rot" to describe slip-skin maceration disorder in dried French prunes. Undocumented reports indicate that this disorder is responsible for serious losses.

Symptoms

Deterioration of dried prunes is characterized by macerated, wet, sticky areas on the fruit surface and by skin that tends to slip with the slightest pressure. Among fungi tested to be causing slip-skin maceration disorder were *Aspergillus japonicus, A. chevalieri, Penicillium expansum, Monilinia fructicola*, and *M. laxa*, all of which are common on prunes. None of these, however, was able to cause a slip-skin maceration disorder of dried prunes after 24 hours of incubation. The main cause of this disorder of fresh fruit is the fungus *Rhizopus stolonifer*, known as bread mold or whiskers disease on fresh market stone fruits and other fruits. On fresh prunes, *Rhizopus* species produce strong pectinolytic enzymes that degrade cell walls and make the fruit skin slip. Infections develop faster and are more severe at 77°F than at lower temperatures, and significant amounts of prunes can decay within 24 hours of incubation. Although symptoms of infection may not be obvious on fresh prunes, after dehydration, prunes with *Rhizopus* infections show areas with disintegrated skin or without skin and they are sticky. Several prunes can stick together and form a large mass of decayed prunes.

Control

To reduce slip-skin maceration disorder, prunes must be dried within 24 hours of harvest. If this is not possible, they should be stored below 41°F to prevent Rhizopus rot. If cold storage is not available, prunes harvested in the afternoon should be dried first and those harvested in the morning (these have a lower temperature) can be held for dehydration later.

Physiological Disorders

Excessive rain, sunlight, wind, or cold can damage a crop or the trees in an orchard and cause significant economic loss. To some degree, these conditions can be managed, but complete control of extreme environmental conditions is not possible.

Fruit Skin Damage

Wet, cool conditions during early fruit formation, wind abrasion, and hail can cause cosmetic damage to the fruit skin and may reduce the fruit grade.

Russet Scab

Prune russet scab, or lacy scab, is a physiological or abiotic condition and not a biotic disease that damages the fruit skin. It develops during extended wet conditions from full bloom through shuck split. Incomplete development of the outer fruit skin (cuticle) due to excessive moisture and shading by the flower shuck reduce the production of cuticle wax (bloom). Without complete development of this protective layer, affected prunes first develop shiny, lacy areas on the stylar end (opposite from the stem end), which later become brown and scabby as the season progresses (fig. 18.8). If the scabby area exceeds ⅜ inch in diameter, the fruit may be down-graded (see the section "Off-grade prediction protocol," below).

Certain fungicides, including captan and chlorothalonil, applied at full bloom can reduce the amount of russet scab in a wet year, although no fungus has been found to cause the disorder. These fungicides are thought to enhance cuticle development and wax production. No effective management practices currently exist after shuck fall.

Off-grade prediction protocol

Not all scab visible in the orchard causes off-grade fruit, and a protocol has been developed for evaluating prunes preharvest and estimating the percentage of off-grade fruit. This protocol incorporates a correction factor that accounts for the loss of visible russet damage as the fruit dries.

The following step-by-step process can predict the percentage of off-grade dried prunes from a fresh fruit sample.

1. Take a 200-fruit sample per block at any time from June until harvest.
2. Looking at the bottom half of a prune (with the stem held away from you), grade the fruit into five separate categories using the images in figure 18.9.
3. Multiply the correction factor for each class to get a final number of prunes that should be off-grade.
4. Calculate a total percentage of off-grade prunes from the estimated number of off-grade prunes and the total number of prunes in the sample (200).

See table 18.1 for an example of how to predict scab damage using fresh prunes.

Figure 18.8 Russeted prunes just prior to harvest. *Photos:* T. J. Michailides.

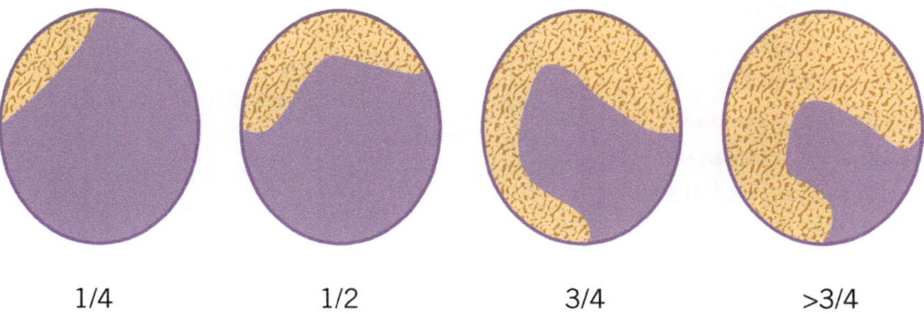

Figure 18.9 Sample illustrations for estimating the proportion of russet scab coverage of the stylar end of prune fruit for use in predicting off-grade prunes.

Table 18.1. Predicting scab damage using fresh prunes

Scab coverage rating	0	¼	½	¾	> ¾	Total
Correction factor	0	0.018	0.124	0.23	0.395	
Number of fresh fruit with scab in each category (see figure 18.9 to define each category; values given are for illustration purposes only)	120	50	20	5	5	200
Corrected number of dried fruit with damage	0	0.9	2.48	1.15	1.97	6.5
Estimated total percentage off-grade due to scab: (Corrected number ÷ total number) × 100						3.25%

Other off-grade categories such as fruit cracks, brown rot, and insect damage contribute to the total quality evaluation. Therefore, russet scab may not by itself downgrade many prunes, but it may significantly contribute to a substandard grading for a particular fruit lot.

Wind Scab

Wind abrasion (wind scab) causes similar damage to russet scab to the skin of prunes. Injuries from rubbing against other prunes, leaves, twigs, or branches can cause russeted areas on the fruit skin that can cause downgrading. Wind scab develops when high winds (more than 32 mph) occur within 3 weeks of full bloom. Prunes on the windward side of the tree (facing the prevailing wind) are more commonly damaged than prunes on the side of the tree away from the wind. Generally, in the prune-growing regions of California, the prevailing winds come often from the northwest, and thus wind scab is most often found on that side of the tree. Scabbed fruit surfaces are more likely to develop cracks into the fruit flesh and trap fungal spores, and are more likely to become infected by brown rot fungi or other fungi, including yeasts, than are healthy fruit. Windbreaks can protect exposed prune orchards from high winds.

Hail

Hail damage cannot be prevented, and the potential damage may be evaluated using the russet scab evaluation protocol described above. If environmental conditions are favorable, hail wounds may be infected by brown rot and other fungi. Manipulating bloom timing (staggering the bloom) and subsequent fruit development is one way to manage the hail damage; this can be done with dormant oil applied in late December through mid-January.

Prune Cracking

Severe fruit cracking can cause a substandard grade for an entire fruit lot and cause extensive economic losses to growers. Cracking often provides easy entry points for brown rot and other fungi, and it further contributes to grower losses (fig. 18.10). There are two different types of fruit cracking in prunes: end cracking (fig. 18.11) and side cracking (fig. 18.12). Each occurs under different conditions when the fruit is rapidly growing (see below). Research has provided a working understanding of how to manage fruit cracking, although its exact mechanism is not understood. End cracking damages the fruit on the stylar end (opposite from the stem), while side cracking damages the fruit in the middle part.

End cracking can occur immediately after irrigating dry trees anytime before late July (when prunes have finished growing in size). The sudden increase in tree water status at least doubles pressure inside the fruit, especially in the tip, which leads to cracking. Managing end cracking is simple: avoid orchard water stress in spring and summer particularly between May and June. The worst end-cracking years are often those with wet, cool springs. Under these conditions, high water uptake causes rapid fruit enlargement that may cause cracking. Regular use of orchard moisture monitoring equipment (such as a pressure bomb and watermark sensors) or programs such as evapotranspiration (ET) estimates or physical evaluation of moisture levels in soil obtained at different depths (using a shovel or soil auger) can help avoid orchard water stress and fruit cracking in the spring.

Side cracking usually occurs when water is present (i.e., dew or rain) on the surface of large exposed prunes when they begin their final growth stage (when the width of the fruit expands). There is a 3-week-long period when side-cracking potential is high, and this stage begins a week

Figure 18.10 Cracked prune infected with brown rot *(Monilinia fructicola). Photo:* T. J. Michailides.

Figure 18.11 Stylar-end cracked prunes. *Photo:* W. H. Krueger.

Figure 18.12 Side-cracked prunes. *Photo:* W. H. Krueger.

after the transverse diameter of the fruit is larger than the suture diameter (fruit length). Historically, the susceptible period for side cracking is usually around July 4, but it can differ by several weeks depending on the year. Extreme day and night temperatures during this period can lead to dew forming on the prunes, an increase in internal fruit flesh pressure as water enters the fruit, and skin rupture (cracking) on the sunny side of exposed prune (the region of the skin that is sun-toughened and less elastic). Summer rain can also lead to side cracking. Managing side cracking is difficult because regional climate cannot be controlled. Some growers do not irrigate during the period when side cracking is a risk, hoping to keep humidity low and dew from forming. Growers with microsprinkler or drip irrigation may have less humidity and side cracking than do growers with flood irrigation. Side cracking varies more from year to year than from orchard to orchard because it is generally related to regional weather patterns rather than management conditions in a particular orchard.

Fruit Sunburn (Blue Prune)

Damage to prune fruit skin (fig. 18.13) and subsequent fruit drop are often associated with sudden increases in the maximum daily temperature, especially in weaker prune orchards (diseased trees, shallow soils, moisture stress, etc.). Leaves in the vicinity of sunburned fruit can be scorched by toxins that move out of sunburned fruit, and the entire shoot from the sunburned fruit out to the tip can be affected (fig. 18.14).

Figure 18.13 Sunburned prune. *Photo:* J. T. Yeager.

Wood (Scaffold or Trunk) Sunburn

Trunks and branches exposed to direct sunlight can overheat, killing the cambium (vascular-generating tissue) and phellogen (bark-generating tissue) just beneath the bark. Trunk sunburn can cause cracks in the protective bark tissue that is the tree's primary barrier against fungal pathogens. Sunburn injuries to trees can decrease water transpiration and cause subsequent defoliation and wounds that in turn provide sites for infections by *Cytospora* spp. and wood decay fungi such as *Schizophyllum commune*. Ultimately, scaffold dieback can reduce orchard production for years. Exposed trunks, especially of young trees, should be protected from sunburn with a coat of inexpensive interior latex paint. Avoid scaffold sunburn by preventing premature defoliation due to potassium deficiency, rust, mites, or water stress and maintain orchard productivity. Clay-based materials applied on the remaining foliage may provide some relief from sunburn damage following defoliation.

Figure 18.14 Leaf scorch associated with sunburned prunes. *Photo:* W. H. Krueger.

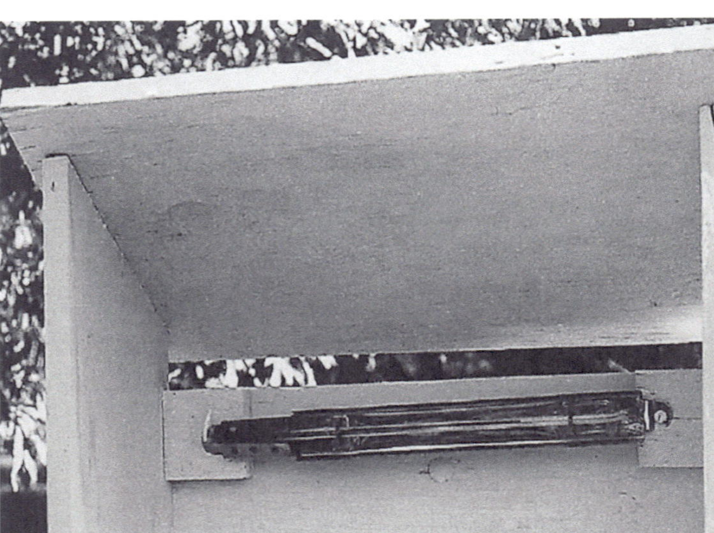

Figure 18.15 Frost thermometer with radiation shield. Photo: F. J. A. Niederholzer.

Frost

Below-freezing temperatures can harm prune flowers as well as developing prunes, and complete crop loss is possible in some orchards in some years. Successful crop protection depends on the stage of flower or fruit development, the length and severity of the cold event, the range of frost management options available in the orchard, and their effective use.

There are two general categories of frost events: mild radiation and advection frosts. Understanding the general characteristics of these events can help in directing management responses to field conditions. Mild radiation frosts occur locally during still, clear, cloudless nights, often with the development of a strong inversion (warmer temperatures in the air above the orchard). Natural radiation from cooling of the soil and plants after sundown following a day with relatively high temperatures causes a rapid decrease in temperature if wind does not mix the surface air with warmer air at higher altitudes. The warmer air traps the cold surface air near the ground. Thus, a frost event can occur under calm (windless), cold, dry conditions.

Advection frosts are characterized by cold air moving into a region from a colder outside area. Cold air is heavy and flows downhill, like water. It can accumulate in low areas or behind obstructions (tall windbreaks, etc.). Advection frosts can occur locally in orchards, for example, at the base of canyons as cold air funnels out of mountains and accumulates in the low areas. Advection frosts can also occur regionally as cold air masses move into a large area from outside the region.

Atmospheric moisture, measured as the dew point, is a key variable in managing frost events. The dew point is the temperature at which moisture condenses (dew forms) from the air. If the air is moist (high dew point) and the temperature drops to the dew point, liquid water condenses out of the air and forms as dew on plant surfaces or the ground. This process adds considerable heat energy to the orchard (when water vapor condenses at the dew point it releases 9,000 BTUs per acre into the orchard).

Frost protection strategies include reducing heat loss, redistributing heat, or directly adding heat to an orchard. The practices most commonly used include

Table 18.2. Critical temperatures for frost damage to prune buds, flowers, or fruit

First white	26°F
First bloom	27°F
Full bloom	28°F
Postbloom	30°F

Note: Temperature endured for 30 minutes or less.

orchard floor preparation and use of undertree sprinklers, wind machines, or helicopters. When bud development begins after dormancy in late winter, orchards should be prepared to reduce the risk of frost damage. Bare soil with a moist surface radiates the most heat energy and maintains the warmest orchard of all orchard floor conditions. If the block has a cover crop, it should be kept lower than 4 inches tall, and preferentially less than 2 inches, to allow sunlight to warm the soil and increase the soil heat bank for release during the night. Dry, bare soil or soil that has been recently cultivated is colder than soil with a 2-inch-tall cover crop. A light irrigation just before a cold event ensures that the orchard soil will hold as much heat energy as possible, and thus the highest possible nighttime minimum temperature. Weed strips in blocks that are not cross-mowed or disked should be treated in the weeks before frost season to avoid the risk to prune flowers and fruit from cooler temperatures on cold nights due to tall weeds. Active addition of heat to an orchard can be used if temperatures approach critical levels (table 18.2), which vary depending on bud and fruit development stage.

Undertree sprinklers can add heat to an orchard as the water cools and freezes or through increased heat radiation from the wetted soil. The higher the volume of water applied, the more heat is added to an orchard, especially if enough water is applied so that it freezes on the trees. Solid-set sprinklers deliver the most water over the widest area, and if angled to put water into the tree canopy, they can add significant heat to an orchard as water freezes on trees. Microsprinklers can provide some frost protection. Flood irrigation provides more heat to an orchard than drip irrigation due to the larger water volumes applied. High-volume microjets (40 gallons per acre per hour) provide more protection than low-volume microsprinklers (15 gallons per acre per hour), as indicated by field research (Snyder 2007).

Drip irrigation provides some frost protection when applied in advance of a predicted frost event. It can increase soil heat storage, especially if the soil was dry before irrigation. Running drip irrigation during a frost event may give some limited benefit. Soil type can greatly influence the potential benefits of drip irrigation in preventing frost damage. If more soil surface is wetted by the drip irrigation, more heat input into the orchard will occur (e.g., heavier soil texture and slower infiltration versus sandy soils with rapid water infiltration). Water should be turned off once the wet bulb temperature upwind of the orchard is above the critical temperature (see table 18.2) or when all ice in the orchard melts. The temperature measured by a wet bulb thermometer using a psychrometer corresponds to the temperature of wet buds as water evaporates from their surfaces. Temperature measurements using a dry bulb thermometer do not correlate to the temperature of wet buds and therefore are not useful for determining the cold damage potential to wet buds.

Redistributing heat by helicopters or wind machines to mix warm and cool air, either from natural inversions or from irrigation water, can provide frost protection. Air movement by helicopters or wind machines can also improve the frost protection provided by widely spaced sprinkler pipelines or flood irrigation.

Accurate thermometers are critical to effective frost management. Thermometers calibrated to read 32°F when immersed in an ice-water mixture should be placed in frost-free shelters (fig. 18.15) in several locations around the orchard. Special attention should be paid to placing thermometers in known cold spots in the orchard. Frost alarms, where used, should be set at 1° to 2° above the critical levels indicated in table 18.2.

References

Adaskaveg, J. E., and J. M. Ogawa. 1990. Wood decay pathology of fruit and nut trees in California. Plant Disease 74:341–352.

———. 1993. Epidemiology and control of brown rot blossom blight and fruit rot of prune caused by *Monilinia* species. In California Prune Board annual report. Pleasanton: California Prune Board. 158–173.

Adaskaveg, J., T. J. Michailides, R. Buchner, and W. Olson. 1994. Management of brown rot of French prunes by detecting quiescent infections and by early and late fungicide-oil applications. In California Prune Board annual report. Pleasanton: California Prune Board. 87–104.

Adaskaveg, J., D. Gubler, T. Michailides, and B. Holtz. 2011. Efficacy and timing of fungicides, bactericides, and biologicals for deciduous tree fruit, nut, strawberry, and vine crops. UC IPM website, http://www.ipm.ucdavis.edu/PDF/PMG/fungicideefficacytiming.pdf. Updated annually.

Batra, L. R. 1991. World species of *Monilinia* (Fungi): Their ecology, biosystematics and control. Mycologia Memoir 16. Berlin: J. Cramer.

Belanger, R. R., W. R. Bushnell, A. J. Dik, and T. L. W. Carver. 2002. The powdery mildews: A comprehensive treatise. St. Paul, MN: American Phytopathological Society.

Bertrand, P. F., and H. English. 1976. Virulence and seasonal activity of *Cytospora leucostoma* and *C. cincta* in French prune trees in California. Plant Disease Reporter 60:106–110.

Bertrand, P. F., H. English, and R. M. Carlson. 1976a. Relation of soil physical and fertility properties to the occurrence of Cytospora canker in French prune orchards. Phytopathology 66:1321–1324.

Bertrand, P. F., H. English, K. Uriu, and F. J. Schick. 1976b. Late season water deficits and development of Cytospora canker in French prune. Phytopathology 66:1318–1320.

Bolkan, H. A., J. M. Ogawa, T. J. Michailides, and P. F. Kable. 1985. Physiological specialization in *Tranzschelia discolor.* Plant Disease 69:485–486.

DeVay, J. E., F. L. Lukezic, H. English, E. E. Trujillo, and W. J. Moller. 1968. Ceratocystis canker of deciduous fruit trees. Phytopathology 58:949–954.

Ellison, P. J., L. McFadyen, and P. F. Kable. 1987. Overwintering of *Tranzschelia discolor* in prune orchards in New South Wales. Australian Journal of Agricultural Research 38:895–905.

English, H., J. DeVay, F. Schick, and B. Lownsberry. 1983. Reducing bacterial canker damage in French prunes. Sun-Diamond Grower 3:19.

English, H., B. F. Lownsberry, F. J. Schick, and T. Burlando. 1982. Effect of ring and pin nematodes on the development of bacterial canker and Cytospora canker in young French prune trees. Plant Disease 66:114–116.

Erwin, D. C., and O. K. Ribeiro. 1996. Phytophthora diseases worldwide. St. Paul, MN: APS Press.

Erwin, D. C., S. Bartnicki-Garcia, and P. H. Tsao, eds. 1983. Phytophthora: Its biology, taxonomy, ecology, and pathology. St. Paul, MN: APS Press.

Gallegly, M. E., and C. Hong. 2008. *Phytophthora*: Identifying species by morphology and DNA fingerprints. St. Paul MN: APS Press.

Holtz, B. A., T. J. Michailides, and C. X. Hong. 1998. Development of apothecia from stone fruit infected and stromatized by *Monilinia fructicola.* Plant Disease 82:1375–1380.

Hong, C. X., and T. J. Michailides. 1998. Effect of temperature on the discharge and germination of ascospores by apothecia of *Monilinia fructicola.* Plant Disease 82:195–202.

Hoy, J. W. 1980. Brownline spreading in prunes. Diamond/Sunsweet News 2:5.

Hoy, J. W., S. M. Mircetich, R. S. Bethell, J. E. DeTar, and D. M. Holmberg. 1983. The cause and control of prune brownline disease. California Agriculture 38:12–13.

Kable, P. F., P. J. Bambach, A. Ellison, A. Watson, and C. J. Kaldor. 1987. Fungicidal control of rust of French prune caused by *Tranzschelia discolor.* Australian Journal of Agricultural Research 38:565–576.

Kable, P. F., B. J. Keen, and R. W. Bambach. 1987. Evaluation of fungicides for curative activity against *Tranzschelia discolor*, cause of the rust disease of French prune (*Prunus domestica* L.). Australian Journal of Agricultural Research 38:577–585.

Luo, Y., T. J. Michailides, D. P. Morgan, W. H. Krueger, and R. P. Buchner. 2005. Inoculum dynamics, fruit infection, and development of brown rot in prune orchards in California. Phytopathology 95:1132–1136.

Michailides, T. J., and D. P. Morgan. 1993. "Wind scab" of French prune: Symptomatology and predisposition to preharvest and postharvest fungal decay. Plant Disease 77:90–95.

———. 1997. Influence of fruit-to-fruit contact on the susceptibility of French prune to infection by *Monilinia fructicola.* Plant Disease 81:1416–1424.

Michailides, T. J., and J. M. Ogawa. 1986. Chemical control of prune leaf rust *(Tranzschelia discolor* f. sp. *domesticae)* in California. Plant Disease 70:307–309.

Michailides, T. J., and B. L. Teviotdale. 1989. Observations on brown rot and studies on the powdery mildew of prune. California Prune Board annual report. 103–173.

Michailides, T. J., J. M. Ogawa, and D. C. Opgenorth. 1987a. Shift of *Monilinia* spp. and distribution of isolates sensitive and resistant to benomyl in California prune and apricot orchards. Plant Disease 71:893–896.

Michailides, T. J., J. M. Ogawa, and P. L. Sholberg. 1987b. Chemical control of decay of fresh prune in storage. Plant Disease 71:14–17.

Michailides, T. J., D. P. Morgan, G. S. Sibbett, and B. L. Teviotdale. 1994. Management of brown rot of French prune by detecting infections in contact surfaces and by early and late summer fungicide applications. In California Prune Board annual report. Pleasanton: California Prune Board. 63–86.

Michailides, T. J., Y. Luo, Z. Ma, and D. P. Morgan. 2007. Brown rot of dried plum in California: New insights in an old disease. APSnet March Feature Story, http://www.apsnet.org/online/feature/prune.

Mircetich, J. S., and G. T. Browne. 1989. Phytophthora root and crown rot of deciduous fruit trees: The influence of different rootstocks and soil moisture conditions on severity of the disease. Compact Fruit Tree 22:115–123.

Mircetich, S. M., and J. W. Hoy. 1981. Brownline of prune trees. Phytopathology 71:30–35.

Mojtahedi, H., B. F. Lownsberry, and E. H. Moody. 1975. Ring nematodes increase development of bacterial cankers in plums. Phytopathology 65:556–559.

Moller, W. J., and J. E. DeVay. 1968. Insect transmission of *Ceratocystis fimbriata* in deciduous fruit orchards. Phytopathology 58:1499–1508.

Moller, W. J., J. E. DeVay, and P. A. Backman. 1969. Effect of some ecological factors on Ceratocystis canker on stone fruits. Phytopathology 59:938–942.

Moore, L. W., and G. Warren. 1979. *Agrobacterium radiobacter* strain 84 and biological control of crown gall. Annual Review of Phytopathology 17:163–179.

Munnecke, D. E., M. J. Kolbezen, W. D. Wilbur, and H. D. Ohr. 1981. Interactions involved in controlling *Armillaria mellea*. Plant Disease 65:384–389.

Ogawa, J. M., and H. English. 1991. Diseases of temperate zone tree fruit and nut crops. Oakland: University of California Division of Agriculture and Natural Resources Publication 3345.

Otta, J. D., and H. English. 1970. Epidemiology of the bacterial canker disease of French prune. Plant Disease Reporter 54:332–336.

Roos, I. M. M., and M. J. Hattingh. 1987. Systemic invasion of plum leaves and shoots by *Pseudomonas syringae* pv. *syringae* introduced into petioles. Phytopathology 77:1253–1257.

Schroth, M. N., and D. C. Hildebrand. 1968. A chemotherapeutic treatment for selectively eradicating crown gall and olive knot neoplasms. Phytopathology 58:848–854.

Schroth, M. N., A. R. Weinhold, A. H. McCain, D. C. Hildebrand, and N. Ross. 1971. Biology and control of *Agrobacterium tumefaciens*. Hilgardia 40:537–552.

Shaw, C. G., and G. A. Kile, eds. 1991. Armillaria root disease. Agriculture Handbook No. 691. Washington, DC: USDA Forest Service.

Sholberg, P. L., and J. M. Ogawa. 1983. Relation of postharvest decay fungi to the slipskin maceration disorder of dried French prunes. Phytopathology 73:708–713.

Snyder, R. L. 2007. Frost protection. In E. J. Mitcham and R. B. Elkins, eds., Pear production manual. Oakland: University of California Division of Agriculture and Natural Resources Publication 3483. 117–123.

Teviotdale, B. L. 1984. Prune brownline. Sun-Diamond Grower 3:11.

Teviotdale, B. L., and D. M. Harper. 1991. Infection of pruning and small bark wounds in almond by *Ceratocystis fimbriata*. Plant Disease 75:1026–1030.

Teviotdale, B. L., D. M. Harper, T. J. Michailides, and G. S. Sibbett. 1994. Lack of effect of stone fruit rust on yield of French prune trees and survival of urediniospores of the pathogen on leaves, shoots and buds. Plant Disease 78:141–145.

———. 1995. Postharvest prune rust does not lower French prune yield. California Agriculture 49:23–26.

Zeller, S. M. 1926. Brown-pocket heartrot of stone fruit trees caused by *Trametes subrosea* Weir. Journal of Agricultural Research (USDA) 33:687–693.

———. 1926. Observations on infections of apple and prune roots by *Armillaria mellea* Vahl. Phytopathology 16:479–484.

19 Plant-Parasitic Nematodes

• Michael V. McKenry and Becky B. Westerdahl

Plant-parasitic nematodes are particularly problematic when growers purchase plants already contaminated with nematodes or when they grow a perennial crop in a field in successive seasons without rotation to a different crop. Since 1960, California has had an effective CDFA-regulated nursery certification program that halts transport of nematode-contaminated nursery stock across county lines. Today, nematode-free nursery stocks are the norm throughout the state. However, rotation of perennial crops is a grower decision that is made even more difficult because of economics and regulations associated with fumigation of soil before replanting. Selection of the best rootstock is another grower decision, and the presence or absence of certain nematode species in a field should play a role in making that decision. This chapter provides information on nematode biology, distribution, and management considerations when replanting prunes. Nematode damage is relatively subtle compared with that of other pests and diseases, but it also cannot be stopped once a field becomes infested. The most common symptoms are smaller prunes, fewer prunes, and reduced ability of the tree to take up nutrients and water from the soil.

Identifying Nematodes and Diagnosing Damage

Certain nematode species derive all their energy from the roots of plants. We call these plant-parasitic nematodes. They are microscopic: seen with a microscope that enlarges 60-fold, plant-parasitic nematodes usually have a visible spear, or stylet, at their head end (fig. 19.1). The presence or absence of a stylet provides a quick method to separate the plant-feeding types from those that feed on bacteria, fungi, algae, insects, or invertebrates, although some nematodes that feed on fungi do possess a stylet. In the surface few inches of orchard soil perhaps only 10% of the nematodes present are plant-parasitic. Deeper in the soil profile, the percentage increases to 50% or more. Plant-parasitic nematodes can migrate and reproduce anywhere the roots of a tree grow, but they tend to be most plentiful from 6 to 36 inches deep in the soil profile.

Increase the magnification of the microscope to 600-fold and it becomes apparent that there are different types of plant-parasitic nematodes (figs. 19.2, 19.3, and 19.4). The biology of these plant-parasitic nematodes can be as different as the biology of one insect type compared with another. In fact, the knowledge of which nematode is present will require diagnosis by someone trained in nematode identification. Several private laboratories in California can extract nematodes from soil or roots, identify them, and count them. The answer from the lab will indicate which plant parasites are present, how many there are, and usually whether the counts are low, medium, or high. Sampling

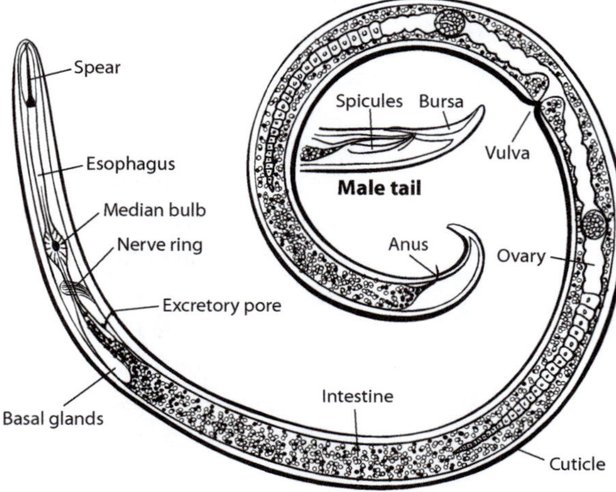

Figure 19.1 Body structure of a typical plant-parasitic nematode.

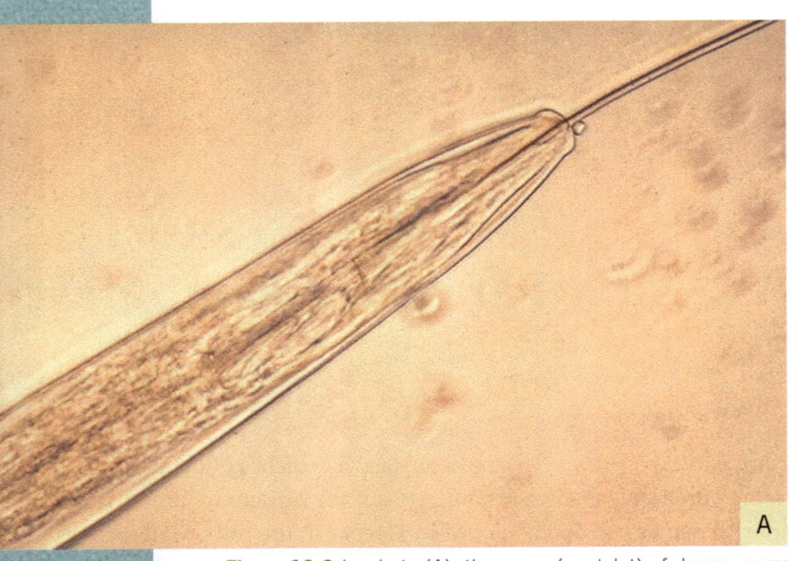

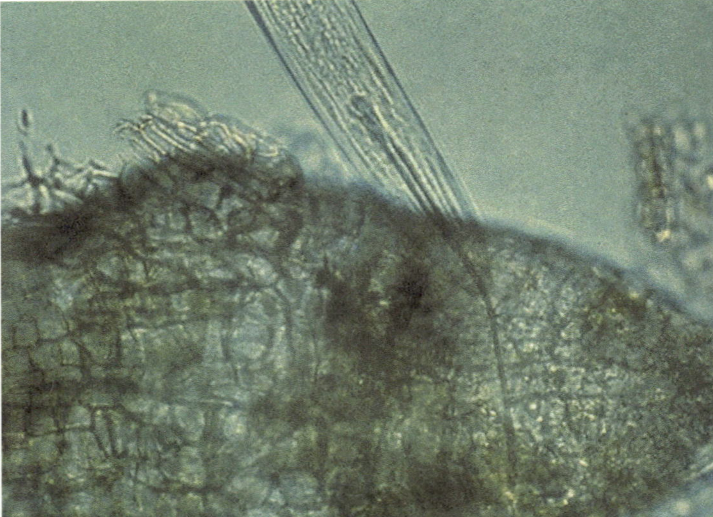

Figure 19.2 In photo (A), the spear (or stylet) of dagger nematode is extended for feeding on plant roots. In photo (B), the spear of a dagger nematode is inserted into the root for feeding. *Photos:* Department of Nematology, UC Davis.

results should also indicate the method the lab used for nematode extraction.

Prune growers in the Sacramento Valley must know whether their soil is abundant with

- ring nematode *(Mesocriconema xenoplax)*
- root lesion nematodes *(Pratylenchus vulnus, P. penetrans, P. thornei, P. neglectus,* or *P. scribneri)*
- dagger nematode *(Xiphinema americanum)*
- pin nematode *(Paratylenchus hamatus)*

Other genera and species of plant parasites may exist in the orchard, but these are the most common. Occasionally, the citrus nematode *(Tylenchulus semipenetrans)* may be found on Marianna 2624 rootstock in prune, but to this date it has not been found at a high population level. Thirty years ago, Ben Lownsberry at UC Davis reported that pin nematode could be found in 67% of prune orchards, dagger nematode in 62%, ring nematode in 38%, and root lesion nematode *(Pratylenchus vulnus)* in 7%. Prune orchards in the Sacramento Valley are usually planted on Marianna 2624 or Myrobalan 29C rootstock, and the prevailing soil is clay loam.

Prune growers in the San Joaquin and adjacent coastal valleys may find all the nematodes listed above, but several of the root lesion nematodes *(P. neglectus, P. penetrans,* and *P. thornei)* are less common, while *P. vulnus* may be found in 30% of prune orchards (McKenry 1999). Root knot nematode (*Meloidogyne* spp.) can be quite damaging on many crops grown in coarse-textured soil, but the four common prune rootstocks (except for Lovell seedling) are resistant to it. Also, some prune orchards planted in sandy soil commonly support population levels of ring nematode that are double those in the Sacramento Valley. In the San Joaquin Valley, the dominant rootstock is Nemaguard, and prevailing soils tend to be loamy sand to sandy loam. Marianna 2624 rootstock planted to loamy sand soils usually succumbs to bacterial canker complex within a few years if the site is a replant. Preplant fumigation increases the chance that trees will survive at least 3 years. Orchards may even become productive on this rootstock if a good postplant nematicide is available and applied each fall once the trees become established. Compared with Marianna 2624, Nemaguard is a slightly poorer host for ring nematode, but bacterial canker complex can kill many trees on it.

Weeds and cover crops growing on the orchard floor have relatively shallow roots that also support plant-parasitic nematodes. *Pratylenchus vulnus*, the most damaging root lesion species of concern, is not known to feed on most cover crops or weeds; it feeds instead primarily on roots of woody perennials (see fig. 19.4). By contrast, *P. thornei* and *P. neglectus* feed almost exclusively on roots of annual plants that grow on the orchard floor. *Pratylenchus penetrans,* and to a lesser extent *P. scribneri,* are known to feed on a diversity of root types, but the damage they cause to prune rootstocks has not been clarified. Several species of ring nematode, including *Mesocriconema ornata* and *M. curvatum,* feed on grasses, but *M. xenoplax* feeds on grasses and the smallest feeder roots of woody perennials (see fig. 19.3). *Paratylenchus hamatus* feeds almost exclusively on the roots of woody perennials, but several attempts have failed to show direct damage to *Prunus* rootstocks because of its feeding. *Xiphinema americanum* is an important vector of certain ring spot viruses that are not prevalent in California. This

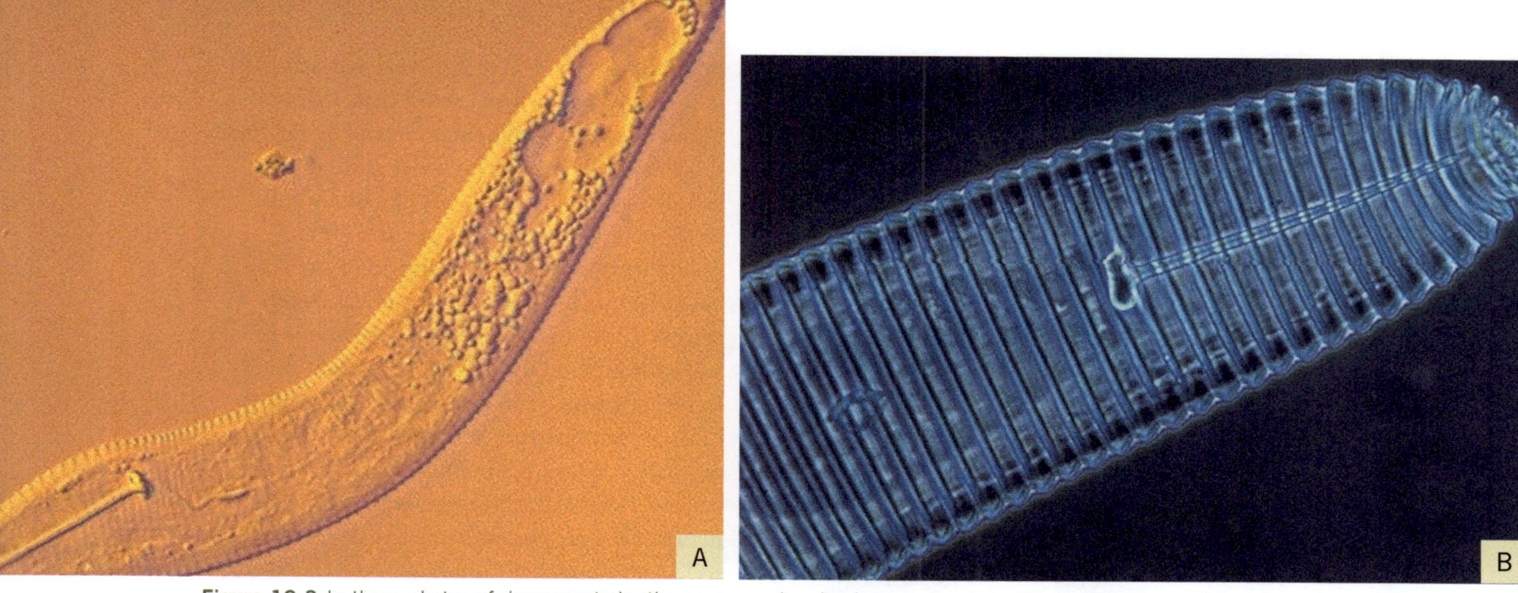

Figure 19.3 In these photos of ring nematode, the spear can be clearly seen. In photo (B), the rings around the nematode cuticle, from which the nematode gets its name, are clearly visible. *Photos:* B. Jaffee.

nematode is common in orchards, particularly where grasses and woody roots grow together, but thus far no particular damage to roots of prune trees has become apparent.

To summarize our current level of knowledge about replanting, prune growers must be concerned about *M. xenoplax* and *P. vulnus* particularly when planting after any perennial crop. It is not clear whether *P. penetrans* damages prunes, but damage does occur at high population levels. Growers must also be concerned about the rejection component of replanting, which appears as poor and uneven growth in first- and second-year plants, particularly when replanting one *Prunus* spp. after another in the San Joaquin Valley (for more information, see the section "Rejection of Replants" at the end of this chapter).

Figure 19.4 This photo of a lesion nematode in a root shows its wormlike nature and microscopic size as it is stretched out across several cells of the root. *Photo:* S. D. Van Gundy.

Nematodes of Greatest Concern

The nematodes of greatest concern in prune or plum orchards are the ring nematode *(Mesocriconema xenoplax)* and the root lesion nematode *(Pratylenchus vulnus).* It is common for these two nematodes to occur together, but the damage they cause is quite different. The preferred feeding site of ring nematode is the smallest feeder roots. Adult stages of this nematode are sedentary, but smaller, younger stages are capable of moving through soil. It builds to its highest population levels on Marianna 2624, followed by Nemaguard, then Lovell rootstock. The ring nematode host status of Myrobalan 29C is unclear, but it is apparently a slightly poorer host than Marianna 2624. This nematode develops its highest population levels in highly porous soils. Sand, loamy sand, and coarse sandy loam soils that infiltrate water quickly provide its best habitat. Fine sandy loam soils that are relatively slow to infiltrate water will support this nematode but not at levels high enough to permit extensive population buildup. Clay loam soils that are well structured, and hence have large pore spaces, also provide adequate opportunity for ring nematode buildup. These are the same soils where Lovell or Marianna 2624 rootstocks are the preferred choice of the grower.

Several years of extensive feeding by the ring nematode can greatly reduce the abundance of young feeder roots and cause abundant populations throughout the soil profile. The nematodes are "lying in wait," and just as fruit production becomes adequate, some trees or tree limbs may suffer springtime death. The nematode predisposes the upper portion of the tree to rampant infections by *Pseudomonas syringae*. The complex disease caused by this bacterium is referred to as bacterial canker complex. In 95% of orchards with bacterial canker, the ring nematode provides the trigger for damage. There are also occasions where the nematode is not present or present at quite low population levels. In those few instances, some other field condition usually limits proliferation of small feeder roots, such as excessive soil saturation due to broken irrigation lines or leaky ditch banks. Ring nematode is also associated with dieback in the tops of trees as they age and lose vigor. Application of an appropriate postplant nematicide can noticeably invigorate such trees for as much as half a year: repeated applications are necessary to maintain the new growth.

Pratylenchus vulnus is an endoparasitic root lesion nematode that may feed and lay eggs outside or within the roots (see fig. 19.4). On occasion, the roots of prune may exhibit dark lesions due to the feeding of this nematode, but by the time this damage is evident considerable damage is under way. This nematode can oftentimes be found together with *M. xenoplax*, but *P. vulnus* by itself is not a predisposing agent of bacterial canker. Adults, juveniles, and egg stages of both of these nematodes survive in soil without food for lengthy periods of time. In settings where roots of old trees have been killed with a systemic herbicide, 5% of the original *P. vulnus* population remained in the soil for 5 years after tree removal. Survival of this nematode is greatest at 3 to 5 feet deep in the soil. Applications of Roundup to old tree trunks kill remnant roots and, coupled with 1 year of fallow, alleviate 85% of rejection. However, even with 2 years of fallow, an abundance of this nematode can remain within dead remnant roots, as well as within the soil. This root lesion nematode may feed at various locations along the young root, but it does the greatest damage if it is present at high population levels when new trees are planted. Excavation of young affected trees can often display an inadequate number of primary roots. Once trees become established, fruit size and numbers per tree will be reduced. In a 15-year evaluation in sandy loam soil, the damage level to fresh plums was 13% on Nemaguard and Lovell but 8% on Marianna 2624 and Myrobalan 29C (McKenry 1988). Although these differences were not significant, damage was visible in locations where other stress events were also present. In trials at Kearney Agricultural Center and at UC Davis, it was the presence of a sand layer located approximately 2 to 3 feet beneath the sandy loam, or clay loam soils, respectively. Water infiltration is reduced where several layers of soil exist; the result can be a limited root system aggravated further by root lesion nematode feeding.

Pratylenchus vulnus may be found in a wide range of soil textures, with population levels largely influenced by the host rather than soil type. Colt cherry rootstock, English walnut, and rose are among the best hosts, commonly supporting 1,000 nematodes per 250 cubic centimeters of soil. These are followed by black walnut (750 per 250 cc of soil); Nemaguard, Marianna 2624, and Myrobalan 29C (250 to 500 per 250 cc of soil); grape (20 to 100 per 250 cc of soil), and *Pistacia atlantica* or UCB1 rootstock

(0 to several per 250 cc of soil). This root lesion nematode is also hosted by beans and tomato but not by most weeds or rotation crops such as California alfalfa types, sudangrass, or sorghum × sudan hybrids.

Managing Nematodes in Prunes

The one best time to manage nematodes is prior to planting. Consider the following points:

- Soil fumigation has become relatively expensive, while economic returns from dried prunes have remained stagnant. A proper preplant fumigation for clay loam soils will cost double and require more care than the same fumigation applied to sandy loam soil. Methyl bromide is the preferred fumigant for clay loam soils, whereas in sand to sandy loam soils 1,3-dichloropropene applied to well-dried soil performs adequately. For more information on fumigation, see the replanting section at the UC Kearney Agricultural Center Nematode website, http://www2.uckac.edu/nematode.
- Purchase only certified nematode-free nursery stock. In California, nursery trees are required to be free of nematodes for sales that cross county lines.
- Choose the best rootstock for your soil conditions with consideration of soil texture first and the prevailing soil pests second (see chapter 6, "Rootstocks").
- Postplant nematicides such as Enzone that perform well in sandier soils may not be useful in clay loam soil. Thus, consider preplant fumigation as one of the few chemical treatments available when replanting orchards into finer-textured soil. Newer postplant nematicides are on the horizon and will appear in the UC IPM Pest Management Guidelines for Prune as they become registered and their performance quantified.
- Research is under way using what is termed a "starve and switch" tactic to replace soil fumigation. A trunk application of glyphosate herbicide followed by 1 full year of waiting can starve and alter prevailing soil ecosystems. Switching to a very different rootstock can further improve first-year tree growth, particularly if the new rootstock is known to be tolerant to rejection and also resistant or tolerant to the prevailing nematode species. The current shortfall for prune growers is the paucity of prune rootstocks with nematode resistance as well as tolerance to rejection.

Postplant nematicides can improve the vigor and yield of prunes that are infested with nematodes. There are, however, serious considerations relative to their performance.

- Nematicides that rapidly degrade within 24 to 48 hours should not be used in clay loam soils. By the time the active ingredient moves adequately through the soil and into soil particles, the concentrations are too low to provide adequate nematode reduction in clay particles. The half-life of Enzone in soil is about 24 hours. Its use in sandy soils can provide dramatic benefit, but there may be minimal benefit in most clay loam soils.
- Currently, the most economic procedure for delivering nematicides is via chemigation. The chemigation process for delivery of nematicides is not the same as it is for delivery of fertilizers. For nematicides, the goal is to reach as many roots as possible without excessively diluting the product. Research is currently under way to identify foliar-applied nematicidal agents, and some progress has been made.
- Postplant nematicides may be relatively specific in their performance. Some perform well against ring nematode but give only short-term relief against root lesion, or vice versa.

Nematode Protection of Current Rootstocks

There have been numerous examples of nematode relief due to the use of rootstocks, but decades often pass before new rootstocks actually receive evaluation against nematodes. This section reviews current rootstocks with an eye toward their nematode management value. Consult table 19.1 for a partial list of prune rootstocks and other planting choices for assistance in planning rotations that might reduce the incidence of certain nematode species.

Nemaguard is useful because of the resistance it provides to all root knot nematode (*Meloidogyne* spp.) populations throughout California. Use of this rootstock is dominant in sandier soils of the San Joaquin Valley. Its sensitivity to zinc deficiency in sandy soil can be overcome, but sensitivity to waterlogging and iron chlorosis limit its planting to the sandier soils. When evaluating a wide selection of *Prunus* species rootstocks, Nemaguard does carry some protection against the root lesion nematode (*Pratylenchus vulnus*), but it is not resistant. Against *M. xenoplax*, it is susceptible and a poor choice, but it is usually the only choice available where soil profiles are highly porous.

Lovell seedlings perform similarly to Nemaguard in the presence of *P. vulnus.* Against *M. xenoplax,* Lovell supports half as many ring nematodes as does Nemaguard, which engenders some relief from bacterial canker complex. Lovell is highly susceptible to *Meloidogyne* species, and in sandy soil this nematode may cause some trees to die as early as the first year. Preplant fumigation is essential if this rootstock is to be planted in sandy soils with *Meloidogyne* species present. In finer-textured soils, Lovell rootstock has an advantage over Nemaguard because it is less sensitive to waterlogging and iron chlorosis.

Marianna 2624 supports *P. vulnus* at population levels similar to those of Nemaguard, but it appears to be more tolerant of its presence. It has resistance to *Meloidogyne* species, is appreciated for its resistance to certain *Armillaria mellea* populations, and does relatively well in clay loam soils. However, this rootstock is highly susceptible to *M. xenoplax.* It also produces suckers at levels that hamper grower acceptance.

Myrobalan 29C is susceptible to *P. vulnus,* but like Marianna 2624 it exhibits some tolerance to its damage. It also has resistance to *Meloidogyne* species. Its host status to *M. xenoplax* has not been adequately tested, but unproven testimonials indicate that it performs better than Marianna 2624 in the presence of bacterial canker complex. This stock grows well in clay loam or sandy loam soils.

There are new rootstocks that need field evaluation in California with prune scions. Among these new stocks is M40. It is known to host *P. vulnus,* but its other attributes remain unknown. A second rootstock worthy of testing in large field settings is Viking. This rootstock is a complex cross of almond, plum, apricot, and Nemaguard, with vigor similar to Nemaguard. It is reported to be compatible with all plum varieties. In recent field evaluations, its resistance to *M. xenoplax* was on par with that of Lovell, except during the warmest months of the year. In the presence of bacterial canker, it performs almost as well as Lovell while providing resistance to *Meloidogyne* species. It supports *P. vulnus* at levels similar to Nemaguard. Its performance against rejection has been improved by the purchase of potted plants rather than bare roots, but it cannot be considered very tolerant.

Krymsk 1 is a dwarfing rootstock compatible with plum, French prune, and Tulare Giant, but it produces a proliferation of suckers, which may be caused by partial incompatibility with certain scions. A disadvantage of this stock is its susceptibility to crown gall, but this subject needs greater study. An advantage is that it is the only *Prunus* species rootstock that offers resistance and tolerance to *P. vulnus* while also limiting development of *Meloidogyne* species to root terminals that are 60 days or less in age. The *Prunus tomentosa* parentage within this stock needs greater attention by plant breeders.

Table 19.1. Nematode-rootstock profile and crop rotation choices

	Relative nematode population levels in first 2 years as compared with Nemaguard (= 100)		
	Mesocriconema xenoplax	***Pratylenchus vulnus***	***Meloidogyne incognita***
Rootstock			
Nemaguard	100	100	0
Lovell	40	100	200
Marianna 2624	150	100	0
Myrobalan 29C	100	100	0
Viking	flux from 40–90	120	0
Guardian SC-17	flux from 50–90	100	on young roots
HBOK 1*	30	30–90	0
Krymsk 1*	150	0	on young roots
Walnut, English	100	300	300
Walnut, black	100	200	150
UCB-1 pistachio	100	0–50	0–100
Crop rotation choices			
winter wheat	20	0	10
sudan grass	50	0	0
velvetbean*	100	0	0
tomato	20	20	30
Soils most suitable to the nematode	prefer highly porous soils	any soil	sandy soils south of Sacramento

Note: *Not commercially available.

HBOK 1, a selection of Harrow Blood peach by Okinawa peach, is of interest because it offers resistance to *M. xenoplax* that is on par with that present in Lovell while also providing resistance to root knot nematode species. It also displays vigor similar to that of Lovell.

Nematode Sampling

Because nematodes present few distinctive diagnostic symptoms or signs, soil and root samples should be collected and sent to a diagnostic laboratory whenever tree vigor seems limited without an apparent cause. To begin the procedure, visually divide the orchard into sampling blocks that represent differences in soil texture, drainage patterns, or cropping history. Take a separate sample from each block so that each block can be managed separately. Because nematodes are usually not uniformly distributed within a field, it is necessary to take a series of subsamples from throughout the area to more accurately determine whether nematodes are present. In a fallow field, collect subsamples from several locations within the sampling block. Be sure to sample to 3 feet deep if the field was previously in woody perennial crops. In an established orchard, collect separate subsamples from the soil around three to five trees that show symptoms and from the soil around three to five adjacent healthy-looking trees for comparison. Subsamples should include feeder roots, when possible, and should be taken in frequently wetted zones at the edge of the tree canopy. Because nematodes feed on roots, they are more prevalent in the rooting zone of the tree, and this is the area from which subsamples should be taken.

Mix the subsamples together and place about 1 quart of the mixed soil and roots into a plastic bag. Seal the bag, place a label on the outside of the bag, keep samples cool (do not freeze), and transport as soon as possible to a diagnostic laboratory. Ask the laboratory to determine whether the nematodes listed in this chapter as pests are present so that they can use the appropriate extraction techniques. Because a number of different species of root lesion nematode (only one of which is known to be pathogenic to prunes) could be found in a sampling site, request a species diagnosis if root lesion nematode is found. For the location of a diagnostic, consult the local UC Cooperative Extension Farm Advisor.

During recent years, increasing emphasis has been placed on the development and use of damage thresholds for making management decisions for aboveground pests. For many reasons, it is difficult to establish damage thresholds for nematodes. These reasons include difficulties in obtaining representative samples and variability in extraction methods and efficiencies of different laboratories, as well as the many biotic and abiotic factors that influence populations. However, a routine soil sampling program can be very helpful in establishing the need for, and the success of, a nematode management program. If the nematodes discussed above as pathogens are present in an orchard with below-normal growth and yield, and no other explanation of the problem can be found, it is likely nematodes are a contributing factor. In order to determine whether the nematode population is increasing or is remaining stable, sample an infested orchard at least once per year and at the same time each year. It is essential if one wants to evaluate the benefits of a management technique to leave some trees in an orchard untreated in order to establish the effectiveness of the technique with respect to nematode control and crop production.

Soil Fumigation

Preplant soil fumigants, listed with highest fuming capacity first, include methyl bromide (MB, registered for orchard replants only), iodomethane (MI, not currently registered), chloropicrin (CP), 1,3-dichloropropene (1,3-D), metam sodium (MS), and metam potassium (MP). California availability and regulatory constraints associated with each is tenuous, so annual updating of regulations and availability is essential.

Prunus rootstocks usually grow poorly following drench applications of MS or MP. If these two products are delivered at only 110 pounds per acre and only into the surface 6 inches of soil, they can provide a useful means of attaining nematode control in the surface 6 inches without using a tarp. These two products are best applied in front of a rotavator device traveling within the 5 inches of field surface. Be sure to obtain the latest regulatory information on each of these soil fumigants.

Applications in excess of 150 pounds per acre of MI to coarse-textured soils can cause marginal necrosis of first-leaf plum and prune scions regardless of the rootstock they are on. The intensity of this necrosis is associated with higher application rates, but even at 200 pounds per acre, the damage has disappeared from second-leaf trees in sandy loam soil. Leaf burn of these same scions in clay loam soils has not been observed at rates as high as 400 pounds per acre of MI. There are also regulatory concerns about this fumigant, so its future use in California is unclear.

Applications of CP at 150 to 350 pounds per acre provide the greatest increased growth response of all the above fumigants. This response occurs even if soil pests are not a problem in the field. The greater the quantity applied, the longer lasting will be the growth spurt. Some of this growth benefit may be associated with protection against rejection. CP is nematicidal but relatively quick to degrade; it can miss nematodes located deep within soil particles and may not kill roots deep within the soil profile. CP is persistent when applied to cold soils; applications made after mid-November can damage trees planted early the following spring. The key to nematode control with this material is that there must be open pore spaces in the soil for it to be effective. This can be accomplished by drying and deep ripping in preparation for the treatment.

Future use of MB is tenuous, but it has been the undeniable product of choice when fumigating finer-textured soils or soils which, even after drying, are in excess of 12% soil moisture within the surface 5 feet of the soil profile.

The alternative to MB is 1,3-D. In California, the prevailing policy is to limit applications to 33.7 gallons per treated acre. This amount can provide adequate nematode control and good relief from rejection if the soil is sandy and has received deep drying throughout the surface 5 feet of depth. At locations where soil moisture content is in excess of 12%, this amount of 1,3-D is inadequate. Clay loam soils commonly hold greater than 12% soil moisture content even after a full year of deep drying. Proper 1,3-D fumigation for soils with up to 15% soil moisture within the profile is 500 pounds per acre of 1,3-D. Soils up to 19% soil moisture content require 670 pounds per acre of 1,3-D. These latter two application rates, on a broadcast basis, are currently unacceptable in California and even if approved are quite expensive to the grower. These higher application rates can be applied as a strip application.

Rejection of Replants

In replanted orchards, poor and uneven growth from tree to tree in the first year is not generally caused by soil pests or diseases, including nematodes. With some of the trees growing poorly and others substantially better, the cause can be overirrigation, waterlogging, and Phytophthora root rot by midsummer of the first year. It is likely that the root development of these trees is being rejected by prevailing soil ecosystems. Within a year or two after replanting, the rejection component is satisfied and trees begin to grow at a rate similar to that achieved with fumigation, unless there are also known nematodes, diseases, or soil profile problems that further limit root development. A limited root system plus the presence of nematodes or diseases can result in a low-vigor orchard that may never be economical.

The intensity of rejection is not similar from tree to tree or from one region to the next. In general, experience in the Sacramento Valley tends to indicate that rejection is less damaging there than in the San Joaquin Valley. Damage by rejection can reduce first-year tree growth to one-seventh of that of fumigated trees. In the second year, the differences may lessen to one-third, but root inactivity comes when trees should have developed far-reaching root systems (for more information, see the UC Kearney Agricultural Center Nematode website, http://www2.uckac.edu/nematode). If growers replanting in the area have observed dramatic first-year growth benefits due to soil fumigation, consider rejection to be the cause rather than nematodes. Rejection may be corrected with strip fumigations at appropriate application rates or with Roundup applications to cut trunks (plus waiting 1 year). Do not expect strip applications of fumigant to protect against nematodes for more than 1 year. If the new trees are on a rootstock with resistance or adequate tolerance to the prevailing nematodes, both rejection and nematodes can be solved with a strip application of fumigant.

References

McKenry, M.V. 1988. Damage and development of several nematode species in a plum orchard. Applied Agricultural Research 4:10–14.

———. 1999. The replant problem and its management. Fresno: Catalina Publishing. UC Kearney Agricultural Research and Extension Center website, www.uckac.edu.

20 Orchard Floor Management and Cover Crops

• Chuck A. Ingels and Fred Thomas

The main goals of prune orchard floor management are to control weed growth, promote optimal tree growth without limiting fruit production or size, improve soil physical characteristics and nutrition, and reduce frost hazard in early spring. Floor management methods include tillage or mowing, herbicides, and/or cover crops. All of these methods influence tree nutrition, soil structure, and water movement into and within the soil profile.

Floor Management Systems and Considerations

Floor Management Options

Complete cultivation

Complete cultivation is disking both down and across the rows for weed control and preparation for irrigation (fig. 20.1). Most clean-cultivated orchards are disked three to six times per year, depending on the number of irrigations and the types and amounts of weeds present. This system is fairly common in the Sacramento Valley.

An advantage of this system is that it can be used with sprinkler, furrow, flood, basin, contour, or strip check irrigation. In areas of high frost hazard, cultivation can provide bare, moist ground, which is several degrees warmer than areas with vegetation.

There are some disadvantages to a complete cultivation system. It generally entails using heavier equipment than other systems. A plow or disk pan may develop because of continual plowing or disking at the same depth, especially if the soil is too moist. Soil aggregation can be destroyed, with the smaller aggregate sizes restricting water movement into and through the soil. Soil density can be increased by compacting soil particles. These can be managed somewhat by deeper tillage (chiseling) and incorporation of organic matter from cover crops. Complete cultivation is usually dustier than other systems and can create additional pest problems, especially spider mites in warm weather.

Cultivation in middles plus herbicide strip

This is a modification of the above complete cultivation with the addition of strip chemical weed control (fig. 20.2). One advantage of this

Figure 20.1 Complete cultivation, with disking both across and down the rows. *Photo:* C. A. Ingels.

Figure 20.2 Disking in the middles and herbicide strip in the tree rows. *Photo:* C. A. Ingels.

system is that the cultivation is done in only one direction, eliminating the cost of cross-disking. It can also be used in orchards with solid-set sprinklers, microsprinklers, or drip irrigation, and in orchards where trees are planted on berms. It has most of the same advantages and disadvantages of complete cultivation.

Mowing in middles plus herbicide strip

The most common floor management method is mowing of a ground cover between the tree rows (middles) and a herbicide strip in the tree row (fig. 20.3). The most common ground cover is resident vegetation (weeds), but a sown perennial sod (grass) cover crop can be used (see fig. 20.21). This system can use sprinkler irrigation or permanent berms in the tree row that form borders for strip check irrigation.

Advantages of this system are that tractor work is in one direction only, preirrigation land preparation is not required, and footing for workers and equipment is better than under tillage. Plant roots of the ground cover aid in maintaining good water penetration. Tree roots can grow closer to the soil surface in the herbicide strip, where nutrients are more readily available than at greater depths. Spider mite problems usually are less severe under no-till than cultivation due to less dust.

No-till floor management does have the disadvantage of using additional water and nutrients, particularly nitrogen. Most of these nutrients are eventually returned to the soil by decomposition of the clippings, but there is a net cost for additional water and nitrogen. No-till orchards generally are cooler, creating a greater spring frost hazard than exists under complete cultivation.

Figure 20.3 Mowed resident vegetation in the middles and herbicide strip in the tree rows. *Photo:* C. A. Ingels.

Figure 20.4 Woollypod vetch produces excellent growth in winter, smothering most winter annual weeds. *Photo:* C. A. Ingels.

Other soil management systems

Variations such as summer mowing with winter cultivation, or summer cultivation with a winter cover crop combine the advantages of two systems to solve specific problems or establish cost advantages. In areas with compaction and poor infiltration, resident vegetation or planted sod or other planted cover crop may be helpful. In some soils, however, orchards under no-till occasionally should be cultivated if water infiltration has been reduced due to surface sealing.

Factors to Consider When Choosing a Floor Management System

No single orchard floor management system is ideal for all orchards. The best system should be tailored to the individual orchard after carefully considering the benefits of each management program. If a method is not working, diagnose and analyze the problem first, then experiment on the best technique to change or improve it.

Economic considerations

Short-term or annual costs are fairly easy to calculate: the number of acres per hour to mow or cultivate, herbicide application costs, cover crop seed costs, water and fertilizer costs, and other related items are easily obtained. However, annual costs or even short-range projections can be misleading if these practices accelerate conditions that are expensive to correct. Resident vegetation management is less expensive per acre per year than cultivated systems, assuming the same type of irrigation systems, because fewer equipment hours are needed even though more water and fertilizer are required. Sown cover crops generally cost more than resident vegetation, but the cost depends on the species or mix used and its management. The added expense of sprinkler irrigation, considering interest and depreciation, makes a sprinkler-irrigated sod system more expensive than a flood-irrigated cultivated system.

A small differential in costs should not be the governing consideration in selecting a floor management system because problems created or solved will more than offset any savings.

Floor management is only one part of the total prune production system that is designed to provide maximum long-term profit and environmental stewardship. If savings in floor management are offset by extra expense in harvesting or pest management, the net result may be undesirable. Likewise, a more expensive soil management system may be desirable if savings in other areas or improved air or water quality more than offset the extra cost.

Effects of floor management on frost protection

Spring frosts can injure flower buds as they swell and open during March and April. The temperature at which flowers are damaged depends on the stage of bloom and duration of the freeze. Critical low temperatures for flower buds and fruit in various stages of development are 25°F at the green tip and popcorn stages, 28°F at full bloom, and 30°F with small green prunes. Sensitivity to frost increases as bloom progresses, so young prunes are more easily damaged than are newly opened flowers. Small prunes directly exposed to the sky radiate heat and are damaged sooner than are prunes protected by foliage.

Frost protection measures typically carried out by California growers include proper orchard floor management and irrigation with well water at 55° to 65°F. During and after bloom, the orchard's floor condition plays an important role. When air temperature cools, an orchard floor that is bare, firm, and moist will yield the warmest air temperatures in the orchard, while a tall cover crop or dry, cultivated soil will yield the coldest. Bare soil absorbs more radiation from the sun than does soil covered with vegetation. A soil with a high moisture content stores more heat than does a dry one. To obtain this condition, it is necessary only to wet the top foot of soil at any time before the frost season. If this optimal soil condition cannot be attained,

Table 20.1. Temperature ranges at 5 feet above the surface resulting from various orchard floor conditions when compared to bare, firm, moist ground

Conditions	Temperature ranges
bare, firm, moist ground	warmest
shredded cover crop, moist ground	0.5°F colder
low-growing cover crop, moist ground	1° to 3°F colder
dry, firm ground	2°F colder
freshly disked, fluffy ground	4°F colder

weeds or cover crops should at least be mowed before bloom, and cultivation should be avoided during the frost season. Table 20.1 compares approximate minimum temperatures of air near soil surfaces of varying conditions with that of the ideal (bare, firm, moist ground).

Some growers furrow or border their orchards after disking just before bloom. This is followed by an irrigation to obtain firm, moist soil and to smooth the furrows or middles in case it becomes necessary to run additional irrigation water during cold nights.

Shallow soils

Special consideration may be required when deciding on proper orchard floor management on shallow soils. The topsoil is generally the most fertile layer in any soil. All surface-applied fertilizers and organic matter that is applied either through plowing or leaf fall impact the surface soil much more than deeper soil layers. Also, irrigation water and oxygen are readily available to the shallow roots on all types of soil. Prune roots generally tend to be shallow, but especially on shallow soils with restricted subsoil. Cultivation kills the roots in the top 4 to 6 inches. Elimination of this top highly fertile zone can reduce the rooting depth even further. A type of noncultivation would tend to encourage roots in this highly fertile surface soil.

Dense stands of competitive weeds such as bermudagrass or johnsongrass can also impact nutrient availability near the surface. Serious attention to their control can also improve the fertility and water availability to the tree.

Potassium

Applications of potassium are generally required to maximize prune production and size, especially in northern California. Potassium is difficult to apply in a way that prune trees can use it before it is absorbed on the clay particles in the soil. Applications made to the soil surface are effective only when

applied to soil that is not cultivated. This can either be under sod culture or to the strip chemical weed control area where middles are cultivated. In either case, maximum effectiveness is obtained when the band is not disturbed.

Band applications shanked into cultivated soils are also effective. For maximum effectiveness, the placement area should be below the cultivated area so that no additional mixing takes place from future cultivation. All floor management decisions should consider how to prevent disturbing the potassium band.

Water infiltration

Water infiltration is another major consideration in selecting an orchard floor management program. What may be the best method on one soil type may not work on another soil type. Increased organic matter from cover crops or resident vegetation helps improve penetration on many soils. Other soils will develop a shallow surface crusting either from the water flow or sprinkler drop impact that seals the surface. Under the latter soil type, periodic cultivation may be needed. Soil or water amendments such as gypsum or sulfur may also be required for improved water infiltration.

Cover Crops

A cover crop is a nonharvested crop that is grown in orchard middles in order to provide specific benefits. A cover crop can be sown from seed, or the resident vegetation (weeds) can simply be managed. Resident vegetation is frequently used because it offers many of the benefits of sown cover crops and is the cheapest way of providing plant cover on the orchard floor. However, the composition of resident vegetation can be variable and may predominantly consist of species that provide little benefit or even include noxious weeds. A sown cover crop is sometimes the only way to obtain specific benefits, such as adding substantial quantities of nitrogen (legumes) or improving year-round water penetration or traction (such as the use of perennial grass sods). This section discusses sown cover crops.

The choice and performance of cover crops often depend on site-specific factors. The main factors to consider when selecting a particular species or mix are cost versus the benefits, tillage practices, irrigation method, nitrogen needs, and frost concerns. Understanding the basic cover crop types and management strategies can greatly improve the chances for success.

Benefits and Drawbacks

Benefits

The benefits of using cover crops are often not easily measured, so it may be difficult to justify the cost of planting and managing them. They also have potential drawbacks, which further limit their utility. The trade-offs of cover cropping are shown on the next page.

An important reason many growers sow cover crops is to improve water infiltration. Plant roots and associated microorganisms produce polysaccharide gums, which increase aggregation of soil particles because the gums act as glues to bind particles together. Aggregation improves soil tilth, aeration, and drainage of water. Cover crops with a higher carbon to nitrogen (C:N) ratio such as grasses usually promote greater polysaccharide production and thus greater aggregate stability than do cover crops with lower C:N ratios such as legumes.

Another important benefit is wheel traction. In Sacramento Valley orchards, heavy rains can limit orchard accessibility on the clay loam soils, so some prune growers have sown perennial grass sod, such as creeping red fescue and perennial rye. These grasses provide excellent traction and reduce wheel rutting, and they take up excess soil moisture. The trade-off is that they require extra water during the summer, which could be a problem in drought years.

Most organic prune growers use legume cover crops as a source of nitrogen, and many often supplement the cover crops with compost to increase potassium. A legume cover crop can potentially satisfy most, and in some cases all, of the nitrogen requirements of prunes. It is not uncommon for a well-managed vetch cover crop to add 150 pounds of nitrogen per planted acre. If the cover crop is sown on only half of the orchard (not in the tree row strip), up to 75 pounds of nitrogen per orchard acre will be added, providing it is incorporated in the spring.

Insect management can be somewhat affected by cover cropping. Cover crops provide a food source (nectar, pollen, or other insects) to many beneficial insect species. These beneficial insects in turn may feed on orchard pests. For example, adult parasitic wasps feed on nectar, while their larvae feed on and kill insect pests such as aphids, which are often a pest in many prune orchards. Predator insects such as lady beetles and green lacewings are frequently attracted to cover crops to feed on soft-bodied insects. However, research examining the effects of cover cropping on pest management of various crops has shown mixed results, with the predator population lagging behind that of the pests (Alins et al. 2007; Daane and Costello 1998). Although some benefits in prune pest management may occur, cover crops should not be counted on to provide substantial control.

Drawbacks

Cover crops require knowledge and management to attain the desired benefit. Therefore, a significant drawback is increased time and monetary cost. As with other production costs, the benefits obtained must outweigh the costs associated with cover cropping in order to justify the expense.

Cover crops also increase water use. It is not unusual for a cover crop to use over 6 inches of water (Grant et al. 2006). In the San Joaquin Valley, water is more limited and expensive than in the Sacramento Valley, so even winter annual cover crops reduce the soil moisture available to trees in the spring (as does resident vegetation), even beyond the increase in the extra infiltration of the winter rainfall. For the least impact, either early disking or close mowing reduces water use, although some species cannot be mowed closely without killing them.

Grass cover crops require nitrogen, and unless sufficient nitrogen is added, the trees may become deficient. Annual grasses that are allowed to mature and then disked in tie up soil nitrogen in their decomposition, also depriving the trees of nitrogen if none is added. For this reason, some growers use a legume-grass blend to increase the fixation of nitrogen and to lower the carbon to nitrogen ratio to prevent nitrogen from being tied up.

As mentioned earlier, cover crops increase the frost hazard in early spring because bare ground is better than living foliage at storing solar heat, which is released at night, warming the orchard. The duration of lower temperatures may also be longer in cover-cropped orchards. The taller the cover crop, the greater the hazard, so frost hazard can be reduced by selecting lower-growing species or covers that can be mowed closely.

Some pests may increase because of cover crops. For example, lygus are attracted to many clovers, and they may move to prune trees or to adjacent cotton or tomato fields after the clover dries or is mowed in the spring. Gophers are attracted to vegetation, especially clovers. Some cover crops may also become weeds requiring control if allowed to go to seed or allowed to creep onto the herbicide berms, such as strawberry clover.

Trade-Offs in Cover Cropping in Prune Orchards

Potential Benefits

- Improved soil structure and water infiltration
- Improved orchard access on wet soils
- Addition or conservation of nitrogen
- Addition of organic matter to soil
- Weed suppression
- Beneficial insect habitat
- Reduced dust and associated spider mite pressure
- Reduced soil erosion and nutrient runoff
- Cooler summer orchard environment

Potential Drawbacks

- Increased water use
- Competition with prune trees for soil moisture and nutrients
- Increased frost hazard
- Increased vertebrate pests
- Increased costs
- Increased management

Cover Crops for Disked Orchards

High-biomass cover crop species are often used in disked orchards. Mixes that produce large amounts of biomass (plant matter) can be used to add organic matter to the soil. The periodic addition of organic matter enhances soil microbial populations, improving soil structure and nutrient cycling. However, the same disking that incorporates plant matter also aerates the soil, causing the microorganisms to quickly break down organic matter.

Seeds of high-biomass mixes are generally large and easy to grow. In general, they are sown each fall and disked in the spring; this is referred to as a green manure cover crop. Where border check flood systems are used, most prune growers disk twice within several days to incorporate the material before the soil dries excessively. Two basic types of high-biomass mixes are often available from seed companies: pure legume and legume-grass blends. In addition, some growers use a monoculture of either a grass or a legume.

Pure legume blends, usually containing bell beans, vetch, and/or field peas, are used to add a large amount of readily available nitrogen in the form of plant proteins to the soil. Bell beans produce vigorous, upright growth and cannot be

mowed closely. Field peas are shallow rooted and therefore subject to drought on sandy soils and likewise cannot be mowed closely. Vetches are frequently used because of their seedling vigor, biomass and nitrogen production, and weed suppression (fig. 20.4, p. 214). Lana woollypod vetch and purple vetch produce excellent cool-season growth. Common vetch has less vigor, but it has extrafloral nectaries on the stipules, which provide nectar to beneficial insects. Languedoc is a variety of common vetch that grows faster during the winter.

Various legume-grass blends are also available. The addition of grasses such as barley, oats, triticale, or cereal rye in a mix imparts several benefits. The fibrous roots of grasses greatly enhance soil tilth and water penetration. Grasses also take up excess nitrogen from the soil, improving the growth and nitrogen-fixing ability of the legumes. Lastly, grasses provide structural support for the twining vetches and peas. Typical blends often consist of bell beans, vetch, peas, and oats or barley (fig. 20.5). Barley-vetch or oat-vetch blends are relatively inexpensive although they have to be sown every year.

Figure 20.5 High-biomass blend of bell beans, peas, common vetch, and oats. *Photo:* C. A. Ingels.

Cover Crops for No-Till Orchards

Winter annuals

Some prune growers who use drip or microsprinkler irrigation systems that favor dry middles sow winter annual species that reseed and die in the spring and regenerate each fall with rainfall. Such species include annual legumes like crimson clover, rose clover, Persian clover, subclovers, and bur medic, or the annual grass Blando bromegrass. Although these species can be planted individually, they are often used in various blends. If not replanted or if neglected, in time these species may simply become minor components of the ground cover. If needed, periodic replanting every 3 or 4 years can ensure dominance by these species.

Bur medic (burclover) is well adapted to California's climate and grows well in neutral- to higher-pH soils. It is often a major component of resident vegetation (see fig. 20.13). It will reseed even under fairly close mowing and, because of its high percentage of hard seed, it usually reestablishes well. Subterranean clover, or subclover, usually performs best in acid to neutral soils. Other self-reseeding clovers, such as crimson rose and Persian clover, are well adapted to drip-irrigated prune orchards on sandier soils.

Perennials

Perennial cover crops are very common in prune orchards, with perhaps 10 to 15% of the northern California prune acreage planted with perennial ryegrass and fescue blends, New Zealand white clover, or managed stands of resident Ladino clover. There are several advantages to having a permanent cover, including reduced mowing and year-round traction for winter operations. Because they use water during the growing season, perennial cover crops are used more in northern California than in the southern San Joaquin Valley.

The perennial clovers, such as white and strawberry, are fairly low growing and add nitrogen, but strawberry clover is invasive, resistant to some herbicides, and may significantly increase rodents. Some perennial grass mixes require very little mowing; these include fine-leaf fescues, such as creeping red, chewings, hard, and sheep fescue, as well as dwarf perennial ryegrass blends. Dwarf varieties often require as few as three or four mowings per year and effectively smother most annual weeds. On clay or clay loam soils, however, the grasses may not receive the frequent irrigation they need, leading to a weak stand after a couple of years. However, some prune growers whose orchards have poor water infiltration and no-till flood irrigation systems find that the best management benefits are obtained from the perennial sod mixtures, New Zealand white clover as a monoculture, or a mixture of grasses and white clover.

Cover Crop Management

Planting

Winter annual and perennial cover crops perform best when sown by mid-October, but they can usually be successfully grown if planted by early November. In general, lower rates can be used for early seeding and higher rates should be used for later seeding. For example, vetch sown in late September or early October may be seeded at 40 pounds per planted acre, whereas an early November seeding would require 60 pounds. Establishing small-seeded cover crops in years with sparse fall rains may be very difficult if not irrigated, especially on sandy soils. Good seedbed preparation, control of noxious perennial weeds such as johnsongrass or bermudagrass, and a well-land-planed, flat, permanent floor are essential for establishing a perennial grass or clover cover crop that will live for 10 to 15 years in a prune orchard.

Legume seeds must be inoculated with nitrogen-fixing *Rhizobium* bacteria to ensure nitrogen fixation. Small-seeded legumes are usually preinoculated, but large-seeded legumes (bell beans, vetch, and peas) must be inoculated by the grower, at least the first time they are planted. Use about 8 to 16 ounces of the proper inoculant per 100 pounds of seed. If a legume-cereal mix is used, up to 50% more inoculant and sticker are required because a greater proportion of the mix adheres to the cereal seed. The most reliable inoculation method is the wet method, in which a slurry of inoculum and adhesive are added to the seed and then allowed to dry before planting. Some growers simply pour dry inoculant into the seed hopper in layers along with the seed; however, the effectiveness of this method has not been tested. Inoculant and inoculated seed should be kept cool and out of direct sunlight, and seed that has been broadcast should be incorporated as soon as possible.

Mowing

Because cover crops can increase frost hazard, they are often mowed in late winter. Bell beans and field peas do not perform well if mowed. If vetch is mowed, it should be mowed high, no lower than 8 to 10 inches. High-biomass blends can be killed in spring if mowed close to the ground. Such a mowing is usually done just before disking, but it can also be done to leave a thick mulch on the soil surface.

No-till clover mixes, which are sometimes used in drip-irrigated prune orchards, should be mowed in late winter to suppress tall weeds and encourage seed setting. Subclover and bur medic can usually reseed even under fairly close mowing, whereas crimson, rose, and Persian clovers flower above the foliage. Therefore, after the early spring mowing for frost protection, these species must not be mowed until about early June to allow for reseeding (fig. 20.6).

Low-growing perennial grasses or white clover can be managed with only three mowings per year. However, it is important to keep the mower height at no less than 4 to 6 inches to keep from stunting the grasses or damaging the crowns that weaken the stand and allow other weeds to become established.

Figure 20.6 No-till blend of crimson, rose, and subterranean clovers and bur medic in an almond orchard. The cover crop on the right was previously mowed for frost protection, but the left side was not mowed. *Photo:* C. A. Ingels.

Nutrition

As with prune trees, soil fertility is important for cover crop production. Legumes fix nitrogen, so nitrogen should not be applied before planting or during their growth. However, legumes do require adequate sulfur and phosphorus for good growth. Annual grasses such as oat and barley may require nitrogen additions if grown alone and respond well with 40 to 50 pounds of supplemental nitrogen per planted acre. Perennial grasses may require even more than this amount once established. In general, grasses predominate on highly fertile sites, whereas legumes usually grow best in soils with low nitrogen content.

Vetches and peas fix more nitrogen than do clovers, and medics with a green manure legume cover crop disked in April can add 150 pounds or more of nitrogen per planted acre. Nitrogen production in orchards may be reduced by shading and the use of wide herbicide strips. When residues are mowed and left on the soil surface, a portion of the nitrogen can volatilize into the atmosphere. No-till clovers and medics may add only about 30 to 40 pounds of nitrogen per planted acre because they are normally allowed to mature their seed and then are left on the orchard surface and not incorporated.

Prune growers whose trees regularly suffer potassium deficiency usually correct the problem with banded applications of potassium sulfate or muriate, and it is most effective to shank the potassium into the ground below the soil surface rather than banding it on top of the cover crop.

Species Selection Guidelines

If disking is used, it may be wise to consider using winter annual species that are disked under, such as high-biomass legumes and/or grasses. In no-till orchards in which the middles are dry or partially dry, no-till annual species or perennial dryland species or a mixture of both may be used. If the no-till orchard is fully irrigated, either no-till annual species or perennial irrigated species should be considered.

Disked Winter Annual Species

The following cover crop species can be used in prune orchards alone or in mixtures at the recommended seeding rate. For more information on these species, see Ingels et al. 1998.

Figure 20.7 Bell bean. *Photo:* C. A. Ingels.

Figure 20.8 Magnus field pea. *Photo:* C. A. Ingels.

Bell Bean
Vicia faba L.

Fig. 20.7

100–150 lb/seeded acre

Although bell bean is a true vetch, it differs greatly from other vetches with its strong, upright growth. It also has a relatively shallow, thick taproot, which may be useful for opening up heavy soils. Bell bean is often used in mixes with vetches, peas, or cereals. Because of its height and because it does not tolerate close mowing, it is often omitted from mixes in frost-prone areas. Bell bean is frequently infested by the bean aphid *(Aphis fabae)*, which seldom affects its use as a cover crop. The aphid, which does not attack prunes, and the presence of extrafloral nectaries may attract beneficial insects into orchards.

Field Pea
Pisum sativum L.

Fig. 20.8

40–70 lb/seeded acre

Numerous field pea cultivars are available and are most often used in mixes. Those most commonly planted as cover crops include Austrian Winter, Magnus, and Miranda. Austrian Winter, which has pink and reddish flowers, is dormant during cold weather and produces most of its biomass during the spring. However, it usually produces as much biomass as most other legumes if allowed to grow through the spring. Magnus can be distinguished by its large, light- and dark-pink flowers and its large tendrils. Unlike Austrian Winter, Magnus grows rapidly through the winter and matures earlier and is therefore a better choice in orchards that are disked early.

Common Vetch
Vicia sativa L.

Fig. 20.9

40–80 lb/seeded acre

Common vetch was once the most important vetch species in California, but now woollypod and purple vetches are also frequently used. Common vetch remains dormant through much of the winter, developing nearly all its biomass in March and April. For this reason, it is not the best choice to grow in orchards that will be disked in March. Some growers include common vetch in mixes

Figure 20.9 Purple vetch (left), Lana woolypod vetch (right), and common vetch (bottom). *Photo:* J. K. Clark.

Figure 20.10 Oats. *Photo:* C. A. Ingels.

because it has extrafloral nectaries on the stipules, which provide a readily available source of nectar for beneficial insects.

Purple Vetch
Vicia benghalensis L.
Fig. 20.9

40–60 lb/seeded acre

Purple vetch has been used commercially since the 1920s for forage, cover crops, and green manures. Like woollypod vetch, it produces excellent cool-season growth, but it blooms and matures later than woollypod vetch. Purple vetch leaves are markedly hairy, giving a silvery, downy appearance to shoot tips in the early spring. It also has reddish flowers. Although purple vetch is among the least–cold hardy cultivated vetches, in most years it will thrive in all but the coldest mountain locations in California.

Woollypod Vetch
Vicia dasycarpa
Fig. 20.9

40–60 lb/seeded acre

Lana woollypod vetch is the earliest-flowering and maturing vetch available; in the warmest regions, it may mature by late April, but it usually matures in mid- to late May. Its flowers are dark purple. Woollypod and purple vetches usually produce similar quantities of biomass and nitrogen and are both quite vigorous.

Oats
Avena sativa L.
Fig. 20.10

100–120 lb/seeded acre

Oats are frequently sown in prunes, often in mixes but also in monocultural stands. They tolerate wet and heavy soils better than barley and can also tolerate a wide range of soil types. Under moderate fertility and drainage, oats can tolerate a lower pH than barley; they tolerate a soil pH as low as 4.5. However, oats are not as tolerant as other cereals of drought, sandy soils, or cold. Dozens of cultivars have been developed, primarily as forage species. Cultivars vary in their period of flowering; Montezuma is the earliest, followed in descending order by Swan, Sierra, California Red, and Cayuse.

Barley
Hordeum vulgare L.

Fig. 20.11

80–100 lb/seeded acre

Barley is an inexpensive, fast-growing cereal that produces substantial biomass and competes well against weeds. It produces more tillers at the base than do cereal rye and oats. It is the most salt-tolerant cereal and is more drought tolerant than rye or oats. Barley is not as tolerant of wet soil conditions as cereal rye or oats; it will not grow well in heavy, poorly drained, or low-permeability soils, especially after periods of heavy rainfall. Many barley cultivars are available. UC 476 is a popular tall-growing cultivar that has good disease resistance but poor self-regeneration; UC 603 is a short-statured cultivar. Care should be given to cultivar selection, avoiding those that are not tolerant or resistant to yellow dwarf virus and rust.

Figure 20.11 Barley. *Photo:* C. A. Ingels.

Triticale
X Triticosecale

100–120 lb/seeded acre

Triticale, a cross between wheat and cereal rye, is similar in productivity to both these species. Many types are available, with widely differing growth habits and maturity dates. It is used less in orchards than other cereals.

No-Till Winter Annual Species

Berseem Clover
Trifolium alexandrium

Fig. 20.12

15–20 lb/seeded acre

Berseem clover is a rapidly growing winter annual that flowers in late spring and early summer, much later than most annual clovers. It is very tolerant of waterlogging and can be used to remove excess soil moisture. It is an excellent forage plant and thus responds well to mowing, exhibiting basal branching and rapid regrowth from the crown. It can be mowed three or four times in late winter and spring. These clippings are rich in nitrogen. Where used, it is often disked in the spring to conserve moisture and reduce the nitrogen contribution.

Figure 20.12 Berseem clover. *Photo:* C. A. Ingels.

Bur Medic
Medicago polymorpha L.

Fig. 20.13

15–20 lb/seeded acre

Bur medic is the most popular of the medics because it usually grows best and reestablishes reliably each year. Although it is frequently referred to as a burclover, it is not a true clover. It can be distinguished by its coiled seedpod and the short stalk extending from the middle leaflet. Bur medic is the most widely adapted of the medics to soils of varying pH. In some areas, Egyptian alfalfa weevil may damage bur medic, and in extreme cases the weevil may skeletonize plants. Bur medic is well adapted to California orchard growing conditions. It lends itself well to drip-irrigated orchards because it germinates readily in fall rains, grows rapidly during the winter, and produces much seed by early May. Seedlings produce early taproot growth and therefore may be better adapted to early-season drought than subterranean clover.

Figure 20.13 Bur medic in a walnut orchard. *Photo:* C. A. Ingels.

Crimson Clover
Trifolium incarnatum L.

Fig. 20.14

15–25 lb/seeded acre

Crimson clover performs well in annual clover mixes. Like other mowable clovers, it can be mowed to 3 to 5 inches tall during the winter and early spring. However, because it produces its flower heads above the foliage, it must be allowed to grow from mid-March or early April onward until the seed mature in late spring to ensure reseeding. Whether used alone or in mixes, crimson clover often produces a brilliant display of red flowers.

Figure 20.14 Crimson clover mixed with rose clover and bur medic. *Photo:* C. A. Ingels.

Subterranean Clover
Trifolium subterraneum L., *T. yanninicum*, or *T. brachycalycinum*

Fig. 20.15

20–25 lb/seeded acre

Subterranean clover is an excellent cover crop species for many orchard sites. It performs well in mowable clover mixes and usually requires periodic mowing to stimulate vigorous growth. In the spring, it often forms a dense mat of stems below the height of mowing, which helps reduce soil erosion and suppress weed seed germination. It is even more tolerant of very close mowing than bur medic due to its low, spreading habit. In addition, the peduncle reflexes and elongates downward after flowering, driving the seed head slightly underground in some cultivars. There are dozens of subclover cultivars, and differences exist among them in the time of flowering and maturity and in soil pH requirements. In general, subclovers are best adapted to acid or moderately acid to neutral

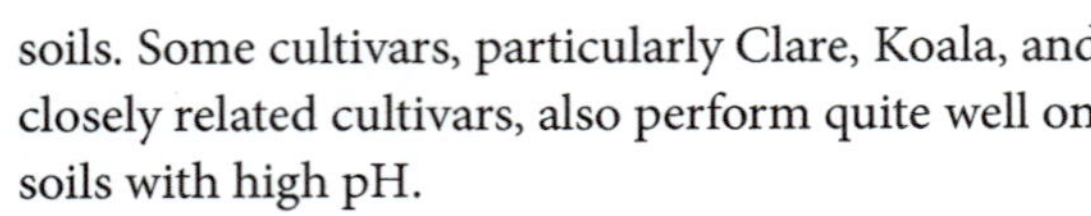

soils. Some cultivars, particularly Clare, Koala, and closely related cultivars, also perform quite well on soils with high pH.

Soft Chess (Blando Brome) *Bromus mollis* L.

Fig. 20.16

10–15 lb/seeded acre

Blando brome is well suited to no-till orchards, especially those that are drip-irrigated, but it is rarely used in prune orchards. It is low growing and mowable and matures early. It has strong seedling vigor, excellent reseeding ability, and dense, fibrous roots. For these reasons, it can reduce soil erosion while not competing excessively with trees. Blando brome is widely adaptable to a range of soils and climates. It can be grown in monocultural stands or mixed with clovers or other low-growing species.

Perennial Dryland Species

Hard Fescue *Festuca ovina*

Fig. 20.17

15–20 lb/seeded acre

Hard fescue establishes more slowly than does sheep fescue and has little spreading tendency. It is more drought tolerant than chewings or creeping red fescue but less so than sheep fescue. It does not tolerate waterlogging. Both hard and sheep fescues perform well on sandy soils. Like other fine-leafed fescues, sheep and hard fescues tolerate certain grass herbicides used to control most weedy grasses.

Figure 20.15 Subterranean clover foliage (top) and foliage lifted back to reveal seed pods, some of which are buried (bottom). *Photos:* C. A. Ingels.

Figure 20.16 Soft chess. *Photo:* C. A. Ingels.

Figure 20.17 Hard fescue *(Festuca ovina)*. *Photo:* C. A. Ingels.

Sheep Fescue
Festuca ovina var. *duriuscula*

Fig. 20.18

15–20 lb/seeded acre

Sheep fescue is a short-statured noncreeping bunchgrass that forms a dense turf. Because it is long lived and relatively summer dormant, it is suited to drip-irrigated orchards. The cultivar Covar is frequently used; it is somewhat slow to establish but is competitive once established. Similar to hard fescue, sheep fescue is very drought tolerant.

Figure 20.18 Sheep fescue. *Photo:* C. A. Ingels.

Perennial Irrigated Species

White Clover
Trifolium repens L.

Fig. 20.19

5–12 lb/seeded acre

White clover is similar in habit and growth to strawberry clover but is less invasive and may attract fewer pocket gophers because of its smaller taproot. It tolerates a wide range of soil conditions and thrives best under cool, moist growing conditions; it is also shade tolerant once established. It performs better in heavy, moist soils than on sandy soils that may be droughty and contain fewer nutrients. White clover cultivars are arbitrarily classified by size of the plants: small, intermediate, and large. The small types often have "wild white" in their names. Intermediate types often include the term "New Zealand" or "common"; most unnamed U.S. cultivars are intermediate types. The large type was introduced from Italy into the United States in the early 1900s. Seed derived from this ecotype were designated Ladino until the early 1950s, when new cultivars were developed.

Figure 20.19 White clover. *Photo:* C. A. Ingels.

Strawberry Clover
Trifolium fragiferum L.

Fig. 20.20

10–15 lb/seeded acre

Strawberry clover is a long-lived perennial that roots at the nodes of stolons and grows year-round. It tolerates saline and alkaline soils, wet or submerged soils, infrequent irrigation, and frequent close mowing. When established, it often outcompetes some weed species and is useful for erosion control. However, it is very invasive, competes with prunes for water, is very resistant to most herbicides, and attracts pocket gophers. The cultivar Salina is well adapted to California conditions; it was developed in California from selections of Palestine, a productive Australian cultivar.

Chewings Fescue
Festuca rubra var. *commutata*

20–25 lb/seeded acre

Chewings fescue is one of the red fescue cultivars that have been developed and introduced for use in turf. It is a noncreeping bunchgrass that produces a firm sod. It is frequently included in fine-leaf fescue mixtures to provide cover until more-sod-forming cultivars such as creeping red fescue can become established.

Creeping Red Fescue
Festuca rubra

20–25 lb/seeded acre

Creeping fescue has also been introduced for use in turf. It spreads by short rhizomes. It has been used extensively throughout California orchards and vineyards for 30 years and provides an excellent sod floor for mowed and irrigated orchards. Creeping red fescue is often mixed with perennial ryegrass for rapid establishment.

Perennial Ryegrass
Lolium perenne

25–35 lb/seeded acre

Perennial ryegrass is a short-lived (3 to 4 years) perennial bunchgrass that is frequently used in lawns. Of the many cultivars, Elka, a short-statured cultivar, had been most frequently used in cover crop mixtures. It grows well in heavy soils but needs a large amount of extra water. Although it is similar to tall fescue in growth and management, it is less aggressive and not as well adapted to poor or submerged soils.

Ryegrass–Fescue Mixtures

Fig. 20.21

25–35 lb/seeded acre

The mixture of grasses most frequently planted in California is a 75-25% mixture of perennial ryegrass and creeping red fescue, respectively. Once this "lawn" is established, it provides a durable, long-term floor for orchard operations in prune production. The additional advantages include a cooler environment, improved water infiltration, and year-round orchard access. Because the grass mixture does use up to 30% more water and nitrogen, it must be managed properly for the prune crop.

Figure 20.20 Strawberry clover. *Photo:* C. A. Ingels.

Figure 20.21 Blend of perennial ryegrass, and creeping red fescue, partially mowed. *Photo:* C. A. Ingels.

References

Alins, G., S. Alegre, and J. Avilla. 2007. Improving sustainability in organic and low input food production systems. Proceedings, 3rd International Congress of the European Integrated Project Quality Low Input Food, University of Hohenheim, Germany, 20–23 March, 2007.

Daane, K. M., and M. J. Costello. 1998. Can cover crops reduce leafhopper abundance in vineyards? California Agriculture 52:5, 27–33.

Grant, J., K. K. Anderson, T. Prichard, J. Hasey, R. L. Bugg, F. Thomas, and T. Johnson. 2006. Cover crops for walnut orchards. Oakland: University of California Division of Agriculture and Natural Resources Publication 21627.

Ingels, C. A, R. L. Bugg, and F. L. Thomas. 1998. Cover crop species and descriptions. In C. A. Ingels et al. ed., Cover cropping in vineyards. Oakland: University of California Division of Agriculture and Natural Resources Publication 3338. 8–26.

21 Weed Management

• Kurt J. Hembree

Managing weed growth on the orchard floor is an important component of prune production. Weeds compete for water and nutrients, and in some cases, for light. If weeds are not controlled in newly planted orchards, tree growth can be reduced and normal production may be delayed a year or more. Once an orchard is established (about 3 years), competition from weeds may only minimally impact growth. However, if not properly managed in established orchards, weeds can reduce irrigation water and nutrient availability to the trees; increase the risk of crown rotting diseases; host insects that can feed directly on leaves and fruit or transmit diseases; harbor voles, squirrels, and other destructive rodents; increase the risk of frost damage by lowering the temperature of the orchard; and hinder harvest efficiency and other cultural operations.

Cultural Considerations

Prunes are grown under a variety of localized conditions, including topography, soil type and texture, climate, planting style (raised berm versus flat), irrigation method (basin flood, furrow, drip, sprinkler, or low-volume sprinklers or misters), and other cultural considerations. Where trees are planted on raised berms, herbicides are normally applied in a strip 4 to 6 feet wide within the tree rows, while the area between trees (middles) is repeatedly mowed or cultivated to destroy weeds. Sometimes the middles are treated four to six times per year with herbicides to control emerged weeds. When trees are planted on flat ground, complete tillage (in two directions) can be used to destroy weeds. However, this practice usually leaves weeds in an area 1 to 2 feet wide around each tree that must be sprayed or removed by hand. Care should be used to avoid mechanical damage to the base of the trees in order to reduce the incidence of canker or other diseases.

The type of irrigation used in prune orchards can influence how weeds are managed. For example, in orchards with drip, microsprinkler, mister, or sprinkler irrigation, some soil-residual (preemergent) herbicides degrade faster or leach more than in fields that are furrow or basin flood irrigated. Herbicides sprayed on raised berms generally last longer under furrow or basin flood irrigation, because water does not normally come into contact with the herbicide sprayed on the soil surface. Additionally, under low-volume frequent irrigation, weed growth can be significant near emitters, requiring repeated applications of foliar (postemergent) herbicides for control during the summer months.

Soil type and texture can also influence selection and performance of herbicides. The application rate of some preemergent herbicides, such as norflurazon, must be adjusted to match the particular soil type, or tree injury could result. Other preemergent herbicides, such as oxyfluorfen, are less prone to leaching and degradation under wet conditions, and rates may not have to be adjusted for the soil texture. Always read and follow all herbicide label recommendations for the particular soil and irrigation condition to maximize weed control and reduce the likelihood of crop injury.

Weed Identification

It is important to recognize the types of weeds present in the orchard before deciding on the type of management strategies to employ. Understanding the type of weed flora present helps maximize control and reduce unnecessary waste of prunes and other resources. Weeds commonly found in prune orchards are classified as annuals, biennials, or perennials. Annual

weeds reproduce by seed only, completing their life cycle in less than 1 year or season. These weeds are further separated into summer annual (emerge in spring to summer and grow and mature during summer to fall) and winter annuals (emerge, grow, and mature during winter to spring). Biennial weeds require more than 1 year or season to complete their life cycle, growing vegetatively the first season and reproductively the next. Perennial weeds live more than 1 year and often can reproduce by seed, above- and belowground roots or stems, bulbs, tubers, or other means. Perennial weeds are the most difficult to control once they become established, so they should be eradicated as soon as they are observed. Several good pictorial publications (including *Weeds of California and Other Western States* and *Weeds of the West*; see the References for more information) and online sources (including the UC IPM website, http://ucipm.ucdavis.edu) are available to help identify weeds. A chemical representative, pest control adviser, or UC Farm Advisor can provide assistance in their identification. Examples of important weeds found in prune orchards are given in table 21.1.

It is equally important to recognize the difference between broadleaf weeds, grassy weeds, and sedges. Some herbicides, such as glyphosate, are nonselective in activity and control a wide variety of weeds in each of these categories. Other herbicides, such as sethoxydim, are selective and control only grassy weeds.

Table 21.1 Annual and perennial weeds commonly found in prune orchards

Common name	Scientific name
Winter annuals	
annual bluegrass	*Poa annua*
annual ryegrass	*Lolium multiflorum*
burning nettle	*Urtica urens*
cheeseweed	*Malva parviflora*
chickweed	*Stellaria media*
common groundsel	*Senecio vulgaris*
filaree	*Erodium* spp.
shepherd's-purse	*Capsella bursa-pastoris*
Summer annuals	
barnyardgrass	*Echinochloa crus-galli*
cupgrass	*Eriochloa* spp.
hairy fleabane	*Conyza bonariensis*
horseweed	*Conyza canadensis*
lambsquarters	*Chenopodium album*
pigweeds	*Amaranthus* spp.
purslane	*Portulaca* spp.
puncturevine	*Tribulus terrestris*
sprangletop	*Leptochloa* spp.
Biennials	
wild carrot	*Daucus carota*
Hooker's evening primrose	*Oenothera hookeri*
common mullein	*Verbascum thapsus*
Perennials	
bermudagrass	*Cynodon dactylon*
dallisgrass	*Paspalum dilatatum*
field bindweed	*Convolvulus arvensis*
johnsongrass	*Sorghum halepense*
purple nutsedge	*Cyperus rotundus*
yellow nutsedge	*Cyperus esculentus*

Weed Scouting

Detecting new weeds and weeds that are escaping control is important in preventing new weed establishment and shifts in weed types and populations. Regularly scout orchards, particularly after each treatment, to help determine whether the current program is working or whether other options should be used. Look for small patches of perennial weeds that may appear after annual weeds are effectively controlled. Control perennial weeds at this point when they are small and easier to kill. Perennial weeds are generally less susceptible to herbicides than are annual weeds and can quickly replace annual weeds. Perennial weeds such as yellow nutsedge, field bindweed, and bermudagrass can quickly become established and are very difficult to control. Scouting fields for weed escapes is also a very useful tool for preventing herbicide resistance from occurring.

When scouting the orchard, pay particular attention to high and low spots (particularly where water may accumulate), the head and tail ends of the field, field borders or edges, fencerows, and irrigation ditches. Areas surrounding the field are good reservoirs for the introduction of new weed seeds. Maintain accurate records of when and what weed control tactics were used to help modify the approach as warranted. A map of the field showing where certain weed escapes have occurred can also be useful.

Weed Management before Planting

It is always best to have a weed management plan before the orchard is planted. This can help reduce the need for additional inputs later and can save money. A weed-free period after planting helps ensure that trees receive proper water and nutrients, encourages growth, and reduces the likelihood of tree injury due to rodents, mechanical damage, insects, or diseases. Take into consideration the types of weeds known to be present in the field before planting an orchard. Avoid planting into fields that have a history of perennial weeds such field bindweed, bermudagrass, and nutsedge.

Take advantage of a fallow period to control as many weeds as possible before trees are planted. Preplant weed management, along with other cultural considerations, will help get the orchard off to a good start. Methods to rid fields of many weeds before planting include repeated tillage of emerged weeds, deep plowing (e.g., using the Kverneland plow) to bury weed seeds deep in the soil profile, solarization with clear plastic to superheat the soil and kill germinating weed seeds, and repeated uses of postemergent herbicides (e.g., glyphosate) on emerged weeds.

Herbicides

Preemergent herbicides are available to help control weeds before trees are planted. Trifluralin and pendimethalin can be used in prune orchards to kill germinating seeds of important grass and broadleaf weeds that can hinder early tree growth, including barnyardgrass, junglerice, crabgrass, foxtail, johnsongrass (from seed), lambsquarters, pigweed, and purslane. These herbicides can be applied in strips where the tree rows will be or across the entire field. Within 24 hours of treatment, thoroughly incorporate the herbicide 1 to 2 inches deep with a harrow or offset disk, making a pass up and back. Pendimethalin should be used at 2 to 3.9 pounds of active ingredient per acre, and trifluralin should be used at 0.5 to 1 pound of active ingredient per acre. When planting trees, be careful not to let treated soil fall into the bottom of the hole where it can come into contact with the tree roots, or injury can occur. If possible, use clean, untreated soil to fill in around the tree roots.

Weed Management after Planting

It is important to maintain weed control after planting, especially for the first 2 to 3 years, because young trees are sensitive to the effects of weed competition. Many weeds (e.g., lambsquarters, horseweed, and summer grasses) can rob developing trees of valuable water and nutrients needed for growth. If weeds are not adequately controlled during this time, maturity may be delayed and production reduced. Several options can be used to manage weeds after planting, including both chemical and nonchemical tools.

Cultivation and Cover Crops

Cultivation can be an important tool for destroying weeds in prune orchards, especially in the middles. Disking three to five times per year is effective for killing young annual weeds but may not provide adequate control of perennials such as field bindweed. In some cases, repeated cultivation can spread weeds such as nutsedge (particularly purple nutsedge) throughout the field. In these cases, postemergent herbicides may need to be applied during the season for control.

Cultivation equipment is also available that can provide effective weed control within the tree row. For this tactic to work effectively, the field should be furrow or basin flood irrigated; low-volume irrigation tubing should be suspended above the soil surface or buried. In-row cultivators are equipped with tripping mechanisms that pivot the cutting arm around the trunk of the tree to avoid damaging the tree. Several companies manufacture specialized in-row cultivators, such as Kimco, Bezzerides, and L&H Manufacturing. When using these types of equipment, cultivate after the field has been irrigated and when weeds are less than 4 inches tall and easy to dislodge from the soil. Weeds that may escape control around the trees can then be spot-treated with postemergent herbicides.

Weeds can also be suppressed in the row middles by using cover crops. Cover crop species and management techniques vary. Select a cover crop that will not become competitive with young trees. Some clover species can encroach into the tree row and are difficult to control with herbicides, so select a cover crop or mix that can be easily managed. To preserve the cover crop, mow at a recommended height. For information regarding cover crop variety selection and management, refer to the UC SAREP Cover Crop Resource website, http://www.sarep.ucdavis.edu/ccrop.

Organic Production

Cultivation, propane flaming, weeder geese, and other nonchemical options can provide acceptable weed control in certified organic orchards. Organic-based herbicides are also available for controlling emerged weeds. The registered products have a variety of active ingredients, including clove oil, citric acid, limonene, and others. These products typically kill weeds by contact activity. Some California-registered products labeled for prunes include Matran EC, Green Match, Green Match EX, Weed Zap, and others. These products must be applied when weeds are very small (less than 2 to 3 inches tall). Repeated treatments will be needed during the season to maintain weed control, as there is typically no systemic or residual activity. Grasses and perennial weeds are not generally controlled with these products.

Herbicides

Several pre- and postemergent herbicides are available for weed control after prune trees are planted. Tables 21.2 to 21.5 list the currently registered herbicide products in California at the time this publication was written. These tables give the chemical names as well as the trade names and also list some of the weeds that are controlled and a brief description of their use. Products, rates, or use patterns listed in this publication are not intended to be a written recommendation for the use of herbicides, merely a suggestion for their use. Because product availability and formulations change, always read and follow recommendations in current herbicide labels before using any product. For information on current registrations, consult your local agricultural commissioner, pest control adviser, or UC Cooperative Extension Farm Advisor.

Some herbicides can be used only before the trees begin producing prunes (nonbearing), some can be used only after the trees have been established for a designated amount of time, and others can be used regardless of tree age. It is important to consider these constraints, along with the costs of the various products and the weed control spectrum, before selecting specific herbicides to use. It is equally important not to rely on a single herbicide program year after year, because weed population shifts (typically to weeds that are tougher to control) or herbicide resistance can occur. In most cases, using combinations of pre- and postemergent herbicides provides the most effective and affordable control and reduces the risk of weed population shifts or herbicide resistance.

Preemergent herbicides

Preemergent herbicides must be applied before weeds emerge. Their efficacy usually is improved when they are applied to a clean soil surface. Where leaves and other debris have accumulated on the soil, blowing off these areas before spraying exposes the soil and improves herbicide contact and performance. Herbicides such as flumioxazin and oryzalin adhere to debris on the soil surface and do not provide effective control. All preemergent herbicides require a means of incorporation to be activated. In most cases, rainfall or irrigation water is used for this purpose (table 21.6). If these herbicides sit on the soil too long before rainfall occurs, a reduction in efficacy is likely to occur due to volatilization or photodegradation. Products such as trifluralin work better if they are mechanically incorporated into the soil within 24 hours of treatment. Use properly calibrated spray equipment and nozzles to deliver the herbicide uniformly over the treated area. When spraying young orchards, direct nozzles to spray as low as possible to avoid contact with green bark or foliage, which can cause unacceptable crop injury. Additionally, using a spray volume of 15 to 20 gallons per acre is usually adequate to provide uniform coverage of the soil surface. In some cases, higher volumes may need to be used if soil debris or dried weeds are present during application. To help reduce the amount of herbicide needed and its cost, the treatment can be limited to a narrow strip (4 to 6 feet) within each tree row.

Several herbicides can be applied immediately after planting to provide residual control of weeds. Herbicide treatments should be applied in late winter (before mid-February) before a rainfall or split into sequential applications (first in the fall after harvest, then again in late winter). Preemergent products can be applied alone or in combination with other pre- or postemergent products (tank-mixed). One such tank mix might include oryzalin (3 to 4 lb a.i./acre) plus oxyfluorfen (1 to 2 lb a.i./acre). This combination controls a wide array of annual grasses and broadleaves. Pendimethalin can also be substituted for oryzalin to provide similar weed control at a lower cost. Other preemergent herbicides and tank-mix combinations are also effective, depending on the targeted weed spectrum. Examples include tank-mixing rimsulfuron plus pendimethalin, flumioxazin plus oryzalin, thiazopyr plus pendimethalin, and others. A postemergent herbicide such as glyphosate can be added to the tank to control weeds that are emerged at the time of spraying.

Postemergent herbicides

Several postemergent herbicides are available for use in prune orchards in California (see tables 21.2 to 21.5). Postemergent materials are intended to control small (usually less than 4 inches tall) actively growing weeds. Weeds that are large or stressed for moisture during application often escape control. For example, junglerice (a shallow-rooted summer annual grass) is normally easily controlled with glyphosate when conditions favor growth. However, when moisture is limited, control can be greatly reduced, even when high rates are used. For improved control, spraying after an irrigation can improve control with many postemergent herbicides.

Some postemergent herbicides, referred to as contact herbicides, act only where the spray contacts the weed foliage. Because these products

Table 21.2. Susceptibility of annual broadleaf weeds to herbicides in prune orchards in California

Annual broadleaves	Herbicides (see key below)																			
	1	2*	3	4	5	6	7	8*	9	10	11*	12*	13	14	15*	16	17†	18	19*	20†
cheeseweed	•	&	&	&	•	&	•	•	N	•	N	•	N	&	N	•	•	&	N	&
chickweed	•	•	&	•	&	•	•	•	•	&	N	•	N	•	•	&	•	•	N	N
cocklebur	—	&	•	N	&	N	&	—	N	&	N	•	N	&	&	•	•	•	N	•
cudweed	—	&	•	N	N	N	—	•	N	—	N	•	N	•	N	&	•	•	N	&
fiddleneck	—	•	&	•	•	•	•	•	•	•	N	•	N	•	N	•	•	&	N	&
filarees	•	•	&	&	•	N	•	•	&	&	N	•	N	&	N	•	&	&	N	•
goosefoot	•	•	&	•	•	•	&	•	•	•	N	•	N	•	N	N	•	•	N	•
groundcherry	•	•	•	N	•	N	•	&	&	•	N	•	N	•	&	•	•	•	N	•
groundsel	•	•	&	N	•	N	•	•	N	—	N	•	N	•	N	•	•	•	N	&
hairy fleabane	&	&	&	N	&	N	•	&	N	&	N	&	N	•	N	&	•	&	N	•
henbit	•	•	&	•	•	•	•	&	&	—	N	•	N	•	•	•	•	•	N	&
horseweed	&	•	&	N	&	N	•	&	N	&	N	&	N	•	N	&	•	&	N	•
knotweed	N	•	&	•	&	•	•	•	•	—	N	&	N	•	N	&	&	•	N	&
lambsquarters	•	•	&	•	•	•	•	•	•	•	N	•	N	•	N	•	&	•	N	•
London rocket	•	•	&	N	•	&	•	&	N	•	N	•	N	•	N	•	•	•	N	•
morning glories	•	•	&	&	&	N	&	—	&	•	N	&	N	&	N	&	&	•	N	&
mullein	—	•	&	N	&	N	—	•	&	&	N	&	N	&	N	&	&	&	N	&
mustards	•	•	&	N	•	&	•	&	N	&	N	•	N	•	N	&	•	•	N	&
nettles	•	•	•	&	•	N	•	•	N	•	N	&	N	N	N	&	•	•	N	&
nightshades	•	•	•	N	•	N	&	&	N	&	N	•	N	•	N	•	•	•	N	•
pigweeds	•	•	&	•	•	•	•	•	•	•	N	•	N	•	N	•	•	•	N	&
prickly lettuce	&	•	&	N	•	N	&	•	N	&	N	•	N	•	N	•	&	•	N	•
primrose	N	&	N	&	&	&	—	•	&	&	N	•	N	•	N	&	•	&	N	—
puncturevine	•	•	•	•	&	&	•	&	&	—	N	•	N	•	P	&	•	•	N	•
purslanes	•	•	•	•	•	•	•	•	•	&	N	•	N	•	N	•	•	•	N	&
Russian thistle	•	&	•	&	&	&	—	&	&	•	N	&	N	•	N	&	&	•	N	•
shepherd's-purse	•	•	&	N	•	&	&	•	N	&	N	•	N	•	N	&	•	•	N	•
sowthistles	&	•	&	N	•	N	&	•	N	&	N	•	N	•	N	•	•	•	N	•
spotted spurge	•	•	&	&	&	&	•	&	&	—	N	•	N	•	N	N	•	•	N	•
wild radish	•	•	&	N	•	N	•	•	N	&	N	•	N	•	N	&	•	•	N	•
willowherb	&	&	&	&	•	—	•	—	—	—	N	&	N	&	—	N	N	&	N	•

Key:
• = Control
& = Partial control
N = No control
— = No information
1. flumioxazin (Chateau SW)
2. isoxaben (Gallery T&V)*
3. norflurazon (Solicam)
4. oryzalin (Surflan, Oryzalin, etc.)
5. oxyfluorfen (preemergent) (Goal 2XL, GoalTender, etc.)
6. pendimethalin (Prowl H_2O)
7. rimsulfuron (Matrix FNV)
8. thiazopyr (Visor)*
9. trifluralin (Treflan, etc.)
10. carfentrazone (Shark)
11. clethodim (SelectMax)*
12. diquat dibromide (Reglone)*
13. fluazifop-p-butyl (Fusilade DX)
14. glyphosate (Roundup, Touchdown, etc.)
15. msma (MSMA 6 Plus, etc.)*
16. oxyfluorfen (postemergent)
17. paraquat (Gramoxone Inteon, etc.)†
18. pyraflufen (Venue)
19. sethoxydim (Poast)*
20. 2,4-D amine (Dri-Clean, etc.)†

Notes:
This table does not show a complete list of herbicides and trade names available. Refer to your local chemical dealer for other herbicides available. This is not a written recommendation for the use of herbicides. Refer to the appropriate pesticide label for recommended uses. Proper weed identification, timing, and accurate application are imperative for effective control. Weed control ratings in this table are based on label recommendations and observations made in California under varied conditions. Read and follow all label recommendations before using any pesticide.
*Herbicide for use only in nonbearing orchards.
†Permit required from county agricultural commissioner for purchase or use.

typically have little to no mobility within the plant, thorough coverage of the weed foliage is critical for control. Paraquat and carfentrazone are contact herbicides. Other herbicides move, or translocate, within the plant for systemic activity. These kinds of herbicides generally take longer to kill the weeds than do contact herbicides. Examples of translocated herbicides include 2,4-D amine, glyphosate, and sethoxydim. Postemergent herbicides can be further divided based on their selectivity. For example, glyphosate is considered nonselective because it can control a wide variety of annual and perennial grasses, broadleaves, and sedges. Clethodim, on the other hand, is a selective grass herbicide and kills only certain grass species without affecting broadleaf weeds or sedges.

Spray volume, nozzle selection, and spray coverage are important factors to consider when applying postemergent herbicides. For contact herbicides to adequately kill weeds, they must be applied uniformly over the entire plant. In general, applying 30 to 40 gallons per acre of spray gives adequate coverage of the weed foliage for most postemergent products. While flat fan or off-center type spray nozzles are adequate for preemergent herbicides, these nozzles do not necessarily distribute postemergent contact herbicides properly. In most cases, twinjet nozzles provide more uniform coverage of weed foliage. Nozzles should be replaced when they no longer perform within 10% of the manufacturer's recommended flow rate discharge for nozzle size and operating pressure. It is always a good idea to check and replace worn nozzles at least once per year for improved efficacy.

When using postemergent herbicides, be extremely careful not to spray green tissue, because trees can be seriously injured. Spraying under minimally windy conditions (3 to 6 mph) with hooded sprayers can significantly reduce the risk of crop injury. Low-volume herbicide shielded applicators (e.g., the Enviromist) mounted on an ATV can provide very good weed control with herbicides like glyphosate at a spray volume of 3 to 5 gallons per acre, saving product and cost. When applying herbicides through low-volume application equipment, always use the appropriate rate as shown on the herbicide label. Reducing the herbicide rate to compensate for a lower spray volume decreases the herbicide concentration delivered to the weeds and will likely reduce effectiveness. Caution should be taken with these types of sprayers to avoid drift, because droplets sprayed through these systems are much smaller than with conventional sprayers and

Table 21.3. Susceptibility of annual grass and perennial weeds to herbicides in prune orchards in California

	Herbicides (see key below)																			
	1	2*	3	4	5	6	7	8*	9	10	11*	12*	13	14	15*	16	17†	18	19*	20
Annual grasses																				
annual bluegrass	•	N	•	•	ꝏ	•	—	•	•	N	•	ꝏ	N	•	N	ꝏ	N	ꝏ	N	N
barnyardgrass	•	N	ꝏ	•	ꝏ	•	•	•	•	N	•	ꝏ	•	•	ꝏ	N	N	ꝏ	•	N
canarygrass	ꝏ	N	•	•	ꝏ	•	—	•	•	N	•	ꝏ	•	•	N	ꝏ	N	ꝏ	•	N
downey brome	ꝏ	N	•	•	ꝏ	•	•	•	•	N	ꝏ	ꝏ	ꝏ	•	ꝏ	N	N	ꝏ	ꝏ	N
fescues	ꝏ	N	•	•	N	•	•	ꝏ	•	N	ꝏ	ꝏ	ꝏ	•	—	N	N	ꝏ	ꝏ	N
foxtails	•	N	ꝏ	•	N	•	•	ꝏ	•	N	•	ꝏ	•	•	—	N	N	•	•	N
junglerice	•	N	ꝏ	•	ꝏ	•	•	•	•	N	•	ꝏ	•	•	ꝏ	N	N	ꝏ	•	N
Italian ryegrass	ꝏ	N	•	•	•	•	ꝏ	•	•	N	•	ꝏ	•	•	ꝏ	N	N	ꝏ	•	N
large crabgrass	•	N	ꝏ	•	N	•	•	•	•	N	•	ꝏ	•	•	•	ꝏ	N	•	•	N
lovegrass	•	N	ꝏ	•	•	•	—	ꝏ	•	N	•	ꝏ	•	•	ꝏ	N	N	ꝏ	•	N
ripgut brome	•	N	•	•	—	•	•	•	•	N	ꝏ	ꝏ	ꝏ	•	ꝏ	N	N	ꝏ	ꝏ	N
sandbur	•	N	•	•	N	•	—	•	•	N	•	ꝏ	•	•	ꝏ	N	N	ꝏ	•	N
sprangletop	ꝏ	N	ꝏ	ꝏ	N	ꝏ	—	•	•	N	•	ꝏ	•	•	N	ꝏ	N	ꝏ	ꝏ	N
wild barley	ꝏ	N	•	•	ꝏ	•	ꝏ	•	•	N	•	ꝏ	•	•	N	N	N	ꝏ	•	N
wild oats	•	N	•	ꝏ	ꝏ	ꝏ	ꝏ	ꝏ	ꝏ	N	ꝏ	ꝏ	•	•	N	N	N	ꝏ	•	N
witchgrass	ꝏ	N	ꝏ	•	ꝏ	•	—	ꝏ	•	N	•	ꝏ	ꝏ	•	ꝏ	N	N	N	•	N
Perennials (seedlings)																				
bermudagrass	N	N	•	•	N	•	N	N	N	N	•	N	•	•	N	N	N	ꝏ	•	N
buckhorn plantain	—	ꝏ	—	—	N	—	N	ꝏ	N	N	N	ꝏ	N	•	N	N	N	—	N	•
dallisgrass	—	N	N	•	N	N	N	•	•	N	•	N	•	•	•	N	N	N	•	N
johnsongrass	•	N	•	•	N	•	ꝏ	•	•	N	•	N	•	•	•	N	N	•	•	N
field bindweed	—	N	ꝏ	ꝏ	N	ꝏ	ꝏ	ꝏ	ꝏ	ꝏ	N	ꝏ	N	•	ꝏ	N	N	N	N	ꝏ
Perennials (established)																				
bermudagrass	N	N	ꝏ	N	N	N	N	N	N	N	ꝏ	N	ꝏ	ꝏ	N	N	N	N	ꝏ	N
buckhorn plantain	N	ꝏ	N	N	N	N	N	N	N	N	N	N	N	ꝏ	N	N	N	—	N	ꝏ
dallisgrass	N	N	N	N	N	N	N	N	N	N	ꝏ	N	ꝏ	ꝏ	•	N	N	N	ꝏ	N
johnsongrass	N	N	ꝏ	N	N	•	N	N	N	N	ꝏ	N	ꝏ	ꝏ	N	N	N	N	ꝏ	N
field bindweed	N	N	N	N	N	ꝏ	N	N	ꝏ	ꝏ	N	ꝏ	N	ꝏ	ꝏ	N	N	N	N	ꝏ
purple nutsedge	N	N	ꝏ	N	N	N	ꝏ	ꝏ	N	N	N	ꝏ	N	ꝏ	•	N	N	N	N	N
yellow nutsedge	N	N	ꝏ	N	N	N	ꝏ	ꝏ	N	N	N	ꝏ	N	ꝏ	•	N	N	ꝏ	N	N

Key:
• = Control
ꝏ = Partial control
N = No control
— = No information
1. flumioxazin (Chateau SW)
2. isoxaben (Gallery T&V)*
3. norflurazon (Solicam)
4. oryzalin (Surflan, Oryzalin, etc.)
5. oxyfluorfen (preemergent) (Goal 2XL, GoalTender, etc.)
6. pendimethalin (Prowl $H_{2}O$)
7. rimsulfuron (Matrix FNV)
8. thiazopyr (Visor)*
9. trifluralin (Treflan, etc.)
10. carfentrazone (Shark)
11. clethodim (SelectMax)*
12. diquat dibromide (Reglone)*
13. fluazifop-p-butyl (Fusilade DX)
14. glyphosate (Roundup, Touchdown, etc.)
15. msma (MSMA 6 Plus, etc.)*
16. oxyfluorfen (postemergent)
17. paraquat (Gramoxone Inteon, etc.)†
18. pyraflufen (Venue)
19. sethoxydim (Poast)*
20. 2,4-D amine (Dri-Clean, etc.)†

Notes:
*Herbicide for use only in nonbearing orchards.
†Permit required from county agricultural commissioner for purchase or us
This table does not show a complete list of herbicides and trade names available. Refer to your local chemical dealer for other herbicides available This is not a written recommendation for the use of herbicides. Refer to th appropriate pesticide label for recommended uses. Proper weed identification, timing, and accurate application are imperative for effective control. Weed control ratings in this table are based on label recommendations anc observations made in California under varied conditions. Read and follow a label recommendations before using any pesticide.

Table 21.4. Preemergent herbicide use rates, timings, and other label uses in prune orchards in California

Preemergent herbicide	Rate/ac/ application	Rate/ac/ season	Postharvest interval	Application information
flumioxazin (Chateau SW)	6–12 oz	24 oz	60 days	Bearing/nonbearing; ¼–½" rain within 28 days; spray volume 20–40 gpa; can tank-mix; leaves and dried weeds are removed before treatment; don't disturb soil after treatment.
isoxaben (Gallery T&V)	0.66–1.33 lb	4 lb	365 days	Nonbearing only; ½" rain in 21 days; spray volume 20–40 gpa; can tank-mix.
norflurazon (Solicam)	1.25–5.00 lb	5 lb	60 days	Bearing/nonbearing at > 2 years old; adjust rate to soil texture; ½" rain within 28 days; spray volume 20–40 gpa; can tank-mix; permit needed if used in a GWPA.
oryzalin (Surflan A.S. , Oryzalin, etc.)	2–6 qt (liquid) 2.4–7.1 lb (dry)	12 qt (liquid) 14 lb (dry)	0 days	Bearing/nonbearing; ½–1" rain within 21 days; spray volume 20–40 gpa; chemigation OK; can tank-mix.
oxyfluorfen (Goal, Galigan, etc.)	5–6 pt (2 lb a.i./gal) 2.5–3 pt (4 lb a.i./gal)	6 pt (2 lb a.i./gal) 3 pt (4 lb a.i./gal)	7 days	Bearing/nonbearing; ¼" rain within 28 days; spray volume 20–40 gpa; chemigation OK; can tank-mix; don't disturb the soil after treatment.
pendimethalin (Prowl H_2O)	2–6.3 qt (H_2O) 2.4–4.8 qt (3.3 EC)	6.3 qt (H_2O) 4.8 qt (3.3 EC)	60 days 365 days	Bearing/nonbearing; ½" rain within 21 days; spray volume 20–40 gpa; chemigation OK; can tank-mix; (3.3 EC)–nonbearing only; preplant/postplant.
rimsulfuron (Matrix FNV)	2–4 oz	4 oz	14 days	Bearing/nonbearing >1 year old; ½" rain within 21 days; spray volume 20–40 gpa; spray pH > 5; chemigation OK; can tank-mix.
thiazopyr (Visor 2E)	2–4 pt	4 pt	365 days	Nonbearing only; ½" rain in 28 days; spray volume 20–40 gpa; two sequential applications of 2 pt/ac for nutsedge control; can tank-mix; permit needed if used in a GWPA.
trifluralin (Treflan 4E, etc.)	1–2 pt (preplant) 2–4 pt (postplant)	6 pt	60 days	Preplant or established planting; adjust rate to soil texture; incorporate mechanically immediately; spray volume 20–40 gpa.

Note: This table does not show a complete list of herbicides and trade names available. Refer to your local chemical dealer for other herbicides available. This is not a written recommendation for the use of herbicides. Refer to the appropriate pesticide label for recommended uses. Proper weed identification, timing, and accurate application are imperative for effective control. Weed control ratings in this table are based on label recommendations and observations made in California under varied conditions. Read and follow all label recommendations before using any pesticide.

Table 21.5. Postemergent herbicide use rates, timings, and other label uses in prune orchards in California

Postemergent herbicide	Rate/ac/ application	Rate/ac/ season	Postharvest interval	Application information
carfentrazone (Shark)	2 fl oz	7.9 fl oz	3 days	Bearing/nonbearing; weeds < 4" tall; COC, NIS, or MSO adjuvant is used; spray volume at least 30 gpa; can tank-mix; spray pH at 5–8.
clethodim (Prism)	9–16 fl oz	64 fl oz	365 days	Nonbearing only; grasses before tillering; repeat treatment on perennial grasses; NIS adjuvant is used; spray volume at least 30 gpa.
diquat dibromide (Reglone)	1.5–2 pt	2 pt	365 days	Nonbearing only; COC or NIS adjuvant is used; spray volume at least 30 gpa; weeds < 4" tall.
fluazifop (Fusilade DX)	16–24 fl oz	72 fl oz	14 days	Bearing/nonbearing; grasses before tillering; repeat treatment on perennial grasses; COC or NIS adjuvant is used; spray volume at least 30 gpa.
glyphosate (Roundup, Glyphomax, etc.)	0.5–1.5 lb a.i.	10.5 lb a.i.	3 days	Bearing/nonbearing; weeds < 6" tall; AMS and/or buffering agents to bring pH to 5–6; spray volume 5–40 gpa; can tank-mix.
msma (MSMA)	2.5 pt	7.5 pt	365 days	Nonbearing > 1 year old; nutsedge with 4" leaves and retreatment necessary; NIS adjuvant used; spray volume 50 or 100 gpa.
oxyfluorfen (Goal, Galigan, etc.)	1–6 pt (2 lb a.i./gal) 0.5–3 pt (4 lb a.i./gal)	6 pt (2 lb a.i./gal) 3 pt (4 lb a.i./gal)	7 days	Bearing/nonbearing; broadleaf weeds < 4" tall; NIS adjuvant is used; at least 30 gpa; can tank-mix.
paraquat (Gramoxone Inteon, etc.)	2.5–4 pt	20 pt	7 days	Bearing/nonbearing; weeds < 6" tall; Restricted Use Permit required; NIS or COC adjuvant is used; spray volume at least 30 gpa; can tank-mix.
pyraflufen (Venue)	0.7–4 fl oz	6.8 fl oz	Prebloom	Bearing/nonbearing; broadleaf weeds < 4" tall; COC or NIS adjuvant is used; spray volume at least 30 gpa; can tank-mix; spray pH 5–7.5.
sethoxydim (Poast)	2.5 pt	10 pt	15 days	Bearing/nonbearing; grass weeds before tillering; repeat treatment on perennial grasses; Dash, MSO, or COC adjuvant is used; spray volume at least 30 gpa.
2,4-D amine (Dri-Clean, etc.)	27 fl oz	54 fl oz	40 days	Bearing/nonbearing at least 12 months old; broadleaf weeds < 6" tall; Restricted Use Permit required; spray volume at least 20 gpa; can tank-mix.

Note: This table does not show a complete list of herbicides and trade names available. Refer to your local chemical dealer for other herbicides available. This is not a written recommendation for the use of herbicides. Refer to the appropriate pesticide label for recommended uses. Proper weed identification, timing, and accurate application are imperative for effective control. Weed control ratings in this table are based on label recommendations and observations made in California under varied conditions. Read and follow all label recommendations before using any pesticide.

Table 21.6. Incorporation requirements for activation of preemergent herbicides

Herbicide	Maximum time to activation	Incorporation needed
flumioxazin (Chateau SW)	21 days	> ¼" rainfall or irrigation water
isoxaben (Gallery T&V)	21 days	> ½" rainfall or irrigation water
norflurazon (Solicam)	28 days	> ½" rainfall or irrigation water
oryzalin (Surflan, etc.)	21 days	½–1" rainfall or irrigation water
oxyfluorfen (Goal, etc.)	21–28 days	> ¼" rainfall or irrigation water
pendimethalin (Prowl H_2O)	21 days	> ½" rainfall or irrigation water
pronamide (Kerb)	24 hours	> ½" rainfall or irrigation water
rimsulfuron (Matrix FNV)	14 days	½" rainfall or irrigation water
thiazopyr (Visor)	28 days	> ½" rainfall or irrigation water
trifluralin (Treflan, etc.)	24 hours	1½–2" deep, mechanically

Note: Times listed reflect the maximum time herbicides can remain on the soil before efficacy is significantly reduced. Incorporation/activation occurring before times indicated can result in improved control.

drift farther. Drift-reducing nozzles (e.g., turbo twinjets or air induction) can be useful for minimizing spray drift while still achieving efficacy. Specific nozzle type should always be considered when switching between preemergent and postemergent (contact versus systemic) products.

Using properly cleaned sprayers before and after applications is also important to prevent crop injury. Herbicides like 2,4-D amine and flumioxazin can adhere to rubber hoses and polyethylene tanks and can cause crop injury in subsequently treated sites unless the equipment is properly cleaned according to label directions.

Combinations of postemergent herbicides can be used to control a broader spectrum of weeds than either used alone. Some common herbicide combinations used in prune orchards include glyphosate (1 to 2 lb a.i./acre) plus 2,4-D amine (1 to 1.4 lb a.i./acre) during the dormant period; glyphosate (1 to 2 lb a.i./acre) plus a low rate of oxyfluorfen (0.0625 to 0.125 lb a.i./acre) in bearing and nonbearing fields; and glyphosate plus flumioxazin (0.0625 to 0.125 lb a.i./acre) in nonbearing orchards. Spray additives, such as ammonium sulfate, spreaders and stickers, citric acid, and other materials can also improve the efficacy of postemergent herbicides. Follow all label recommendations when tank-mixing herbicide products or adding spray additives to the tank. Regardless of the herbicide, combinations of herbicides used, or whether additives are used, be sure to time the application to when the weeds are small and growing vigorously in order to improve control.

Traditionally, most prune growers apply broadcast or strip sprays to kill emerged weeds. However, equipment is available that can help reduce herbicide inputs based on the amount of weed cover during application. The smart sprayer, for example, uses a system of solenoids and nozzles to deliver the spray (usually a systemic product such as glyphosate) to the orchard floor only when it "sees" green weed foliage. This type of equipment works well after preemergent herbicide applications or where the population of emerged weeds is sparse or appears patchy throughout the field. With this system, postemergent herbicide use can be reduced by as much as 85% compared with standard broadcast in-row applications.

A Word of Caution When Using Herbicides

Read and follow all label instructions carefully before using any herbicide. The information presented in this chapter is designed to supplement herbicide label recommendations. Information concerning limitations on use is also available from regulatory officials, through agricultural chemical trade channels, and from University of California Cooperative Extension offices. Observe all precautions and restrictions on the herbicide label regarding use of protective clothing and equipment, handling and storage, and protection of pets, domestic animals, wildlife, fish, and desired vegetation. A permit is required from the county agricultural commissioner for purchase and use of restricted herbicides such as paraquat and 2,4-D amine. Additionally, a permit is required from the county agricultural commissioner for use of certain preemergent herbicides (e.g., norflurazon) in regulated groundwater protection areas (GWPA).

Since preemergent herbicides can persist in the soil for a relatively long time after application, discontinue their use at least 1 year before orchard removal. Plan ahead and cultivate or use postemergent herbicides for weed control before orchard removal. Follow label recommendations regarding plant-back restrictions after any herbicide treatment. If it is necessary to replant trees in the field and preemergent herbicides have already been applied, use clean, untreated soil to backfill around the tree's roots. To avoid applying an excessive amount of herbicide, shut the sprayer off when stopping or turning at the ends of rows; apply at a constant speed, since

Table 21.7. Herbicide-resistant weeds in California

Weed common name	Scientific name	Situation	Herbicide mode of action	Year of resistance
common groundsel	*Senecio vulgaris*	asparagus	PS II inhibitors	1981
perennial ryegrass	*Lolium perenne*	railways, roadsides	ALS inhibitors	1989
smallflower umbrella sedge	*Cyperus difformis*	rice	ALS inhibitors	1993
California arrowhead	*Sagitaria montevidensis*	rice	ALS inhibitors	1993
Russian thistle	*Salsola iberica*	roadsides	ALS inhibitors	1994
wild oat	*Avena fatua*	barley, wheat	pyrazoliums	1996
redstem	*Ammania auriculata*	rice	ALS inhibitors	1997
ricefield bulrush	*Scirpus mucronatus*	rice	ALS inhibitors	1997
late watergrass	*Echinochloa phyllopogon*	rice	ACCase inhibitors	1998
late watergrass	*Echinochloa phyllopogon*	rice	thiocarbamates	1998
rigid ryegrass	*Lolium rigidum*	almonds	glycines	1998
long-leaved loosestrife	*Ammania coccinea*	rice	ALS inhibitors	2000
barnyardgrass	*Echinochloa crus-galli*	rice	thiocarbamates	2000
barnyardgrass	*Echinochloa crus-galli*	rice	ACCase inhibitors	2000
early watergrass	*Echinochloa oryzicola*	rice	ACCase inhibitors	2000
early watergrass	*Echinochloa oryzicola*	rice	thiocarbamates	2000
littleseed canarygrass	*Phalaris minor*	onions	synthetic auxins	2001
smooth crabgrass	*Digitaria ischaemum*	rice	synthetic auxins	2002
horseweed	*Conyza canadensis*	roadsides, orchards, and vineyards	glycines	2005
hairy fleabane	*Conyza bonariensis*	roadsides, orchards, and vineyards	glycines	2007

slowing down increases the amount of herbicide applied.

Always adjust the amount of water and herbicide needed depending on whether the application is broadcast, strip, or spot. The herbicides and rates discussed in this chapter are within current label recommendations at the time this publication was written. They also represent those generally found to be safe and effective by University of California weed scientists. Rates may need to be adjusted for specific regions, soil types, and conditions.

Herbicide-Resistant Weeds and Management

The development of herbicide-resistant weeds is not a new phenomenon. It is a naturally occurring evolutionary process in response to a selection pressure exerted by herbicides. Currently, 328 different weed biotypes have developed herbicide resistance globally, of which 20 have been identified in California (see table 21.7). Glyphosate-resistant horseweed and hairy fleabane have been difficult to control in crop and noncrop situations, including prune orchards. Other important weeds that have developed herbicide resistance and are common to prune orchards in California include lambsquarters, redroot pigweed, and prickly lettuce.

Herbicide resistance has been defined by the Weed Science Society of America (WSSA) as a plant's inherited ability to survive and reproduce after being exposed to a dose of herbicide that normally would be lethal to the wild type of the plant. Herbicide resistance should not be confused with herbicide tolerance, which according to WSSA, is a plant's inherent ability to survive the effects of a given herbicide. Resistance may be naturally occurring or induced by such techniques as genetic engineering. It is also important to note that herbicides do not cause mutations that lead to resistance. Rather, they select for biotypes of the plant (weed) that already has genes that confers insensitivity to the herbicide.

Several factors lead to or accelerate the development of herbicide resistance. These factors can be classified into three groups: chemical properties, weed characteristics, and cultural practices. The chemical properties of some herbicides can lead to more rapid development of resistance among weeds than the properties of other herbicides. Herbicides with a single site of action are more likely to lead to resistance than those with multiple sites of action; herbicides that are detoxified by a common metabolic pathway in weeds are more likely to lead to resistance than those detoxified by an uncommon metabolic pathway.

Characteristics in weed growth also contribute to resistance. For example, the top 10 most common weeds in the world that have developed herbicide resistance have an annual growth habit, are widely distributed, and proliferate in a

variety of climatic and environmental conditions. This suggests that genetically diverse species are most likely to contain a gene that confers resistance to a particular herbicide. For example, glyphosate-resistant horseweed is an annual species, produces at least 250,000 seeds per plant, is disseminated by wind, and grows under a variety of habitats, including wet or dry soils, deep or shallow soils, gravely roadways, and shade or full sunlight.

Cultural practices that increase the selection pressure on a weed population increase the likelihood of resistance. For example, reliance solely on herbicides for weed control increases the risk of resistance. Repeated use of the same herbicide or a herbicide with the same mode of action greatly increases the risk of developing weed resistance. Furthermore, the likelihood of developing resistance is increased by using increasingly higher rates of the same herbicide over time.

Some weeds found in prune orchards in California have been previously shown to have herbicide resistance. Horseweed and hairy fleabane have developed resistance to glyphosate, a commonly used herbicide in prunes. Once a weed develops resistance to glyphosate, it is resistant to all products containing the same active ingredient.

It is important to recognize resistant weeds early so that strategies can be implemented to control them. In most cases, growers observe about 30% failure in the control of a particular species before suspecting resistance. Herbicide-resistant weeds are not always easy to identify. They occur naturally within a given population but cannot be visually distinguished from susceptible weeds until the population increases over time. Resistant weeds generally begin to appear in clumps or patches, unless their seed is windborne, which makes this appearance unlikely. Thus, scouting for the initial evolution of resistance patterns of weed escapes is a critical aspect of management and prevention.

Herbicide resistance is rarely signaled by the complete failure of a single herbicide application. Even isolated escapes are seldom associated with resistant weeds, although it is certainly possible. Weed escapes appearing in single-line patterns in a field are usually associated with a plugged spray nozzle or some other application error. Other causes for weeds to escape herbicide treatments include misidentification of the weeds present, inappropriate herbicide rates, weed emergence after application, degradation of preemergent herbicides, and application at an insensitive stage of weed development.

The weed may be herbicide resistant if all of the following conditions are met:

- most of the weeds in a field have been controlled but a few weeds have escaped control
- the application consisted of the label rate of a herbicide registered for control of the weed
- the herbicide was applied at the recommended timing
- the application equipment was calibrated and working properly
- all other label recommendations were followed

As soon as weed resistance is suspected, alternative management efforts must be employed in order to achieve effective weed control and preserve the herbicide's effectiveness. Preventing weed resistance from occurring is best achieved with an integrated weed management approach. Killing the weeds before they produce seeds has the best long-term effect in reducing the incidence of resistance. To prevent and reduce the level of resistance, do not use the same herbicide (or herbicides with a similar mode of action) year after year. Rather, rotate or tank-mix herbicides with different modes of actions. Mechanical cultivation is also an excellent means of controlling these weeds. Finally, rotating chemical and mechani-

Figure 21.1 Hairy fleabane. *Photo:* J. M. DiTomaso.

cal options whenever possible reduces the risk of weed resistance. Contact the local UC Cooperative Extension Farm Advisor or chemical representative if weed resistance may be occurring, so an alternative approach to weed control can be developed.

Damaging Weeds in Prune Orchards

Hairy Fleabane
Conyza bonariensis

Fig. 21.1

Hairy fleabane is a summer annual that emerges in February, but it can also emerge from September through the winter if the temperature is warmer than the seasonal normal. The bushlike plants have woody stems and sparse narrow or crinkled leaves, and usually grow 2 to 3 feet tall. Each plant can produce more than 10,000 seeds that spread by wind, readily infesting orchards, particularly those growing along frequently traveled roadsides.

Shallow cultivation is effective when weeds are shorter than 4 inches. Mowing is not effective and does not prevent seed production. Lightly disturbing the soil in the late fall can discourage seeds from sprouting. Of the preemergent herbicides, rimsulfuron gives the most consistent control, while flumioxazin also provides control at the higher label rates. Preemergent herbicides such as isoxaben and thiazopyr can be used only in nonbearing orchards, but they can provide effective control. Paraquat, glyphosate, and 2,4-D amine give effective control if hairy fleabane has fewer than 8 to 10 leaves; control is reduced as plants grow older. Combinations of glyphosate (2 lb a.i./acre) plus 2,4-D amine (1 to 1.4 lb a.i./acre) work well during the winter dormant season. Paraquat also gives good control of small plants when used at the highest label rate, and also if a crop oil concentrate or nonionic surfactant is added and a spray volume of at least 30 gallons per acre is used. Low rates of flumioxazin (0.0265 to 0.125 lb a.i./acre) or oxyfluorfen (0.0625 to 0.125 lb a.i./acre), mixed with glyphosate (2 lb a.i./acre), can also provide control of young plants. The addition of spray-grade ammonium sulfate (5 to 17 lb/100 gal of spray solution) often increases the activity of postemergent herbicides, but it should be added to the spray tank before herbicides and other additives are added. Turbo twinjet nozzles provide better spray coverage on these weeds than do flat fan nozzles and usually improve control when contact herbicides are used as part of the tank mix. They also help minimize spray drift.

Figure 21.2 Horseweed. *Photo:* J. M. DiTomaso.

Because repeated use of glyphosate alone over several years has resulted in glyphosate-resistant hairy fleabane biotypes, this type of spray program should be avoided. If any weed may be herbicide resistant, contact the local pest control adviser, chemical representative, or UC Cooperative Extension Farm Advisor immediately to determine whether resistance has occurred, so a plan can be implemented to control the weed and prevent its spread to other areas.

Horseweed
Conyza canadensis

Fig. 21.2

Horseweed is a summer annual weed closely related to hairy fleabane. It has wide leaves alternating around a single woody stem and can grow to 10 feet tall, even in mature orchards. It produces more than 200,000 seeds per plant that are readily dispersed in the air well over a mile or more. Horseweed plants growing in a prune orchard can reduce harvest efficiency and host several important insects, including the glassy-winged sharpshooter. The growth habit, control measures, and susceptibility to herbicide resistance are similar to that of hairy fleabane. While horseweed grows upright with a single stalk, if it is mowed or treated late with postemergent herbicides, it usually produces lateral branching, appearing similar to hairy fleabane. Once this type of regrowth occurs after a herbicide or mowing treatment, it is nearly impossible to control without using hand weeding or cultivation.

Figure 21.3 Common purslane. *Photo:* J. M. DiTomaso.

Common Purslane
Portulaca oleracea

Fig. 21.3

Common purslane is a summer annual that reproduces from seed. It begins emerging in April to early May and can continue to emerge throughout summer. It is found mainly in the sunny parts of orchards, growing mostly prostrate in large clumps, and has very fleshy stems and leaves. Root development can initiate along the aboveground stems and can continue even after cultivation, especially when moisture is present.

While shallow cultivation of young weeds (less than 4 inches wide) can sometimes be an effective tool in drier soils, herbicides offer the best means of control. In nonbearing orchards, isoxaben (0.66 lb/acre), pendimethalin (1.9 to 3.8 lb a.i./acre), thiazopyr (0.5 lb a.i./acre), or flumioxazin (0.375 lb a.i./acre) in January can give good preemergent control. In bearing or nonbearing orchards, treating emerged purslane in April with oryzalin (2 lb a.i./acre) and glyphosate gives postemergent control plus extended residual activity and can eliminate the need for a preharvest application. Preharvest postemergent applications of glyphosate (2 lb a.i./acre) or paraquat (0.75 lb a.i./acre) provide effective control. Adding citric acid to reduce the water pH to between 5 and 6 before adding glyphosate often improves control. The addition of spray-grade ammonium sulfate at 5 to 17 pounds per 100 gallons of spray solution often increases the activity of postemergent herbicides.

Puncturevine
Tribulus terrestris

Fig. 21.4

Puncturevine, or goathead, is a summer annual weed that reproduces from seed that are borne within spiny seedpods and spread by sticking to clothing, shoes, and tires of equipment. It emerges from the soil in April or May and forms prostrate clumps that can be 5 feet or more in diameter. Woody stems make control with postemergent herbicides difficult as plants mature.

Cultivation provides effective control when plants are less than 6 inches in diameter. In nonbearing orchards, pendimethalin applied at a high label rate (3.8 lb a.i./acre) can effectively suppress growth from seed. In bearing and nonbearing orchards, a tank mix of oxyfluorfen (1.5 to 2 lb a.i./acre) plus oryzalin (3 lb a.i./acre) can give excellent season-long control. When using postemergent herbicides for control, make the application when the weed is about 4 to 6 inches in diameter; the foliage should be thoroughly wet, particularly when using paraquat. Glyphosate and 2,4-D amine also provide good control when weeds are small.

A stem-boring weevil *(Microlarinus lypriformis)* and a seed-boring weevil *(Microlarinus lareynii)* were released into California in 1961. These weevils, which have a narrow host range, have populated the state and can provide some biological control of the puncturevine. Weevils can be purchased from various insectories, but they can also be collected by hand, sav-

Figure 21.4 Puncturevine. *Photo:* J. M. DiTomaso.

Figure 21.5 Field bindweed. *Photo:* J. K. Clark.

Figure 21.6 Yellow nutsedge. *Photo:* D. Rosen.

ing money. To collect weevils, find an area (particularly fencerows or other noncrop areas) where the puncturevine population is high; look for evidence of scarring on the underside of stems and holes in the stem or holes appearing in the seedpods; place a pan or plastic tarp under the plant and shake the plant until the adult weevils drop from the plant; place the weevils (250 per acre for moderate infestations) in a paper bag; and allow the weevils to migrate from the open bag to nearby puncturevine plants in the orchard. Collect entire puncturevine plants infested with the weevils and place the plants in the orchard, then allow the insects to migrate from the bag on their own.

Field Bindweed *Convolvulus arvensis*

Fig. 21.5

Field bindweed is a vigorous summer-growing perennial that grows in large patches. It can reproduce from seed that can last in the soil for 30 years, but it more commonly grows from underground rhizomes or extensive roots and stems 15 feet or deeper in the soil. It begins emerging from the soil in about March and can continue to emerge from seed or root and stems throughout summer. As the temperature decreases in the fall, the plants move carbohydrates to the roots for storage during winter dormancy. Stored food reserves are used in the spring during emergence. Field bindweed can persist for many years once established. The plants are usually spread by soil cultivation and the movement of root segments. Its wirelike stems can wrap around the trunks of young trees and eventually the foliage, and it can also entangle irrigation emitters and tubing.

To eradicate this weed, plants must be killed when they first appear, before they produce seed or an extensive root system. Cultivation is effective when plants have emerged from seed. Although cultivation will not likely control established plants in a single pass, repeated cultivations at 2- to 3-week intervals can deplete the food reserves in the roots and starve the plants. Unfortunately, few herbicides labeled for use in prune orchards are effective on this field bindweed. If it appears in the orchard, spot-treat it with glyphosate (2 lb a.i./acre). Adding a good nonionic surfactant improves control. Irrigation prior to glyphosate treatment may also be necessary, because field bindweed is not controlled effectively when stressed for moisture. Established plants may require several applications per year to be effective. Additionally, tank-mixing glyphosate with 2,4-D amine after harvest can be effective before the weed enters dormancy.

Yellow Nutsedge *Cyperus esculentus*

Fig. 21.6

Yellow nutsedge is a summer-growing perennial in the sedge family. While it looks similar to grassy species, it should not be treated the same. It reproduces primarily from underground tubers, or nutlets, that appear at the end of roots or rhizomes and can survive in the soil for 2 to 5 years. Yellow nutsedge

Figure 21.7 Bermudagrass. *Photo:* J. K. Clark.

thrives in lighter soils that are frequently wetted. The tubers are readily spread by disks and other cultivation equipment. Each tuber contains several buds that can produce individual plants. Even if one or two of the plants arising from a tuber are killed, additional plants can sprout from the tuber.

Do not use cultivation as a means of controlling yellow nutsedge, because it can spread the weed. When nutsedge first appears, spot-treat it with repeated applications of glyphosate or paraquat before the plants have four leaves, which is about the time when new tubers begin to form. Spraying newly emerged plants every 21 to 28 days can eliminate the new infestation in as few as 2 years. Do not wait to spray these weeds until most of the plants have emerged. Although the aboveground plants can be controlled, new tubers will already have formed and the population may actually increase. Light to moderate infestations of nutsedge can be suppressed with a preemergence application of norflurazon or rimsulfuron in late winter before tubers break dormancy. Norflurazon can injure prune trees growing on sandy soils irrigated with frequent low-volume irrigation. Thiazopyr can provide good nutsedge suppression in nonbearing orchards as long as two sequential applications are made. For control, apply 2 pints per acre in the fall and again in late winter (January or early February). Thiazopyr works best if adequate rainfall occurs during the winter to move the herbicide into the top 2 to 3 inches of the soil, where the tubers are likely to sprout.

Bermudagrass *Cynodon dactylon*

Fig. 21.7

Bermudagrass is a vigorous spring- and summer-growing perennial grass. It can reproduce from seed, but more commonly it grows from extensive roots and aboveground stems or stolons. It is easily spread through the field on cultivation equipment. In fields where it is present in the middles and is repeatedly mowed, it can become a significant problem because mowing increases the amount of light the stolons receive, stimulating their growth. Bermudagrass is very competitive for water and nutrients, which can greatly reduce prune tree vigor and health.

If bermudagrass becomes established in localized areas in the field, immediately spot-treat it with postemergent herbicides such as glyphosate or sethoxydim in bearing orchards and fluazifop or clethodim in nonbearing orchards. Allowing about 8 to 12 inches of regrowth to occur before treating helps starve the plant of stored food reserves. As many as three sequential applications may be required to control the weed. Applications of preemergent herbicides such as norflurazon, oryzalin, and pendimethalin in late winter control bermudagrass emerging from seed.

References

Ball, D. A., D. W. Cudney, S. A. Dewey, C. L. Elmore, et al. 2003. Weeds of the West. 9th ed. Newark, CA: Western Society of Weed Science.

DiTomaso, J. M., and E. A. Healy. 2007. Weeds of California and other western states. 2 vols.; includes CD-ROM. Oakland: University of California Agriculture and Natural Resources Publication 3488.

Heap, I. 2010. International survey of herbicide resistant weeds. Weed Science website, http://www.weedscience.com.

Lanini, W. T., F. Niederholzer, and A. Shrestha. 2009. UC IPM Pest Management Guidelines for Prune. Oakland: University of California Agriculture and Natural Resources Publication 3464. UC IPM website, http://ucipm.ucdavis.edu/pmg/selectnewpest.prune.html

d and Buff Corsi, © California Academy iences, www.scarysquirrel.org.

22 Vertebrate Pest Management

• Rex E. Marsh and Terrell P. Salmon

Bird and mammal pests of several kinds are found in and around virtually every prune orchard in California, although they may not always cause significant damage. Damage by depredating birds may destroy a portion of the current crop, but it may not affect future production. Tree injury by ground squirrels, pocket gophers, meadow voles, rabbits, or deer is often more serious, killing trees outright or causing permanent damage that can lower yields for years after the initial feeding.

A number of birds can damage prunes, including house finches (linnets), European starlings, scrub jays, and crows. Crowned sparrows, house sparrows, and to a greater extent, house finches often disbud the trees as the buds begin to swell, since these birds can readily distinguish fruit buds from leaf buds. Other damage includes pecking ripening prunes or removing nearly ripe prunes, as is often done by crows. On rare occasions, sapsuckers peck holes in the bark of the trunk or major limbs, or a flock of band-tailed pigeons may strip the trees of green, immature prunes.

Several rodents and rabbits eat roots, small branches, or bark and can kill young trees outright. Pocket gophers and ground squirrels dig burrows and mounds that can interfere with orchard maintenance and also gnaw on drip irrigation lines. Deer strip young trees of foliage and can stunt or even kill saplings.

A program incorporating the following points reduces vertebrate pest damage and makes control more economical.

- Correctly identify the species causing the damage.
- Alter the habitat, when feasible, to make the area less favorable to the pest species.
- Take early action and use control methods appropriate for the orchard and time of year, with due consideration for the environment.
- Establish a monitoring system to detect initial infestations and reinfestations to help determine when action or additional corrective measures or controls are necessary.

Most vertebrate pest control equipment and supplies (baits, fumigants, traps, etc.) are available at local retail outlets such as farm supply and hardware stores. In addition, many county agricultural commissioners' offices make certain rodent pesticides available to growers. More-specialized equipment may have to be ordered from the manufacturer or from outlets serving the needs of vertebrate pest control operators. For further information or sources of special control materials, consult the local UC Cooperative Extension Farm Advisor or county agricultural commissioner.

Endangered Species Guidelines

Where prune orchards are located within the range of threatened or endangered species protected by federal or state regulations, special precautions and guidelines come into play. Concerns may relate to the use of traps, burrow fumigants, or toxic baits, depending on the species occupying the area and the pest to be controlled. The county agricultural commissioner can provide the latest distribution maps of the ranges of endangered species and can also provide current information on restrictions that apply to proposed vertebrate pest control activities. More information on endangered species regulations is available from the California Department of Pesticide Regulation (DPR) website, http://www.cdpr.ca.gov/docs/es/index.htm.

Major Vertebrate Pests of Prunes

Ground Squirrels

The California ground squirrel (*Ostospermophilus beecheyi,* formerly

Spermophilus beecheyi) (fig. 22.1), is a medium-sized rodent that is 14 to 20 inches long from its head to the tip of its long, slightly bushy tail. This species, represented by several subspecies, is responsible for major damage in prune orchards throughout the state. Ground squirrels live in underground burrows and form colonies of 2 to 20 or more animals. Although this native rodent lives in a variety of natural habitats, oak or grassland areas are among their favorites. They adapt well to human activities and are found along roads, ditches, and fencerows, near buildings, and in or around many orchard and agricultural crops. In natural habitats, they tend to avoid thick chaparral, dense woods, and very moist areas. Ground squirrels are active during the day and are large enough to be easily seen. During winter months, most ground squirrels hibernate, but those less than a year old may be active on warm, sunny winter days. Many adults go into a temporary summer sleep (estivation) during the hottest parts of the year. Ground squirrels reproduce once yearly in the early spring; litters average seven or eight young. The young are nursed in the burrow for about 6 weeks before they come aboveground to forage.

Ground squirrels are primarily herbivorous. During early spring, they consume a variety of green grasses and broadleaf plants. When these begin to dry and form seeds, the squirrels switch to feeding mainly on seeds, grains, and acorns where available. They have a great liking for nuts, such as almonds, walnuts, and pistachios, but they also feed on fruits.

Figure 22.1 California ground squirrel. *Photo:* K. Keatley-Garvey.

Figure 22.2 Ground squirrel feeding. *Photo:* Gerald and Buff Corsi, © California Academy of Sciences, www.scarysquirrel.org.

Damage

Ground squirrels can readily infest prune orchards, especially around the perimeters. They climb trees and feed on the prunes. The seeds of some kinds of fruits are more desirable than the flesh; evidence suggests that for ground squirrels to become numerous in prune orchards they must have ample supplies of other foods nearby. Adult squirrels often cache seeds of various kinds in their burrows or in shallow depressions, especially in the late summer and early fall (fig. 22.2).

Ground squirrels, when digging burrows, bring soil and rock to the surface and deposit it in mounds near burrow openings. They enlarge burrow systems each year by constructing new interconnecting tunnels, so the longer the squirrels occupy the burrow, the more extensive and complex it becomes (fig. 22.3). They create new entrances to serve a growing population. The many large burrow openings and soil mounds can damage orchard equipment and make mechanical harvesting more

Figure 22.3 Ground squirrels can burrow beneath concrete slab foundations.

difficult. Ground squirrels frequently burrow around trees and damage the root systems; they can even kill trees. Gnawing on the trunks of young trees and on limbs of older trees is relatively infrequent but sometimes occurs and can be serious. Ground squirrels are not intimidated by people, and their burrows are common beneath buildings and other structures. They are particularly fond of burrowing beneath concrete slabs, causing cracks and settling.

Ground squirrels and pocket gophers, through their extensive burrowing, may significantly increase the loss of irrigation water. This is especially true with basin, border strip, and furrow irrigation. Rodent burrows often divert the water from where it is intended or cause excessive soil erosion. Ground squirrels are notorious for gnawing plastic lines used for drip or microsprinkler irrigation, leading to serious damage and labor-intensive repairs.

Monitoring guidelines

Establish a plan for periodic monitoring of areas where ground squirrels are likely to invade, such as along ditch banks or roadsides or in crops adjacent to the orchard. To monitor, observe squirrel haunts in midmorning, when squirrels feed most actively. If ground squirrels are a major problem, keep annual records of when squirrels emerge from hibernation, when the first young are seen aboveground, and changes in the general number of squirrels, as well as the controls used, dates, and effectiveness. Use these records as the basis for future management decisions.

Management guidelines

If even one or two ground squirrels are present in or immediately adjacent to a prune orchard, control them, otherwise damage is inevitable. Fencing is practically useless against squirrels, and no feasible habitat modification in the orchard expels them once established. Unfortunately, ground squirrels are not responsive to chemical or physical repellent methods. Burrow fumigants, poison baits, and traps are the current means of control.

Habitat modification. In natural habitat, ground squirrels generally feed in open areas where visibility is good (presumably to avoid their natural enemies), although they adapt to other situations. In orchards, ground squirrels often burrow beneath long-standing piles of orchard prunings, wood, or rock or use them as harborage. Removing such piles may make the area somewhat less desirable to them, but the base of trees, fencerows, and ditch banks still offer burrowing sites. Peripheral orchard cleanup may somewhat reduce their numbers; it also makes burrow detection and population monitoring easier and improves access to burrows during control operations.

Predation. Animals that prey on ground squirrels include coyotes, foxes, badgers, and other mammalian carnivores, along with several hawk species. Predation, however, is not a significant factor in keeping ground squirrel populations below the level that causes damage.

Trapping. Because trapping is time consuming, it is most practical for small infestations. Several types of kill traps, including a modified pocket gopher box (fig. 22.4), tube or tunnel traps (fig. 22.5), and Conibear

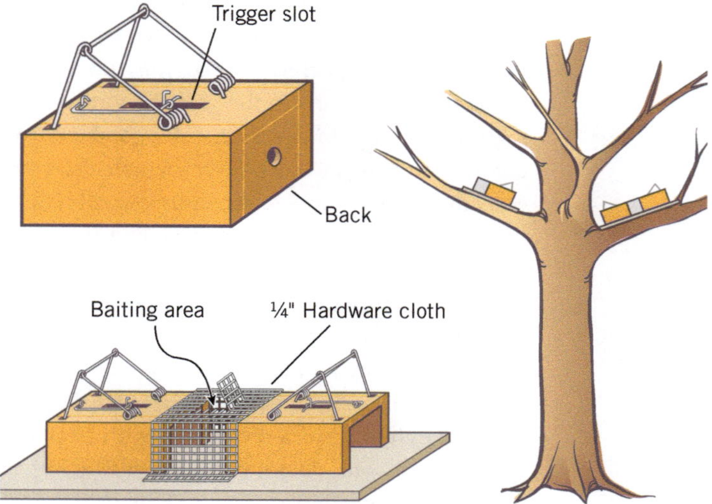

Figure 22.4 An excellent ground squirrel trap can be made by fastening two modified box-type gopher traps back to back.

Figure 22.5 The tunnel-type trap kills animals that pass through it. *Photo:* R. E. Marsh.

(No. 110) traps (fig. 22.6), are effective. Place tube and box traps on the ground near squirrel burrows or runways. For several days, bait them but do not set them; let the squirrels become accustomed to the traps. After the squirrels are taking the bait, rebait and set the trap. Walnuts, almonds, oats, barley, and melon rinds are effective baits. Set unbaited Conibear traps directly in burrow openings so squirrels must pass through them and trip the trigger. Specially designed trap boxes are sometimes used with baited Conibear traps to permit the traps to be placed anywhere squirrels are active (fig. 22.7). As with all traps, take precautions to minimize accidental injury to nontarget wildlife and pets.

Fumigation. Treating ground squirrel burrows with toxic gases (including smoke-generating cartridges and aluminum phosphide) is an effective control method, especially when used within a few weeks after the squirrels emerge from hibernation. Using fumigants early in the year has a great biological advantage in that it provides good control before the young are born. Every female destroyed early means that 7 or 8 fewer squirrels will have to be killed if control were delayed. Fumigants are most effective in the spring when the soil contains enough moisture to retain the toxic gas in high concentrations within the burrow.

Fumigants are ineffective when ground squirrels are hibernating or estivating because they seal themselves into their nest chamber with a soil plug. For safety reasons, do not use fumigants in burrows that extend beneath occupied buildings.

Burrow exploders inject oxygen and propane into the burrow and ignite the mixture, causing a significant explosion. There is very little reported evidence that these devices are effective in controlling ground squirrels.

Poison baits. Poison grain baits are registered for ground squirrel control. The timing of bait application is relatively critical. Baits tend to be poorly accepted and therefore ineffective just after squirrels emerge from hibernation through late spring. During this period, ground squirrels feed extensively on green vegetation and do not readily consume toxic grain baits. They switch from eating green vegetation to eating seeds, grains, and acorns in the latter part of May. Therefore, in central California, May and June are usually the best months for applying baits containing zinc phosphide. In the more southern parts of the state, the squirrels switch to eating seeds earlier in the year.

If a percentage of the adult population goes into estivation during the hottest part of the summer, suspend baiting until about mid-September. If baits were not applied during May to June, the period from mid-September through the end of October offers another opportunity. During this time, many squirrels readily feed on grain baits until they go into winter hibernation.

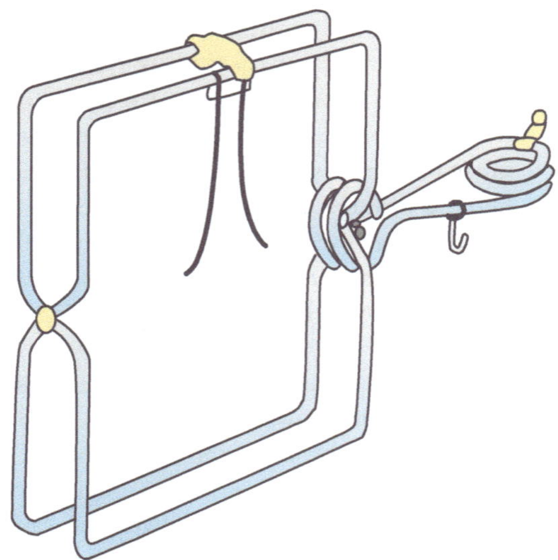

Figure 22.6 Body-grip Conibear trap in set position.

Figure 22.7 Conibear trap set in a trap box near a ground squirrel burrow entrance. *Photo:* R. E. Marsh.

Zinc phosphide poison baits are the most cost effective for ground squirrels and generally produce mortality within 48 hours. Distribute bait by spot baiting (scattering bait by hand on bare ground to cover 2 to 3 square feet at the side or behind each active burrow) or by broadcasting (scattering bait relatively uniformly over the entire infested area). Broadcasting can be done by hand with a belly grinder–type seeder or with a vehicle equipped with a tailgate-type seeder. Consult the product label for recommended application methods and rates.

The major disadvantage of zinc phosphide ground squirrel baits is that if they are applied when the squirrels are not regularly feeding on seeds, poor bait acceptance is apt to occur and, as a result, poor control is inevitable. Any attempt to reapply the zinc phosphide bait will also be doomed to failure because of bait shyness, caused when a squirrel consumes a little bait from the first application and becomes sick but does not die. For this reason, zinc phosphide bait should be applied only once per year, since good grain acceptance is critical to its success.

Baits of multiple-dose anticoagulant rodenticides such as diphacinone provide effective control when squirrels ingest them in multiple feedings for 6 days or more. Death generally occurs from about 6 to 14 days after the first feeding. Consuming anticoagulant bait does not immediately affect the squirrel's feeding activity. To be effective, multiple-dose baits must remain available; effectiveness may be greatly reduced if 48 to 72 hours pass between feedings. The multiple feedings usually required for a fatal dose, the slow action of the anticoagulants, and the availability of an antidote (vitamin K_1) make anticoagulant rodenticides safer than some other rodenticides for use around livestock, pets, and children.

Use anticoagulant baits in bait boxes or, if the label permits, spread them by repeated spot or broadcast baiting. Bait boxes or stations are small structures designed to hold enough bait to provide multiple feedings and to allow the squirrel to enter and feed (fig. 22.8). They also safeguard larger non-target species by excluding them from the bait.

Zinc phosphide and anticoagulant baits are available from commercial distributors and some county agricultural commissioners' offices.

Figure 22.8 Bait stations for ground squirrels can be made from plastic drainpipe. The half cap at the entrance confines the bait but allows squirrels to enter and feed. *Photo:* J. P. Clark.

Figure 22.9 Botta's pocket gopher. *Photo:* T. Borland.

Pocket Gophers

Pocket gophers (*Thomomys* spp.) are stout-bodied, short-legged rodents (fig. 22.9). External fur-lined cheek pouches that open outside the lips, one on each side of the mouth, are used extensively for carrying food. The head and body measure 6 to 8 inches; they have a short, scantily haired tail. In orchards and other irrigated lands, females may produce two litters per year, with litters averaging about five young.

Pocket gophers are most common in areas of abundant plant growth. They feed primarily on succulent underground parts of herbaceous plants; they can pull a 2-foot-tall plant underground to consume it.

This species lives almost entirely underground. Pocket gophers are antisocial and live a solitary life except during breeding and when the young are being raised. Burrow systems may be extensive and include deep main burrows, shallow feeding tunnels, and side tunnels to push out dirt. They create characteristic soil mounds aboveground. Main tunnels are normally 10 to 12 inches beneath the surface, but some can lead to deeper nests or food storage chambers (fig. 22.10). Pocket gophers

plug their burrow openings with soil, so the tunnel system is completely enclosed. As a result, the temperature and humidity in the burrow are stable and close to optimal.

Damage

Pocket gophers can frequently be found living in orchards. They are active throughout the year, and if uncontrolled and food is plentiful, the population can increase to 30 to 40 gophers per acre. Pocket gophers are relatively slow reproducers, however, and population builds up gradually over a period of years. They cause tree damage or death by girdling the roots or crown at or below the soil level.

Monitoring guidelines

Because pocket gopher damage is frequently invisible, it often goes undetected until a tree exhibits stress. By that time the tree may be beyond saving. Gopher activity is readily detected, however, by looking for fresh mounds of soil. The animals produce these soil mounds in greatest numbers in the spring and fall, when the soil is amply moist.

Management guidelines

Persistent efforts can control pocket gophers and even eliminate them from an orchard. The preferred control methods are baiting, trapping, and fumigation. No chemical or mechanical repellents have been shown to be effective against pocket gophers in orchards. Gophers are usually detected when fresh mounds of soil are seen on the orchard floor. After implementing a control program, destroy (flatten) the mounds and resurvey the area in 24 to 48 hours. If new mounds occur, a gopher has probably evaded control.

Poison baits. Single-feeding acute-type poison baits (such as strychnine) placed in burrow tunnels are widely used and are effective for controlling gophers in orchards. Follow bait label directions for application methods and amounts. Two methods of hand baiting are available; there is also a tractor-drawn burrow builder for large, heavily infested orchards.

Hand baiting usually requires using a metal probe to locate the gopher's tunnels. Use a pointed 38-inch steel rod to probe near the fresh mounds or between two recent mounds to find the burrow. Enlarge the probe opening with a larger rod or broomstick and place a small, measured amount of grain-type gopher bait in the burrow. (fig. 22.11).

Hand-operated bait dispensers have a bait reservoir and bait release mechanism that permit probing and bait dispensing in one operation. These devices are substantially faster than stoop-and-pour hand baiting. After the bait is placed, cover the probe hole to prevent light from entering the burrow.

A tractor-drawn mechanical burrow builder creates an artificial burrow and deposits poison grain bait within it at preset intervals and quantities. It is an excellent way to control a serious gopher infestation over a large area. Use the machine to build an artificial burrow at a depth of about 10 inches between the tree rows, where it can intercept some of the gopher's natural burrows. The gophers will soon discover the artificial burrow and consume the deposited bait. During operation, inspect bur-

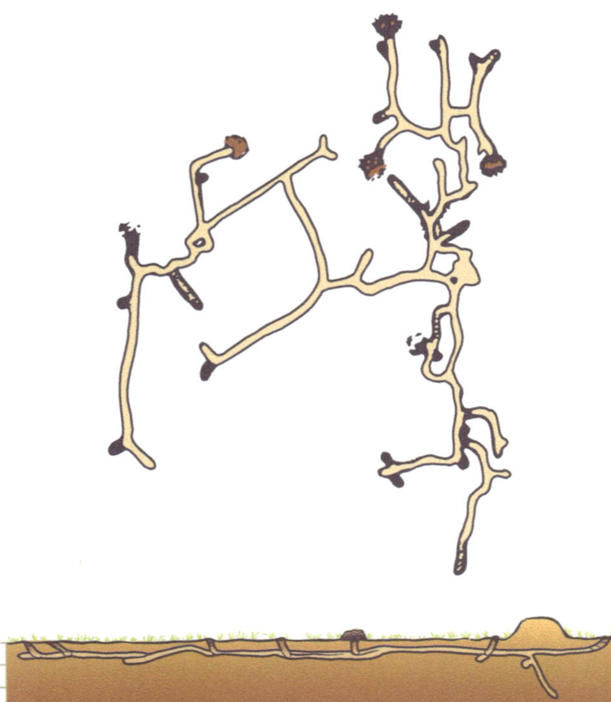

Figure 22.10 Aerial view and cross-section of typical pocket gopher burrow.

Figure 22.11 After bait has been placed, cover hole to prevent light from entering. *Photo:* J. K. Clark.

row depth and condition by occasionally opening a small section with a shovel. The soil must be moist in order to produce a well-formed, smooth, artificial burrow. Follow the instructions that come with the machine. A mechanical burrow builder can apply bait at a rate of 4 to 5 acres per hour. Use a burrow builder only where gophers are active; avoid unnecessary treatments. The bait lasts only a few days, but the artificial burrows can remain for some time and can be used by invading gophers.

Trapping. Pincher or choker traps can effectively control pocket gophers (fig. 22.12). Because trapping is labor intensive, it is most commonly used by organic growers or where only a few pocket gophers are present. To locate the main runway or tunnel, probe with a steel rod a short distance in front of the low side of a fresh mound or between two fresh mounds. After finding the main tunnel, dig a hole to intercept it (the hole will probably be 10 to 12 inches deep). Clean out the burrow and set two traps in the runway, one facing each direction. Whether pincher traps or choker traps are used, wire each pair to a stake so a coyote or fox does not drag them away. Replace the soil after traps are removed.

Fumigation. Pocket gophers create extensive burrows. Portions of them are relatively near the surface; therefore, it is difficult to maintain a lethal concentration of most burrow fumigants (poison gases). Fumigation with smoke or gas cartridges is not effective because gophers quickly seal off their burrows when they detect smoke or gas. However, aluminum phosphide fumigation (a restricted-use material) can be effective if applied when there is enough soil moisture to retain the toxic gas. Follow label instructions and all safety precautions. To use aluminum phosphide, first probe to find the main burrow as with hand application of bait, then insert the number of tablets prescribed by the label into the burrow and seal the probe hole. As with other control methods, monitor for signs of renewed gopher activity. Re-treat the area if new mounds appear after 24 to 48 hours.

Mechanical burrow exploders inject oxygen and propane into a burrow and ignite the mixture, causing a significant explosion. There is very little reported evidence that these devices are effective in controlling pocket gophers.

Habitat modification. Permanent ground covers of certain herbaceous and grassy plant species favor gophers. Removal of permanent ground cover will do much to reduce their numbers, because food is generally the factor that limits population. Deep disking destroys some burrows, making the orchard less habitable; clean cultivation removes cover for gophers and allows new mounds to be detected easily. Controlling gophers along fencerows and roadways adjacent to orchards is important because young gophers disperse from these places into the orchard.

Figure 22.12 Traps used to control pocket gophers. *Photo:* J. P. Clark.

If an orchard is flood irrigated, very young pocket gophers may be drowned and adults may be forced to the surface, where they are temporarily exposed to natural avian predators. By comparison, sprinkler or drip irrigation systems generally favor an increase in pocket gophers.

Predation. A variety of predators feed on pocket gophers; however, their presence does not usually keep pocket gopher populations in orchards low enough to prevent economic damage. There is strong interest in using barn owls for controlling gophers and other small rodents, and some growers encourage them by placing nest boxes in and around the orchard. Little is known about the effect of these avian predators on rodent damage in orchards. It is likely that they feed outside the orchard in grassland areas.

Meadow Voles

Meadow voles *(Microtus californicus)*, also called meadow mice or field mice (fig. 22.13), can severely damage prune orchards by feeding on the bark at the base of trees. Dense grass is their preferred habitat. Vole populations often develop in orchards or on orchard borders, roadsides, and fencerows where grass or other permanent vegetative cover remains year-round. Orchards that have cover crops or those in which grass and herbaceous plants are left to grow next to tree trunks are most susceptible to damage.

Meadow voles are small, blunt-nosed, stocky rodents with small ears and eyes, short legs, and short tails. Their coarse fur is usually dark gray or grayish brown. When full grown, they are larger than a house mouse but smaller than a rat. Females may produce from 5 to 10 litters per year. They may breed year-round, but the principal breeding time is during late winter and spring. Because voles mature rapidly and can bear multiple litters yearly, vole populations can increase quickly. Typically, the numbers peak every 6 to 8 years, when the population can be as high as hundreds of voles per acre. Meadow voles' home ranges are relatively small; they usually travel less than 10 feet from the nest to feed. They may, however, move much farther in search of new food resources.

Figure 22.13 Meadow vole. *Photo:* J. P. Clark.

Figure 22.14 Meadow vole runways in turf. *Photo:* T. P. Salmon.

Damage

Characteristic damage by meadow voles is complete or partial girdling of tree trunks from just below the soil line to as far as they can reach on the trunk, usually no more than 5 or 6 inches. In rare situations, voles climb higher on young trees. The animals attack both young and old trees, but young trees generally sustain greater damage because any bark removal is more damaging to a tree of small diameter.

Monitoring guidelines

Meadow voles are usually found in localized areas marked by numerous 1- to 2-inch-wide surface runways through dense or matted grass, and silver dollar-sized holes to their burrows (fig. 22.14). They are active year-round, irrespective of weather, but they do the most damage to trees in winter or early spring when other plants cannot satisfy their nutritional needs. Meadow voles feed during the night and day. Deposits of small, soft, brownish feces and short 1- to 2-inch pieces of grass stems along the runways are evidence of their presence. Burrows frequently have numerous openings to the surface, which are almost always left open and not filled with soil. The burrows are relatively shallow and contain stored food and nesting chambers.

Starting in midwinter, inspect orchards and surrounding fields for vole activity and population increases. Especially check areas with heavy vegetation; look for new vole runways, fresh droppings, burrow openings, and evidence of bark or grass feeding. Monitoring for voles often requires pulling apart the dense grass cover for close inspection of the ground.

Management guidelines

Vegetative cover provides food and protection from the elements and predators, so management of natural cover or maintained cover crops is essential to meadow vole control. Fencerows or properties adjacent to the orchard may harbor voles. As the vole population increases, young adults begin to disperse into new areas where the habitat is favorable. Eliminating vegetative cover in adjacent areas or providing a buffer 30 to 40 feet wide between vegetated areas and the orchard reduces the frequency and number of voles invading an orchard. Once voles appear in the orchard, the most effective ways to deal with them are clean cultivation of the entire orchard, removal of all vegetation from immediately around the trees, or application poison baits.

Habitat modification. Cultural practices can significantly affect meadow vole populations. Clean cultivation or banded weed control kills vegetation next to the tree, making the immediate habitat unsuitable and thus preventing damage. Maintaining weed-free fencerows, roadsides, and ditch banks is also an important preventive measure. Because voles do not feed more than a few yards from their burrows, any significant destruction of their food and cover causes increased mortality and lowers their reproductive potential. The voles may also seek out new and more favorable habitat.

Predation. Predators such as coyotes, foxes, badgers, weasels, owls, and hawks feed on meadow voles; however, predation is rarely, if ever, a major factor in controlling a rapidly increasing vole population. The meadow voles' high reproductive rate compensates, to a great degree, for losses to predators. There is strong interest in using barn owls for controlling gophers and other small rodents, and some growers encourage them by placing nest boxes in and around the orchard. Little is known about the effect of these avian predators on rodent damage in the orchard. It is likely that they feed outside the orchard in grassland areas.

Trunk guards. Cylindrical wire or plastic trunk guards can protect young trees from voles. To hinder burrowing, guards must extend at least 6 inches below the soil surface, but even then voles may do damage below them. Meadow voles rarely climb over guards. They often work beneath them and continue to gnaw in seclusion behind them, where this early damage may go undetected. Because of this possibility, the use of trunk guards often gives the grower a false sense of security.

Ch**emical repellents**. Chemical repellents have not been shown to be effective in protecting orchard trees from voles.

Poison baits. Zinc phosphide rodenticide grain baits are very effective in reducing meadow vole populations. For the most effective control, apply this acute, single-dose bait in the voles' runways, where most feeding occurs. Spot-apply or broadcast bait over the entire infested area. For broadcasting, use a belly grinder–type seeder, a vehicle with a tailgate seeder, or, in some situations, an airplane. Broadcast application rates depend on the type of toxicant. Multiple-dose anticoagulant poisons are also registered for meadow vole control. Consult the product label for application methods and rates.

Rabbits

Young trees are particularly susceptible to debarking and limb clipping by rabbits. Jackrabbits *(Lepus californicus)* are the major rabbit pest (fig. 22.15), although cottontails (*Sylvilagus* spp.) may cause damage in some areas (fig. 22.16). Jackrabbits breed from early spring to late summer. Females may produce more than one litter per year, especially where irrigated crops are available. The average litter contains four young. Rabbits are active from early evening to early morning year-round. Rabbit populations can fluctuate dramatically and often reach high levels every 5 to 10 years.

Damage

Rabbit damage to prunes is almost always limited to orchards younger than 5 years old. Rabbits may debark trunks during the first winter following planting, and clipping of small branches and leaves may be evident as the tree breaks dormancy. Rabbits tend to partially girdle tree trunks rather than completely girdle them (fig. 22.17). Girdling caused by rabbits is usually higher (6 to 30 inches) on the trunk than that caused by meadow voles.

Monitoring guidelines

Rabbits may breed, bear young, and live in or outside prune orchards. If they move into the orchard only to feed, they may not be seen during daylight

Figure 22.15 Jackrabbits. *Photo:* W. P. Gorenzel.

Figure 22.16 Cottontail rabbit in front of its burrow. *Photo:* R. O'Connell.

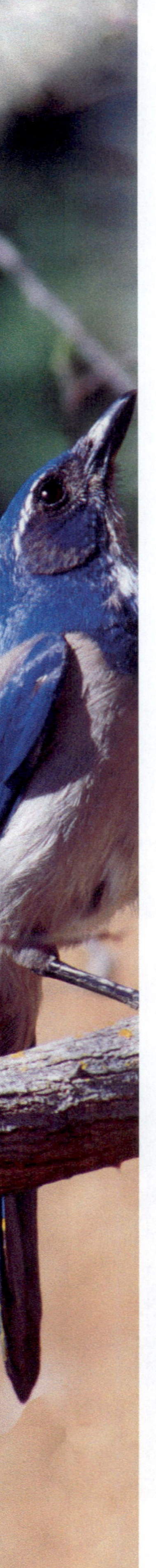

hours. Inspect young trees periodically for debarking to catch a rabbit presence early. Tour the orchard in early morning, late evening, or at night, looking for rabbits with a spotlight.

Management guidelines

Rabbit control in prune orchards includes exclusion, repellents, shooting, and poisoning. The choice of method should depend on the severity and urgency of the problem and the particular situation. Manage rabbit populations before severe damage occurs. Habitat modification to reduce damage in the orchard is rarely possible.

Individual tree guards offer protection from rabbit damage, although rabbits can sometimes reach above a guard and clip lower branches. Guards may be solid or made of net or mesh materials such as metal, hardware cloth, plastic, or sturdy fibrous sheets.

Exclusion. If jackrabbits or cottontails are a constant and continuing threat to young trees, fencing the entire orchard may be the best management approach. To make an effective barrier, build a fence 36 inches high of woven wire or poultry netting using 1½-inch-diameter mesh. After digging a trench 6 inches deep and 6 inches wide along the fence, bury 48-inch-wide wire 6 inches deep, leaving a 6-inch lip turned outward at a right angle at the bottom. When backfilled with soil this will leave a 36-inch-high fence, which rabbits will normally not jump; burying the bottom 6 inches deep prevents them from digging beneath it. For a deer-proof fence, the additional expense of a bottom roll of the smaller-mesh wire, properly buried to also exclude rabbits, may be worthwhile (fig. 22.18).

Figure 22.17 Jackrabbit damage to the trunk of a young peach tree. *Photo:* T. P. Salmon.

Repellents. Chemical repellents may provide temporary relief from rabbits. Spray or paint the chosen repellent on the trunks or foliage, following label instructions. Repeat applications may be required to renew repellency lost through rain or sprinkler irrigation or to protect new growth.

Shooting. Under certain conditions, shooting can be an effective control. Patrol systematically in early morning and late evening.

Poison baits. Poison baits offer a practical and economical way to control large numbers of jackrabbits in large areas, although results are sometimes erratic. Grains such as rolled barley and crimped oat groats make effective baits.

Only multiple-dose poisons are registered for use against rabbits. Place multiple-dose diphacinone baits in open, self-dispensing feeders, shallow trays, or nursery flats. Position the feeders in areas frequented by jackrabbits, such as trails and resting and feeding areas. If rabbits fail to feed after a few days, move feeders to where bait is readily accepted. Keep bait available until all feeding ceases, which may be from 1 to 4 weeks.

Place poisoned bait where livestock and humans—especially children—cannot pick it up. Be aware of all wildlife in the area, such as doves or pheasants, and take precautions to protect them from poisoning. Protect diurnal seed-eating birds by removing or covering the bait during daylight hours, exposing the bait only at night.

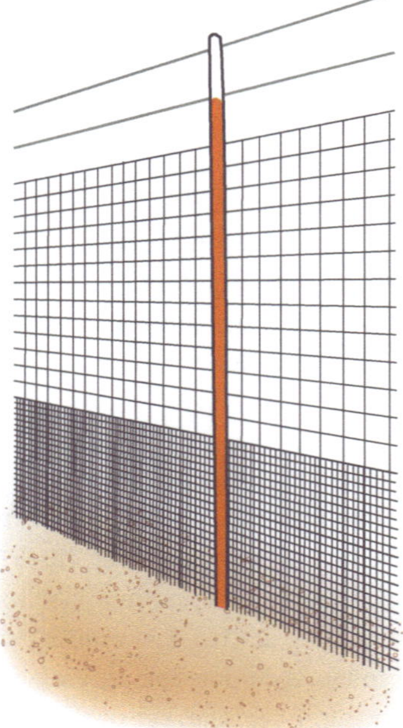

Figure 22.18 A 7-foot deer fence can be made rabbit-proof and even meadow vole–proof by adding ½-inch wire mesh at the base and burying the mesh securely underground.

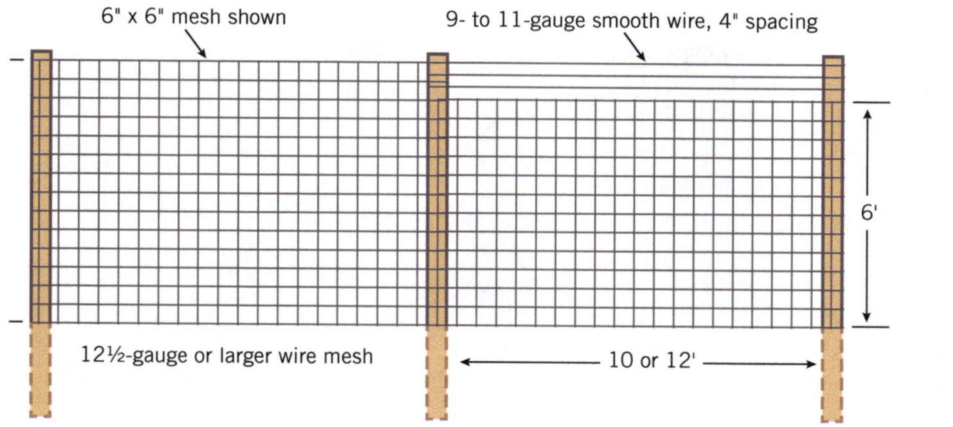

Figure 22.19 Deer fencing.

Habitat modification. Rabbits usually invade orchards from adjacent fields; unless the land is under your direct management, modification of the outlying habitat is usually impractical. Removing orchard cover crops and weeds that serve as rabbit food may decrease the number of rabbits that routinely visit the orchard. It may temporarily increase tree damage, however, because the trees would be all the rabbits have left to feed on. In any case, cover removal does make rabbit detection easier.

Deer

Where the habitat adjacent to a prune orchard supports moderate to high populations of deer *(Odocoileus hemionus)*, they can cause significant damage, especially to a young orchard. Foothill and coastal districts with nearby brush or woodlands that provide cover for deer usually experience the most frequent depredation. Some valley orchards near stream bottoms may also suffer. State game management laws limit the control methods available to growers and make combating deer damage expensive.

Damage

Deer can completely defoliate young trees. They can stunt, distort, or kill a tree by repeated browsing. Buck deer sometimes rub their antlers on trunks and lower limbs, causing severe bark scarring. This, however, is not common. Deer may browse on older trees, but the damage to them is usually less severe than that to young trees.

Monitoring guidelines

In bushy areas, deer usually stay out of sight during the daytime and move into orchards at night to browse. Feeding evidence, droppings, and hoof prints in the orchard indicate their presence. Evening or early morning sightings are likely where deer are numerous.

Management guidelines

Exclusion. Fencing is the most effective method of excluding deer from an orchard. A 7-foot-high wire fence usually works (fig. 22.19). A 6-foot-high mesh fence can be heightened to 7 feet by adding two or three strands of barbed or smooth wire on top. Deer may occasionally clear a 7-foot fence when being chased if the fence is on steep, sloping ground. Electric fences have been used successfully in some areas. Check deer fences periodically to be sure they remain intact. Damaged wire, broken gates, soil washout beneath fences, and the like permit access and must be repaired immediately. Deer that manage to circumvent the fence and get inside may have to be removed by shooting (with a depredation permit) if they cannot be driven out.

Shooting. In some circumstances, depredation permits for shooting deer may be obtained from the Department of Fish and Game. Shooting deer may buy some time to install a fence, but it is rarely a satisfactory, long-term solution to a significant deer problem. Encouraging people to hunt deer in the area may lower the overall deer population and help reduce deer depredation.

Repellents. Growers have tried many odor-producing repellents, such as heavily scented bar soaps, garlic, and blood meal, but deer usually adjust to them rapidly, especially when hungry. When deer populations increase, they compete for food, and odor-producing repellents become totally ineffective. Combined taste and odor repellents such as BGR (Big Game Repellent) applied to the foliage can be somewhat effective, at least for a short time. Re-treatments are needed as new foliage develops or after rain or sprinkler irrigation washes the repellent

away. There are wide differences in the efficacy of commercial deer repellents.

Noise-making devices, such as propane exploders and electronic alarms, have not been effective for more than a day or so because deer rapidly habituate, or become accustomed, to them.

Habitat modification. Eliminating suitable cover for bedding and other survival needs is impractical. Deer are highly mobile, and many travel half a mile or more to reach an orchard, especially when they have become accustomed to feeding there.

Other Vertebrate Pests

Orchards near wooded areas, riparian vegetation along streams, or suburban residential areas or parks may be troubled by the introduced eastern fox squirrel (tree squirrel) *(Sciurus niger)*, which is adept at raiding fruit and nut orchards. Wood rats (*Neotoma* spp.), living in similar habitats, also sometimes clip the small shoots of young prune trees for building their nests. Tree squirrels or wood rats rarely become pests for prune growers.

Birds

Birds that damage prunes and other stone fruits in California include European starling *(Sturnus vulgaris)* (fig. 22.20), house finch *(Carpodacus mexicanus)* (fig. 22.21), goldfinch (*Carduelis* spp.), crowned sparrow (*Zonotrichia* spp.) (fig. 22.22), house sparrow *(Passer domesticus)*, scrub jay *(Aphelocoma californica)* (fig. 22.23), yellow-billed magpie *(Pica nuttalli)* (fig. 22.24), and crow *(Corvus brachyrhynchos)*. Other birds may occasionally be implicated in damage; as a rule, only one or two species may be important to any one grower, and many growers never have any significant bird damage, at least for some years.

Whether birds damage the orchard is often based on the type of habitat adjacent to or surrounding the orchard. Brushy wildlands, woodlands, and riparian habitats that provide ample nest sites and cover for loafing or roosting contribute to greater bird damage.

Because solutions to most bird damage are not species-specific and rely extensively on frightening methods and techniques, the bird pests of prunes are discussed as a group. Less-used pest bird management techniques include habitat modification, shooting, and trapping.

Damage

Bird damage may start when the trees are in the bud stage and continue through harvest. Birds can cause substantial crop loss in bearing orchards,

Figure 22.20 European starling. *Photo:* J. P. Clark.

Figure 22.21 House finch, also called a linnet. *Photo:* W. P. Gorenzel.

Figure 22.22 White-crowned sparrow. *Photo:* J. P. Clark.

often because the losses frequently go undetected for a time and are difficult to quantify once discovered. Even when bird damage is identified, curtailing it may take some time. In the case of disbudding, its discovery may be after the fact, when it is too late to take action.

House finches, crowned sparrows, goldfinches, and house sparrows sometimes disbud prune trees prior to bloom. Of these, the house finch is the worst offender and is overall the most numerous of disbudding birds. The birds commonly move from cover, such as brushy wildlands or trees along a watercourse, to feed on the nearest prune trees. The resulting damage, therefore, is nearly always heaviest on the margins (the first two or three rows) of the orchard adjacent to suitable bird habitat or near power lines the birds use for perching. Disbudding is generally localized, and orchards surrounded only by other orchards with no adjacent wildlands may suffer little or no measurable damage.

Unless the birds are observed in the act, or unless the trees are inspected very closely during the dormancy breaking period, damage may go undetected. Birds can readily differentiate between fruit and leaf buds. House finches may go one step further and remove flower petals of the bloom and feed on the embryonic fruit. Disbudding becomes most apparent when the trees are in full bloom. Fed-upon trees have noticeably fewer blooms in the upper canopy. Disbudding may be the most serious bird damage sustained by some growers.

Scrub jays and yellow-billed magpies feed on ripening prunes. Scrub jays feed singly or in pairs, while yellow-billed magpies feed in small flocks of a few to several dozen birds. Crows are more gregarious and often feed in large numbers, moving from orchard to orchard. Starlings also generally feed in sizable flocks.

On rare occasions, flocks of band-tailed pigeons *(Columba fasciata)* may suddenly move into an orchard and consume large numbers of green, immature prunes. Sapsuckers (*Sphyrapicus* spp.), members of the woodpecker family, sometimes spend time in orchards, where they peck parallel rows of holes into the larger limbs or trunks of trees. Such holes may weaken a limb by creating entry points for disease organisms and insects. Visits by sapsuckers are relatively uncommon.

Monitoring guidelines

Table 22.1 provides brief descriptions of the most common bird pests along with information about the legal aspects of their control. The best monitoring approach uses field personnel to watch for pest species. To detect disbudding, monitoring in the winter months is necessary.

Figure 22.23 Scrub jay. *Photo:* W. P. Gorenzel.

Figure 22.24 Yellow-billed magpie. *Photo:* W. P. Gorenzel.

As prunes set, systematic monitoring is essential. Set up a schedule of biweekly orchard drive-throughs during the time of day when the most bird activity is expected and count the birds. Certain birds often enter the orchard from one point or from one direction, depending on the adjoining habitat; therefore, counts of birds entering or leaving the orchard can also be made from stationary observation points. From these counts, the severity of the damage and its rate of increase can be estimated over time. Hand-powered mechanical counters can be helpful when tabulating bird numbers.

Management guidelines

Habitat modification. In very few situations can the habitat be modified to reduce damage caused by pest birds. However, if brushy areas can be cleared to eliminate bird nesting and loafing sites, it should be done. Occasionally, cutting down a few evergreen

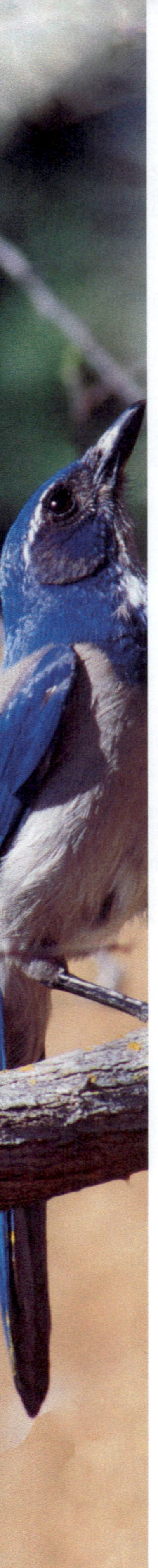

Table 22.1. Common bird pests in California orchards

	Description	Behavior	Damage
crow	*Corvus brachyrhynchos.* A large bird, 17 to 20 inches (43 to 51 cm) long. Coal-black plumage with a rounded tail. Large nests made of sticks are constructed high in trees. Usually found in flocks except when nesting.	Omnivorous. Flocks range from a few birds to a hundred or more when feeding in fields. Congregate in huge winter roosts, often in urban settings.	Droppings contaminate everything below urban roosts. Crop damage includes fruits such as cherries and all kinds of nuts. Corn, milo, and other cereals are often heavily damaged.
crowned sparrow	Two species, from 5¾ to 7 inches (14.5 to 18 cm) long. Typical sparrow coloration, brownish on back, dull grayish breast. Adult white-crowned sparrow (*Zonotrichia leucophrys*): three white and four black alternating stripes on crown. Adult golden-crowned sparrow (*Z. atricapilla*): dull gold crown margined with black.	Forages on ground in grassy and open areas near brush, fencerows, and other such cover.	Feeds on vegetable and fruit crops, especially lettuce, grapes, melons, almonds, and strawberries. Disbuds fruit and nut trees; damages young seedlings in fall and winter.
house finch	*Carpodacus mexicanus.* 5 to 5¾ inches (12.5 to 14.5 cm) long. Male has rosy-red head, rump, and breast, brownish back and wings, sides streaked with brown; female lacks red, has brownish body with heavily streaked breast and abdomen.	Well adapted to human environments; often nests in vines on buildings. Sings and chirps from trees, antennas, or posts. Found in variety of habitats, from deserts and open woods to farmlands, suburbs, and farms.	Eats fruits and berries in orchard and garden; attacks seed crops. Disbuds and deflowers fruit and nut trees.
house sparrow	*Passer domesticus.* 5¾ to 6¼ inches (14.5 to 16 cm) long. Male has black bib and bill, white cheeks, and gray cap; female is dull brown above and dingy whitish below without black bib, bill, or gray cap.	Abundant in farmland, cities, and suburbs; lives in loose flocks. Often nests in eaves, vents, or other openings and cavities in buildings.	Eats emerging seedlings, fruits, and buds; damages flowers, newly seeded lawns, and ripening fruit. Droppings deface and contaminate buildings.
magpie, yellow billed	*Pica nuttali.* 16 to 20 inches (40.5 to 51 cm) long; black and white body with long, streaming tail.	Lives in farming areas of California valleys and nearby foothills. Gregarious, nests in colonies. Builds large nest of sticks high in trees near open grasslands or fields.	Feeds on fruits, nuts, grain, and garbage.
scrub jay	*Aphelocoma californica.* 10 to 12 inches (25.5 to 30.5 cm) long. Head, wings, and tail blue; under parts and back gray; white throat; no crest.	Found throughout California except in deserts and high mountains. Very vocal and noisy. Makes short flights ending in sweeping glide.	Eats fruits and nuts.
starling	*Sturnus vulgaris.* 7½ to 8½ inches (19 to 21.5 cm) long. Short tail; long, slender, yellow bill in spring and summer, dark bill in winter; plumage black to purplish black; heavily speckled in winter.	Abundant in city parks, suburbs, and on farms. Gregarious; uses large communal roosts from late summer until spring. Flies swiftly and directly. Primarily a ground feeder.	Pulls small plants; damages fruit (grapes, cherries, strawberries, and others). Nests in building eaves and other openings; droppings deface buildings.

trees that serve as bird cover reduces disbudding along that side of the orchard. Likewise, cutting down a few trees along a driveway that support crow or magpie nests may help. Always look at habitat modification as a potential management option, even if infrequently used.

Frightening. Most bird control in prune and other stone fruit orchards currently relies on frightening—dispersing birds with sound made by live ammunition or cracker shells, gas cannons (propane exploders), or electronic distress-call noisemakers. Visual frightening devices sometimes augment sound-producing methods. Mylar flags and large "eye" balloons attached to poles above treetops (one device per 3 to 4 acres) may offer temporary repellency (fig. 22.25). Some visual devices produce motion, such as a pop-up scarecrow.

The effectiveness of frightening methods depends on a number of factors:

- Are there nearby alternative feeding sites for birds?

- How great is the birds' affinity for the immediate area in terms of nearby nesting, traditionally high numbers in the area, and so on?
- Have the birds already become accustomed to feeding in the orchard?
- Have the most appropriate frightening methods or combination of methods been selected?
- Are the frightening methods being employed to maximum effectiveness?

Factors cited in the first two questions are not manageable, but those cited in the last three are.

Frightening should begin before the birds have established a regular feeding pattern. Once feeding habits are well established, birds are much more difficult to drive from the orchard.

Not surprisingly, shooting to produce noise is the most frequently used frightening technique because it uses commonly available equipment and gives immediate results. What is not commonly known, however, is that shooting—with live ammunition, cracker shells, or whistle bombs—consists of two elements of harassment: periodic loud blasts and the presence of humans (and their vehicle, if used). After the shooting technique has been employed for a while, some birds, such as crows, often disperse as soon as they see the shooter or the shooter's vehicle. They may return as soon as they see the human and the vehicle leave.

The major disadvantage with all frightening techniques is that most birds habituate to them when used day in and day out. Their effectiveness diminishes with time and as more growers in the same general area use the same techniques. The grower with the most innovative frightening strategies has the advantage.

To prolong the effectiveness of the frightening methods, introduce as many variations as practical. For example, if a shooter normally rides around the orchard on an ATV, he or she should occasionally abandon the vehicle and walk through the orchard. Instead of using normal shotgun ammunition, occasionally use cracker shells, which produce a second blast above the birds.

Employ variations intermittently. Rather than using exploders and an electronic frightening system simultaneously, employ them in a rotational scheme. Use the electronic system for 5 to 7 days, and then use the gas exploders for the next 5 to 7 days. Compared with simultaneous use, rotation usually results in superior and longer-lasting bird dispersal, reduced operating costs, and extended equipment life. Determine the number of days in each sequence by observing the birds' response. In terms of the preceding example, some type of visual stimulus along with the electronic frightening system may improve control, such as suspending large, bright balloons above the treetops. This combination should be used only with the electronic system and taken down when the propane exploders are in operation.

Figure 22.25 Flashing Mylar flags can frighten birds. *Photo:* J. P. Clark.

Growers often make the mistake of using too few gas cannons or electronic alarms. The minimum for effectiveness is one cannon per 5 to 8 acres. Because birds cause damage only during daylight hours, operate gas exploders or other noise-producing devices only from sunup to sundown. Propane exploders should be mounted on tripods or elevated high off the ground so the sound will carry farther. The use of sound-making devices may not be possible where residences are nearby or local ordinances prohibit them.

Although this discussion has emphasized the effectiveness of using frighteners on a rotational basis, do not hesitate to intermittently employ appropriate combinations. Combine a sound-producing device with a visual device (flags, balloons, hawk kits, etc.) or a motion-producing device (e.g., pop-up scarecrow).

Loud electric air horns mounted on ATVs or pickups can be effective if blasted periodically, cutting down on the amount of cracker shells or live ammunition needed. If bird feeding occurs at specific periods of the day, patrol only during those periods, but be aware that birds are quick to alter their feeding patterns to avoid patrols.

Shooting to kill. Scrub jays and magpies are generally found in relatively small numbers, so they can often be controlled by shooting. A depredation permit is required from the U.S. Fish and Wildlife Service to kill scrub jays. If doing damage, mag-

pies and crows can be shot without a permit, but check first with your local Department of Fish and Game officer because regulations change frequently. When permissible, shooting a few crows or any pest species often enhances the effectiveness of ongoing frightening programs. Crows, being numerous, require greater efforts to reduce their populations to acceptable numbers. Sportsmen have long hunted crows and often use crow calls and owl decoys on tall poles to draw crows to the shooting site. Growers can do the same.

Trapping. As a control, trapping can be relatively effective for certain bird pests, and trap plans are available for many species. Trapping can be labor intensive and expensive, however. To produce the desired results at the lowest cost, a trapping program must be well planned and instigated. A cooperative program involving several neighboring growers works best. Once a program is in place, it can be quite efficient. Since trapping depends on drawing birds into the area, this approach is not compatible with frightening methods.

House finches can be trapped in significant numbers by using a modified crow trap similar to those for trapping starlings. A combination of canary grass and rapeseed makes an excellent bait. Once a few birds are caught, they serve as decoys to attract others. Therefore, keep a few decoy birds in the trap with ample food and water. Destroy excess birds humanely. Finch and crowned sparrow trapping must be supervised by the county agricultural commissioner.

Starlings, crowned sparrows, and house sparrows can also be trapped in effective numbers. Crows and magpies are so difficult to trap that few growers attempt to trap them.

Predation. A few raptor species prey regularly on some of the smaller pest bird species—crowned sparrows, house finches, starlings, and house sparrows. In and around orchards, however, avian predators are not numerous enough to have a measurable effect on pest populations. Occasional mammalian predators capture birds when feeding on the ground, but these kills mean little in terms of significant population reduction.

References

Clark, J. P., ed. 1994. Vertebrate pest control handbook. 4th ed. Sacramento: California Department of Food and Agriculture, Division of Plant Industry.

Hygnstrom, S. E., R. M. Timm, and G. E. Larson, eds. 1994. Prevention and control of wildlife damage. Lincoln: University of Nebraska, USDI, and Great Plains Agricultural Council Wildlife Committee.

Micke, W. C., ed. 1996. Almond production manual. Oakland: University of California Division of Agriculture and Natural Resources Publication 3364.

Salmon, T. P., D. A. Whisson, and R. E. Marsh. 2006. Wildlife pest control around gardens and homes. 2nd ed. Oakland: University of California Division of Agriculture and Natural Resources Publication 21385.

University of California Statewide IPM Project. 2002. Integrated pest management for almonds. 2nd ed. Oakland: University of California Division of Agriculture and Natural Resources Publication 3308.

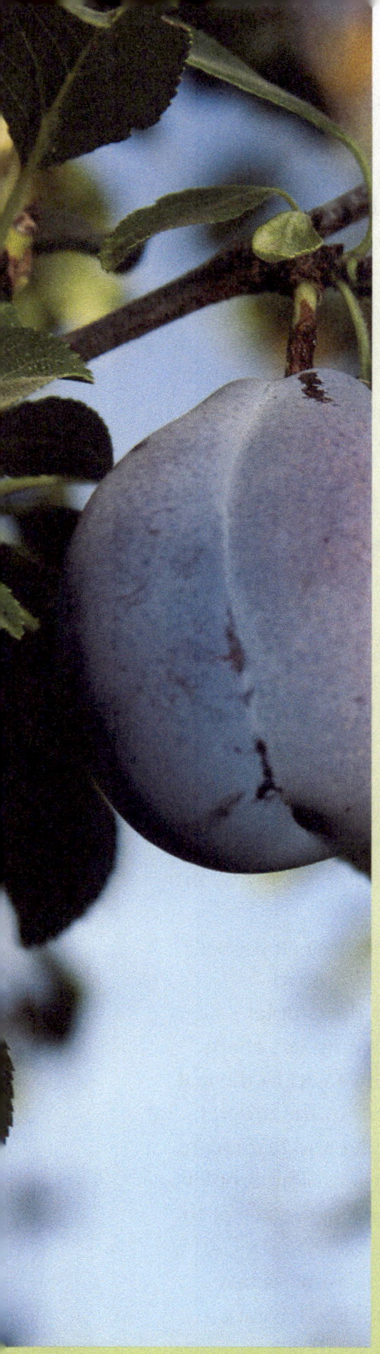

23 Spray Application of Liquid Pesticides, Growth Regulators, and Fertilizers

• Franz J. A. Niederholzer and William E. Steinke

Spray application of pesticides, growth regulators, and fertilizers benefit the growers and the greater community when application protects or strengthens the crop, is cost effective, and does not threaten the health or safety of humans, plants, or animals. Many factors contribute to achieving these three goals, including application equipment, spray volume per acre, adjuvant selection, spray material rate and formulation, atmospheric conditions, and the physical location and structure of the spray site. In this chapter, we will review how these factors contribute to achieving the goals of efficacy, efficiency, and safety. A companion publication to this chapter is the UC ANR publication *The Safe and Effective Use of Pesticides*, 2nd edition (O'Connor-Marer 2000).

Pesticide Application

For analysis and understanding, it is useful to think of pesticide application as a continuous process with the desired result, safe and efficient pest control, as the final step. Pest control is achieved only if each and every step in the process is completed successfully and sequentially. The process consists of the following steps: mixing, atomization, transport, collection, deposit formation and migration, interaction with pest, and biological effect.

Mixing

The active ingredient of the pesticide must be thoroughly mixed into the carriers to make the formulated product, which is usually diluted with water in the sprayer tank. Other pesticides or adjuvants may be added in the sprayer to create a tank mix for a specific purpose or specific conditions. Common adjuvants act as spreader-stickers, drift control agents, wetting agents, sun and ultraviolet light screens, or antifoaming agents. All of the components of the tank mix must be thoroughly dispersed into each other for uniform application to the target. Combinations of materials should be added to the tank in this order:

1. Wettable powders
2. Flowables
3. Water-soluble concentrates
4. Emulsifiable concentrates

Check the label for specific directions for each material being mixed.

Atomization

The liquid to be applied must be broken up into discrete droplets before being carried to the tree. Hydraulic nozzles are commonly used for liquid atomization, although other devices, such as air-shear nozzles, rotary atomizers, and air-liquid mixing nozzles, are also found on orchard sprayers. Atomizers typically perform three functions: they regulate flow, create droplets from the liquid, and disperse those droplets into an initial pattern. All currently available commercial devices produce a range of droplet sizes from very small droplets that are nearly invisible and whose movement is controlled entirely by wind and air currents to droplets as large as raindrops.

In orchard air-blast sprayers, atomization is accomplished by the combination of the nozzle or atomizer and the moving airstream. Hydraulic nozzles create droplets by the expenditure of hydraulic pressure; rotary atomizers use the centrifugal force generated by a rotating cage or disc to create droplets. Subsequently, the high-velocity airstream further breaks up the droplets, making them significantly smaller than they would be if produced by the same nozzle without the airstream. Air-shear nozzles require the moving air to create a venturi effect, which pulls the liquid into the airstream and forms droplets

from the sheet of liquid. Other atomizers appearing on orchard air-blast sprayers include bi-fluid, or twin-fluid, atomizers. These mix pressurized air and water within the nozzle to create droplets. The size distribution of droplets devices emit depends on the internal geometry of the atomizer, the flow rates of air and liquid, the pressures of air and liquid, and the physical properties of the liquid.

Transport

Moving air is the primary agent transporting droplets from the atomizer of an air-blast sprayer to the target tree. Droplets become entrained in the moving airstream generated by the sprayer fan even before they are completely formed. Gravity is the other significant force acting on the droplets. To understand this, visualize a droplet traveling from the nozzle to the target tree. As the airstream carrying the droplet moves up and away from the sprayer, the velocity of the airstream decreases as calm air mixes with the airstream. The result is a larger mass of moving air moving at a reduced velocity. At the same time, gravity is pulling the droplet toward the ground. As the air velocity decreases, the energy of the airstream decreases; the gravitational force becomes relatively more important and the vertical ascent of the droplet slows. This scenario also demonstrates the importance of producing small droplets that minimize the effect of gravity (table 23.1). Unfortunately, the same characteristic that allows small droplets to be efficiently transported to the top of the trees also makes them susceptible to drift, which will be discussed later.

Collection

Droplets are deposited, or collected, on a leaf or branch by one or more of several mechanisms. These modes of collection are interception, impaction, gravitational settling, diffusion, and electrostatic attraction. All these modes operate to varying degrees during a pesticide application. Each may predominate at a different time or in a different portion of the tree canopy during a single pesticide application.

In general, collection efficiency is influenced by the diameter of the droplet; the diameter or projected area of the leaf, twig, or branch; the velocity of droplets; and the proximity of branches, twigs, fruit, or leaves. Diffusion is most important for very small droplets at an extremely slow (near zero) velocity. Impaction is most important for large droplets moving at a high rate of speed. Interception is significant for larger droplets attempting to move around small twigs at high velocities. Gravitational settling is of primary importance for larger droplets and at low velocities.

The electrostatic force between a droplet and the tree is primarily a function of the square of the distance between them, the charge to mass ratio of the droplet, and the geometry of the tree. The magnitude of this force is relatively small, which implies that it is most effective on small droplets that are at or very nearly at rest when they come close to the target. The mechanics of the process are as follows.

1. A charge is placed on the droplet as it is formed and leaves the atomizer.
2. The droplet is transported to the canopy by the moving airstream.
3. The droplet very nearly comes to a stop very close to the target, which is grounded.
4. The charge contained within the droplet induces an opposite, or mirror image, charge on the target as the like charges within the leaf are repelled by the charge within the droplet.
5. The opposite charges of droplet and leaf create an attractive force between them. The droplet is drawn to and deposited on the surface of the leaf.

Note that steps 3, 4, and 5 of this process take place within a fraction of a second. Electrostatic force is most effective over a very small distance and for small droplets with low velocity and inertia. Thus, to maximize effectiveness, the droplets must still be carried to the target leaf, fruit, or other element by moving air. At that time, electrostatic attraction may supplement the other modes of collection.

Table 23.1. Influence of droplet size on potential distance of drift

Droplet diameter (microns)	Type of droplet	Time required to fall 10 ft	Lateral distance droplets travel in falling 10 ft in a 3 mph wind
5	fog	66 min	3 mi
20	very fine spray	4.2 min	1,100 ft
100	fine spray	10 sec	44 ft
240	medium spray	6 sec	28 ft
400	coarse spray	2 sec	8.5 ft
1,000	fine rain	1 sec	4.7 ft

Source: Dexter 1993.

Deposit Formation and Migration

After the droplet comes to rest on the target tree, the diluent and carriers evaporate. What remains is called a deposit. In some applications, the deposit must migrate within or on the surface of the target; in others, the deposit does not move after forming on the target. Captan is an example of a pesticide that does not migrate through a plant, but rather is intended to protect the surface of the plant from disease infection. Foliar nutrients and systemic pesticides are examples of products that are intended to move into and within the target.

Interaction with the Pest

The deposit must then have some sort of interaction with the pest. Common interactions include simple contact, ingestion, absorption, or creation of a suffocating or barrier layer.

Biological Effect

After successful completion of all previous steps, the desired result is a biological effect. If it is achieved, the proper pesticide, growth regulator, or fertilizer has been delivered in the proper quantities, in the proper form, to the proper portion of the target, and at the proper time. Failure in any of the previous steps can prevent this step, which is the entire purpose of the application.

Sprayers

The equipment used to apply pesticides, liquid fertilizers, or growth regulators and other liquid materials to prune trees comes in a variety of types, brands, sizes, and levels of complexity. This section briefly describes the characteristics of each type of sprayer or component but focuses on the designs most commonly used in prune orchards at the time of publication. Before purchasing or leasing a sprayer, test it in the orchard and complete a performance evaluation.

Ground-Based Sprayers

Most ground-based sprayers for use in prune orchards are either pulled or towed behind a tractor; they are powered by the tractor power take off (PTO). A few are engine-drive units with an engine mounted on the sprayer to provide power.

Air-Carrier Sprayers

This type may also be called an air-blast sprayer, an orchard sprayer, a fan unit, or something similar. The defining characteristic is a single large fan that assists with atomization and provides the primary transport mechanism for the droplets. Within this category, sprayers are often further described by atomizer system used (hydraulic, air-shear, etc.).

Figure 23.1 Hydraulic sprayer. *Photo:* F. J. A. Niederholzer.

Hydraulic Sprayers

The hydraulic sprayer (fig. 23.1) is the classic air-blast sprayer. In prune orchards, a machine with a single axial fan pulls air in from the rear of the sprayer, turns it 90 degrees, and expels the air out toward the target trees in an arc greater than 180 degrees. Fan cowlings, deflectors, and volutes are used to straighten and direct the airflow to the tree and improve reach to the top of the trees. Larger sprayers, often used in orchards with tall trees (walnuts, pecans, etc.) can have two fans and air intakes located before and after the air outlet. However, single-fan PTO sprayers are the most common in California prune orchards.

Nozzles are placed radially in the output stream of air. The hydraulic nozzles can be operated at pressures of up to 30 bar, or atmospheres, which is about 450 pounds per square inch (psi), but pressures in the range of 6 to 10 bar (90 to 150 psi) are more common. The nozzles, arranged symmetrically on the sprayer, usually form an arc of 180 degrees or more. Normally each side is on a separate manifold and can be turned off or on independently of the other side. Each position may have more than one nozzle, particularly for high-volume applications. Nozzles are usually of a hollow cone pattern, although the initial pattern is quickly modified by the airstream.

Liquid application rates for hydraulic sprayers in prune orchards range from 50 to 400 gallons per

acre (gpa). This variation in delivery rate is achieved by changing the number of nozzles operating at each position, altering the orifice size of each nozzle, changing travel speed, or changing all three.

Air-Shear Sprayers

An air-shear sprayer (fig. 23.2) usually uses a squirrel cage or rotary fan to generate airflow, which is typically smaller in volume than from a hydraulic sprayer. This type of sprayer may have a fan for each side, or the flow may be split so that a single fan supplies the air to both sides. A series of low-pressure (20 to 30 psi) nozzles or emitters is placed in the airstream so that the liquid is "pulled" from the nozzles by a venturi effect and sheared into small droplets. The liquid is metered by various means upstream of the nozzles. Generally the flow to all nozzles on a side is uniform. It is metered for each manifold and then fed to the nozzles through a series of small tubes without further restrictions.

Air-shear sprayers are generally operated at 40 to 100 gallons per acre and produce a smaller-diameter droplet than hydraulic sprayers operating at more than 100 gallons per acre. Because of this spray volume per acre difference, air-shear sprayers usually deliver a more concentrated spray solution than do conventional hydraulic sprayers, given that pesticides are usually labeled on a per acre basis. Most air-shear sprayers are powered by tractor PTO.

Rotary Sprayers

Rotary atomizers, powered either by propellers in the airstream or by hydraulic or electric motors, can be used on orchard sprayers. The atomizers can be fitted with a metering orifice for each nozzle. The fan used to transport the droplets to and through the canopy may be similar to what would be found in a conventional orchard sprayer, or it may be a series of discreet fan units arranged on a tower (see below). Generally, rotary atomizers produce a narrower droplet spectrum than does a conventional sprayer; in other words, there is less size difference between the largest and smallest drops produced. Droplet size can be modified by changing the rotational speed or altering the type or geometry of the spinning disc.

Tower Sprayers

Although not often used in prune orchards because of the trees' canopy structure, tower (or mast) sprayers are useful in orchards. These devices use a raised mast on a sprayer that travels between the rows of trees and allows horizontal spray delivery. Tower operators must watch for power lines and avoid damaging the sprayer or trees by contact with large branches.

Hydraulic or air-shear tower sprayers

Some hydraulic or air-shear sprayers can be fitted with a hollow tower to direct air from the fan laterally into the canopy (fig. 23.3). Nozzles are located at set distances along the tower. Maximum height of these units is usually 12 to 15 feet, depending on the manufacturer. These tower units may not be able to travel through a prune orchard, depending on how much canopy overhang

Figure 23.2 Air-shear sprayer. *Photo:* J. K. Clark.

Figure 23.3 Air-shear sprayer with tower. *Photo:* F. J. A. Niederholzer.

occurs in the orchard. In addition, the increased exit area reduces the airstream exit velocity and relative volume as it leaves the fan housing or tower compared with a conventional hydraulic unit. This may result in less canopy penetration than with a conventional hydraulic unit at the same ground speeds, especially in dense canopy situations close to harvest. Canopy penetration should be carefully tested in the orchard.

Remote sprayers with multiple fans

Another type of orchard tower sprayer has a solid mast that is raised between the rows and holds six or more rotary spray heads (fig. 23.4). Atomization is either by rotary atomizer or conventional hydraulic pressure. The fans are placed so spray is directed horizontally into the tree. The location of the fans means that the point of atomization (the point of emission of the spray-laden air) is relatively close to the target. Deposition can be more uniform with a tower sprayer than with a conventional air-blast sprayer with a single fan operated from the orchard floor. The arrangement overcomes many of the geometric limitations of a radial fan near the ground. However, to raise the mast there must be an opening in the canopy between the trees. The output airstream from the fans does expand and merge after exiting the fans, but the mast height must be similar to the height of the canopy to be sprayed. Sprayers with remote discrete fans can be used at very low volumes if the atomizers are capable of operating at low flow rates. Tower sprayers using multiple rotary spray heads can deliver effective dormant prune aphid control, depending on the pesticide used, at 30 gallons per acre.

Electrostatic Sprayers

Electrostatic sprayers are currently being used in orchards. This type of device charges previously atomized droplets by passing them through an electric field generated by the nozzle. The droplets carry the charge as they travel to the target site. There, as described earlier in this chapter, the electrostatic force enhances collection.

The electrical properties of the liquid determine its ability to hold a charge. Liquids require the presence of ions (found in all tap, well, and surface water) to hold a charge, but some liquids can hold more charge than others. This capacity is a function of the surfactants used, the formulation of the product, and other additives in the tank mix. The surface tension of the droplet also helps to determine the amount of charge held within the droplet. If all the liquid molecules in the droplet have a like charge, they try to repel each other. The only force available to counteract this force is the surface tension of the liquid. If too much charge is placed on a droplet, the electrical repulsion forces shatter the droplet into smaller droplets, which then have a smaller charge that can be contained by the surface tension of the droplet.

Figure 23.4 Rotary atomizer, multiple-fan sprayer with tower. *Photo:* F. J. A. Niederholzer.

The critical parameter for charge-assisted spraying is the ratio of droplet charge to droplet mass. If this ratio is too large, the droplets shatter into droplets so small that they are subject to wind-driven drift or are too small to contain a lethal dose of pesticide. If the charge to mass ratio is too small, the electrostatic forces will not be strong enough to enhance deposition in the tree. Atomization and charging must work in concert to allow electrostatic sprayers to enhance deposit efficacy.

The same balancing of forces must also occur in conventional spraying. Collection of uncharged droplets is primarily a function of droplet size, droplet velocity, and the dimensions of the collector (leaf, fruit, or twig). Thus, droplets of many different sizes are not equally well collected if all are moving at the same speed. Similarly, turbulence is necessary to expose all leaf and fruit surfaces to moving spray droplets, but too much turbulence can reduce transport to the farther portions of the canopy.

Aerial Spraying

Fixed- and rotary-wing aircraft are used for aerial spraying. The advantages of using aircraft for applying liquids to prune orchards include the high speed of application and the ability to apply materials regardless of soil conditions. Potential disadvantages include poor canopy penetration and possible drift to neighboring sites. The application volume is usually 10 to 30 gallons per acre in prune orchards.

Fixed-wing configurations

Converting a fixed-wing airplane into a sprayer involves placing 60 to 70 hydraulic nozzles along a boom suspended under the wing and fuselage or suspending 4 to 8 rotary atomizers from the wings. Discussions of specific setups and operation are beyond the scope of this manual. For information, consult a pest control adviser and a licensed agricultural pilot.

Rotary-wing configurations

A helicopter, a rotary-wing aircraft, can be used for spraying when outfitted with rotary or hydraulic atomizers. Rotary atomizers used on helicopters are often powered by hydraulic or electric motors; the helicopter engine is the source of power. Compared to fixed-wing airplanes, helicopters can land and refill in more restricted spaces, but they carry a smaller payload, a fact that can reduce operating efficiency. Again, to determine the optimal setup and operating parameters, consult a pest control adviser and a licensed agricultural pilot.

Components of Application Equipment

For a complete description of components generally used in pesticide application equipment, see *The Safe and Effective Use of Pesticides*, 2nd edition (O'Connor-Marer 2000). This sections discusses components specific to orchard sprayers and controllers, which are changing rapidly due to technological advances.

Monitors and Controllers

Two broad categories of sprayer controllers are currently used. The first category includes controllers that simply adjust the flow rate in response to changes in tractor speed. The second category includes sprayers with sensors to detect the presence or absence of trees or portions of trees. Sensor-controlled sprayers are often referred to as smart sprayers.

Rate controllers use either a radar unit or a wheel rotation counter to measure ground speed. Flow rate is adjusted by changing system pressure or liquid flow using pulse width modulation (PWM). (PWM units are currently available only in sprayers for ground application, not canopy spraying.) Pressure modification has a limited effective range, because flow rate varies with the square of the pressure change. Thus, to increase flow rate by 10% when the base is 300 psi, increase the pressure to 363 psi. Similarly, from a base at 300 psi, a 10% decrease in flow rate requires a drop in pressure to 243 psi. This great a pressure change cannot be made quickly, and it affects atomization from hydraulic nozzles. Some sprayer pumps, such as the commonly used centrifugal pumps, are not capable of so great a change. So, the best use of a controller is to handle minor speed variations due to changes in terrain or traction while using a sprayer calibrated for the desired application rate and travel speed.

Smart sprayers are air-blast sprayers equipped with multiple sensors that trigger nozzle flow when a tree part is detected in the target area of each sensor. Each sensor controls several nozzles on that side of the sprayer. Thus, these sensor systems can detect a tree and its overall shape and apply spray adjusted to the height and width of the tree. This reduces chemical use per acre, drift, and ground deposit. These units are either factory installed or retrofitted to any sprayer. Sensor systems can be shut off and the sprayer operated as a conventional unit.

Smart sprayers can significantly reduce pesticide use per acre. A 39% reduction in spray material applied per acre was obtained in a dormant prune orchard using a sensor-equipped sprayer compared with the same sprayer with the sensor turned off (Giles et al. 2011). Growers commonly report a 20 to 25% reduction in pesticide use and cost per acre in mature prunes when using a smart sprayer. These systems have been commercially available for more than a decade and are especially helpful in cost reduction when used in nonbearing orchards or mature orchards with many replants.

There are at least two different sensor designs. One commercial unit uses ultrasound to detect and range a tree. Another design employs a laser scanning technique. Both types of controllers monitor and record information regarding the application, such as total area covered, gallons used in the application, and total area covered since the monitor was reset.

A new, less-expensive sensor sprayer system has recently become available. This system, available from a number of sprayer manufacturers, is primarily designed for use in young orchards of uniform height with significant gaps between the trees. These units use a single sensor on each side of the sprayer that "sees" the tree trunk. One entire sprayer boom is programmed to be on for a given number of seconds following triggering of the sensor's signal. These units are a compromise between the multiple-sensor smart sprayers and the conventional units with no automatic nozzle control. They are less flexible in their response to different-sized trees in the same orchard than are multiple-sensor smart sprayers, but they do offer significant savings in very uniform blocks. Use of single-sensor units in nonuniform blocks may cause poor spray coverage.

Any controller must be tested and evaluated in the orchards where it is to be used, as must the sprayer. Ensure that the material is being deposited in critical areas of the tree and in a quantity sufficient to obtain the biological effect.

Fans

Fan performance and function is one of the key factors in effective and efficient spray application. Fan pitch may be nonadjustable or adjustable depending on the manufacturer and sprayer model. Sprayers may have a gearbox that allows two settings, high and low, for fan speed. Fan speed and air delivery can also be adjusted by changing the idle speed on an engine-drive sprayer.

For commercial air-carrier sprayers, fans are either purchased from fan manufacturers or constructed by the sprayer manufacturer. In either case, they are balanced when new and must be rebalanced if repairs such as welding or brazing are ever required.

Fan performance is often described in terms of airspeed delivery or volume of air delivered. While rate of air volume delivery (cubic feet per minute, etc.) is a much more valid measurement than air velocity to use when evaluating a sprayer for a particular orchard or group of orchards (see the section "Calibration," below), all fan performance data are based on lab measurements and generated under conditions far different from the field. Regardless of the performance rating reported by the manufacturer, testing the sprayer in the field for its ability to do the required spray application is the only way to determine whether it is suitable for use.

Application Process in the Field

The previous sections have described the process of chemical application, the various factors that influence the task, and the characteristics of the necessary sprayer components. This section discusses considerations that are particularly critical to success of spraying in prune orchards.

Physical Description

In simplest terms, if all surfaces of the tree are to receive pesticide deposits, the entire mass of air within the tree must be replaced with spray-laden air moving at sufficient velocity to cause collection by the leaves, fruit, stems, twigs, branches, and scaffold. In addition, spraying must be of sufficient duration to apply the mass of spray necessary to provide the desired deposit characteristics. The following factors affect this process.

Sprayer-Related Factors

Moving air generated by the sprayer fan (or fans) carries pesticide into the tree. The travel speed of the sprayer must be slow enough so the fan can replace the air in the tree with spray-laden air. The output of the fan and the size and density of the canopy also influence spray coverage due to this air replacement process. The number of droplets per volume of liquid applied is a function of the nozzle type, flow rate, and size; the operating pressure of the liquid supply system; and the air speed at the nozzle. The volume of liquid to be sprayed per acre is a variable determined by the sprayer operator after the operator reads the pesticide label.

The effect of travel speed and the energy output from the fan can be thought of as building a moving column of air that, in effect, pushes the spray droplets up to the top of the trees. If a sprayer is stationary, the spray cloud reaches relatively high into and perhaps even out the top of the tree. At the other extreme, if the sprayer is traveling too fast, the spray cloud does not reach very far up into the tree. A successful spray application requires the proper balance between travel speed and the energy of sprayer output.

In engineering terms, the output from the sprayer fan can be thought of as a rectangular jet of air wrapped around the radius of the fan. The rate of energy decay for a jet of air is determined by its initial velocity, the initial level of turbulence within the jet, and the distance from the exit, or the amount of entrainment of still air around the edges and the subsequent reduction in overall jet velocity. If a forward velocity is superimposed on a jet, the entrainment at the leading edge is increased, while the trailing edge loses the energy pushing it. The rate of decay decreases significantly because of the addition of forward travel. An example of this is that a stationary desk fan, pointed at a candle across the room, will blow out the candle, while the same fan in oscillation mode (moving back and forth in an arc across the room) will not push enough air directly at the candle to blow it out.

Canopy density also reduces the strength of this jet of air: the canopy absorbs some of the energy into the leaves and other structures. A portion of the momentum of the air is transferred to the canopy elements, reducing the momentum available to continue replacing the air enclosed by the canopy.

An essential point is that spray volume applied per acre is not related to the spray coverage quality. Air movement provided by the fan affects where

in the tree canopy the spray reaches. "Poor coverage is possible at any spray volume," says Tim Smith, Washington State University Horticulture Extension Agent.

Canopy Development

Leaves, branches, and fruit impede the ability of spray equipment to push pesticide-laden air into the tree. In spring, as leaves are just forming and the canopy is in its early developmental stages, absence of a full leaf cover facilitates spray penetration. Conversely, at midsummer, a full leaf canopy and full-sized fruit require equipment operators to drive slowly so the sprayer remains in front of each tree long enough to overcome canopy interference and degradation of the air jet. Equipment must be adjusted according to canopy development throughout the seasons if optimal spray applications are to be achieved. Indeed, the same sprayer may not be capable of season-long use.

Alternate-Row Application

Spraying every other row became popular as material and labor costs increased. Besides the obvious cost and time savings, this practice was seen as a way to protect natural enemies from disruptive broad-spectrum pesticides, especially insecticides. Now that fungicides with a single-site mode of action, which are at high risk for resistance development, are the primary disease management tools available to prune growers, spraying every other row has become incompatible with pesticide (fungicide) resistance management programs. Spraying every other row causes low levels of pesticide coverage and fungicide concentration in the target tree tissue on the side of the tree away from the sprayer. This low level promotes pesticide resistance.

Spraying every other row is not an effective pesticide resistance management practice, with one exception. Early in the growing season, when flowers are not fully opened, spraying every other row delivers similar fungicide deposits on both sides of a given tree. So, green bud fungicide sprays can be applied to every other row with limited risk of poor coverage and promotion of fungicide resistance. Beginning at full bloom, fungicide sprays should be applied to every row. Where good coverage is essential for effective control, every row should be sprayed at all times. Spraying every other row for prune aphids results in control on every other row. The most effective pest control is delivered by carefully calibrated equipment driving every row.

Alternate-row application of potassium foliar fertilizers is commonly practiced in prune orchards. This practice is usually part of a season-long program where the unsprayed rows are treated on the next spray in the program. This practice has not been reported to harm tree health. Where a single spray of an immobile nutrient, such as zinc, is planned, application to every row should provide the best results.

Sprayer Selection

When planning to purchase or lease a new sprayer, test the sprayer in the most difficult stand, that is, in the tallest and most dense trees prior to harvest when the canopy is most dense. Ask the manufacturer or sales representative to set up the sprayer as he or she would recommend and place water-sensitive paper in critical portions of the canopy. Spray a few rows with water only, retrieve the water-sensitive paper, and assess the deposit characteristics. Note all operating parameters of the sprayer in order to recreate successful applications and accurately set up and calibrate the sprayer. A similar testing technique involves putting a tracer, such as food coloring or Surround WP (a white clay product registered for use in many crops, including prunes), in the water, spraying, cutting branches from critical tree locations, and examining the deposit pattern.

The techniques of sprayer evaluation just mentioned are biased in favor of high-volume machines against low-volume machines. Use these techniques to compare machines operating at the same volume. To compare machines operating at different spray volumes, such as an ultra-low-volume (ULV) sprayer and a conventional high-volume sprayer, requires more care. Begin by field-testing both, using water-sensitive paper to monitor coverage. Arrange the exposed papers in pairs, each pair consisting of one paper from each test; make sure that both papers in one pair were exposed in the same relative position in the tree. Using a microscope, magnifying glass, or hand lens, count the number of droplets that hit it per square centimeter or per square inch on each piece of paper. Compare the droplet count for each pair to determine which machine produced better coverage.

Relative differences in actual material deposition can be measured by spraying fluorescent dyes or micronutrient fertilizers into the canopy, collecting foliage or fruit from critical areas of the tree, washing them, and quantifying the deposit. By describing the deposit in terms of mass per unit of leaf surface area, a more useful and unbiased assessment can often be made. This approach is more time consuming and expensive and may require equipment not readily available. This method also does not evaluate potential differences in the pattern of spray deposition in the canopy. For example, the same amount of deposit, presented on a per leaf surface area basis, can be caused by one large drop or many smaller drops collecting in that area. The large drop that lands in one corner of the area may not provide the same pest protection as many smaller drops that equal the same amount of deposit on a per leaf area basis as the single large drop.

One other significant source of information is other growers and their impressions of equipment that they have used, seen, or heard about. This information can be most helpful in specialized or localized applications, but it also reflects the biases of the observer. In addition, the source may not be up to date regarding new sprayers. After many years of relatively little change, many innovations are making spraying more efficient and effective. These new sprayers deserve a close look and an unbiased evaluation. The cost of the products applied and the cost of the application itself, in addition to the capital outlay, make time spent evaluating sprayers time well spent.

An orchard sprayer is a major investment. The most informed decision can be made by using all three of these methods to evaluate new sprayers: visual evaluation of coverage on water-sensitive paper or on tree parts from tracer materials, actual coverage per unit leaf area, and experiences from trusted friends or business partners. Given the importance of the job done by orchard sprayers, it is worth the extra effort to arrive at the best machine available for a particular operation and personnel.

Custom Applicators

A custom applicator is an independent contractor who can be hired to do the spraying. Ask custom applicators to demonstrate their equipment and the skills of the operators. Have them perform the tests outlined above for selecting a sprayer. Make sure that the custom applicator has sufficient incentive to make a proper application. Payment on a per-acre basis may not be wise, since it encourages high travel speeds and low volumes of spray per acre. Before spraying begins, agree on some objective assessment of spray deposit that will allow the applicator's performance in the orchard to be verified.

Effective and Efficient Spraying

A pesticide, growth regulator, or fertilizer application that provides the best value to the grower requires that a specific spray volume be applied per acre to ensure that label requirements are met, and it also requires that the material is properly applied to the target. To meet these requirements, the sprayer must be set up to make sure the pesticide is delivered uniformly to the entire target and the sprayer must be calibrated to ensure that the application matches the intended spray volume per acre so that proper chemical rate per acre is applied. Both of these steps must be accurate for the most effective and efficient application to be made.

The continued use of pesticides is influenced by current use or misuse of those products. Professional stewardship of agricultural chemicals and the equipment used to apply them is essential to the future of cost-effective prune orchard management in California.

Orchard Sprayer Setup

Nozzle placement and ground speed largely determine the quality of spray coverage delivered by most air-blast sprayers currently operating in California prune orchards.

Matching nozzle placement and relative size is the first step in orchard sprayer setup. Park the sprayer in the orchard where it will be working. Stand behind it and visualize the portion of the canopy targeted by different nozzle locations. This can also be done by tying 2 to 3 feet of bright surveyor tape behind each nozzle and running the fan at normal operating speed. The tape will move with the airflow and show where the material from each fan will be directed. The relative flow from each nozzle should match the amount of canopy (wood or leaves or both) that the nozzle targets. A general rule is that two-thirds of the spray volume should be delivered to the top half of the tree. This is done to counter gravity, the distance the spray must travel from the sprayer to the top of the tree, and the fact that a great deal of pest damage occurs in the top half of the canopy. Select specific nozzle sizes to fit on the different outlet locations on the sprayer from the nozzle manufacturer's flow rate charts.

The next step is to determine gearing and engine settings that deliver best spray coverage. Adequate air displacement by the sprayer fan is responsible for delivering spray to all portions of the target tree. Decreasing or increasing sprayer ground speed is the most commonly available practice for adjusting air displacement. There is no easy way to determine the optimum tractor speed, that is,

the speed that most efficiently moves the sprayer through the orchard while ensuring good spray coverage. It takes a process of trial and error. Apply tracer (water, Surround WP, etc.) at different combinations of tractor gearing and RPM (or tractor gearing and RPM plus engine settings for engine-drive sprayers). Use the nozzles selected through the process described in the previous paragraph. Manually check coverage using a ladder to climb into the tree or a pole pruner to clip branches while standing on the ground. Another option is to position water-sensitive paper (WSP) in the tree before spraying and check the coverage on the WSP after spraying. Choose the gearing-RPM setting that provides the fastest ground speed and good spray coverage in the top center of the tree, where many pests may occur. Don't use the tractor manufacturer's figure for tractor speed at given settings. Tire size, wheel slip, and characteristics of the load being pulled affect the travel speed, so the actual speed may be different from that estimated by the manufacturer.

Determine the actual tractor speed by timing the tractor or sprayer combination at the decided gearing and throttle setting over a measured distance with the sprayer half full of water, the sprayer going (if PTO rig), and running in the orchard to be treated. Many growers use a distance that includes at least four or five trees in a row. The tractor should be moving at the desired throttle setting in the desired gear as it enters and exits the course. Repeat the test three times and average the results to get a representative value for travel speed in feet per second (see figure 23.8 for speed calculation formulas). If a tractor with a hydrostatic transmission is used, make a mark on the transmission lever to facilitate finding the same speed again. An accurate hand-held GPS unit may also be used for ground speed measurement.

The tractor speed that delivers the best coverage possible will decrease through the growing season as leaves and shoots grow and produce a denser canopy. Since tractor speed should be matched to the tree canopy density, calibration components (tractor speed and spray output rates) should change through the season.

Weed Sprayer Setup

The same principles involved in setting up an air blast sprayer are used in setting up a weed sprayer intended to deliver pesticides—usually herbicides—downward onto the orchard floor. Setting up a weed sprayer should be somewhat easier than setting up an air-blast (canopy) sprayer, as the target is more two-dimensional than a tree canopy. In most weed spraying, coverage is closely influenced by spray volume. Nozzle spray tip angle and height of the spray boom to the target influence proper nozzle spacing along the spray boom. Key components of weed sprayer setup include width of spray swath, nozzle spacing along the boom, and spray pattern angle. Spray swath width is affected by tree age (wider for young trees), distance between rows, and mower size. Nozzle placement is affected by nozzle tip angle and boom height (see the nozzle manufacturer's charts for nozzle spacing on the boom). A tape measure and a dry, hard surface can be used to measure the actual width of the spray band delivered while the tractor is not moving. The uniformity of spray delivered to the orchard floor can be evaluated using tools such as the Tee Jet Pattern Check (from Spray Systems, Inc.). This tool can be used to spot-check nozzle wear as well. Ensuring that an herbicide application is applied uniformly to the orchard floor means that the intended rate per acre of pesticide is actually what is applied by any given sprayer. This eliminates herbicide over-application (with potential risk to crop) and under-application (with risk of incomplete weed control).

Aerial Sprayer Setup

Calibrating a sprayer for use in the air is more difficult than calibrating a ground-based sprayer, but the principles remain the same. Travel speed can be estimated by flying both with and against the wind and averaging the two values of indicated airspeed. Swath width is determined by flying over a line of cards and examining the results, or relying on the past experience of the pilot. Flow rates are usually based on a new set of nozzles and the operating pressure of the spray system. More sophisticated methods can be used, involving radar guns, pressurizing the spray system, and collecting the liquid from each nozzle, or analyzing a string laid perpendicular to the flight path to determine the exact application parameters.

Calibration

The process of matching the desired spray volume per acre (grower choice from label options) and sprayer speed (determined during sprayer setup) with an appropriate sprayer output rate is called calibration. Recalibrate whenever any application parameters change. Such changes could include a new sprayer; a different number of nozzles; a new or different type of nozzle; a new tractor, tree spacing, new spray volume per acre; a change from tilled to mowed soil; or any of many other possibilities. The two basic parameters that must be assessed in calibration are spray output or flow rate, and travel speed or land rate.

Some pesticide labels state that the product should be applied only through properly calibrated sprayers; thus, if the sprayer has been calibrated improperly, applying the product is a violation of law. To help establish compliance, keep calibration calculations on file with documentation regarding pesticide or growth regulator application. Also, it is the professional responsibility of anyone qualified as a pest control operator (PCO), or has a qualified applicator's license (QAL) or qualified applicator's certificate (QAC), to make sure that the sprayer is performing as intended and to avoid misuse.

How to calibrate

The flow rate (gallons per minute) and land rate (land area sprayed per minute) are the key measurements. Tools needed include a watch with a second hand (an inexpensive stopwatch works very well) and a calibrated bucket or pitcher (or hose with a flow meter).

Flow rate is measured by determining spray flow per minute from the entire sprayer or the sum of the output from each individual nozzle. The latter practice is more accurate, as it can be used to identify worn or damaged nozzles, saving money in wasted pesticide and delivering effective pest management.

To measure entire sprayer output, park the tractor and sprayer on a flat spot, fill with clean water until water overflows from the top hatch, and run the sprayer at the settings determined for the proper land rate for several minutes. Refill the tank using a calibrated bucket or a hose with a flow meter until the sprayer is just about to overflow to determine how much spray was actually delivered. Divide the gallons actually sprayed by the time sprayed to determine flow rate in gallons per minute.

To determine flow rate for each nozzle of an air-blast sprayer, attach flexible nylon tubing to each nozzle you plan to use in a particular orchard using small stainless steel hose clamps to hold the tubing on the nozzle. Run the tubing from each nozzle into a separate clean jug to catch the water flow during testing. Fill the clean sprayer with clean water. Set the fan gear to neutral. Run the sprayer (with the tractor and sprayer parked) at the settings established during sprayer setup (pressure, PTO or engine speed, etc.) and capture the volume sprayed from each nozzle for a minute or longer in a jug or bucket. Measure the amount of liquid applied in a set amount of time. Sum the volume delivered by all nozzles to determine sprayer output per minute. Replace nozzles that are not within 10% of the output listed in the manufacturer's catalog.

To catch the output of individual nozzles on a weed sprayer, run the clean sprayer (filled with clean water) with all the nozzles open that you plan to use, and catch the output from each nozzle in a calibrated jug or pitcher. Again, sum the total flow from all nozzles to determine output per minute. Replace nozzles that deviate ±10% from manufacturer's specifications.

Divide the flow rate (gallons per minute) by the land rate sprayed (acres per minute) to determine the actual spray volume per acre (gallons per acre) applied by the sprayer at the proper setup settings determined above.

Land rate is readily calculated using the sprayer ground speed (determined in sprayer setup) and the spray swath width. The spray swath width is equal to the across-row spacing in the orchard width for air-carrier sprayers (fig. 23.5) and from the inside edge of the swath to the treeline for weed sprayers (fig. 23.6). If two booms are used in weed spraying, calculate the width of both spray swaths.

With the sprayer calibrated to apply a known volume per acre, the proper amount of pesticide or nutrient can be added along with the volume of water in the spray tank to meet label requirements and deliver the desired outcome of the spray application.

For help calibrating sprayers for specific blocks at specific times of the year, and to keep accurate records of these practices for future reference, use the worksheet in figure 23.7 and the formulas in figure 23.8. Nozzle wear affects spray flow, so sprayers should be recalibrated at least once per year to make sure the flow rate has not changed.

Drift Management

The third component of an effective and efficient spray application is the safety of nontarget organisms (humans, fish, pets, nonlabeled crops, etc.) outside the sprayed field. Although applicator and handler safety is an important part of a safe application, this topic is beyond the scope of this manual; see *The Safe and Effective Use of Pesticides* (O'Connor-Marer 2000). This section discusses off-site movement of pesticide, particularly from direct spray drift as droplets or vapor.

Attitude and training

Complete elimination of off-farm movement of spray materials is virtually impossible, given that spraying is, at best, the semicontrolled aerial release of small droplets some distance from the target. It is

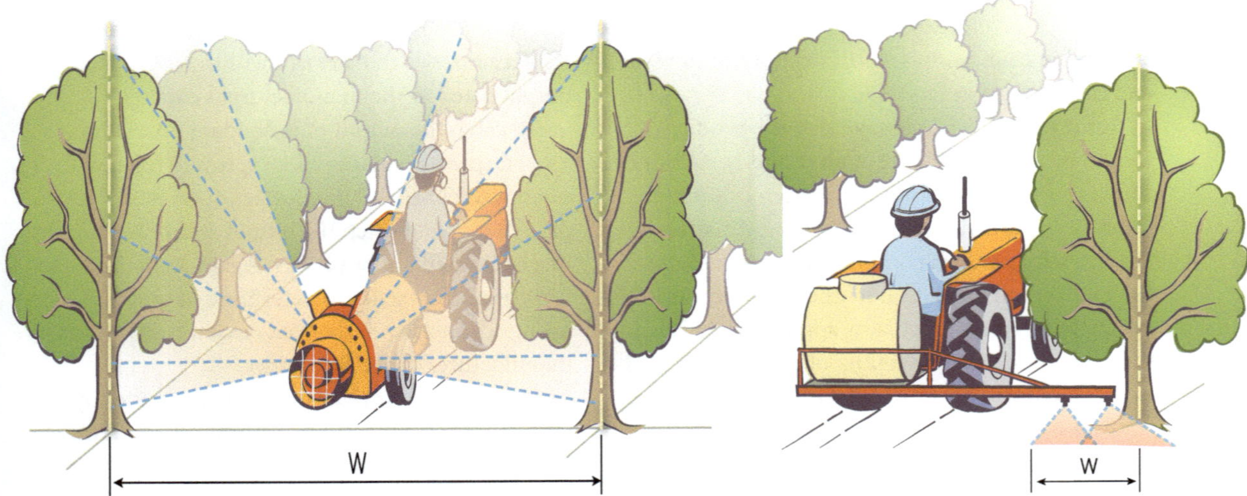

Figure 23.5 Spray swath measurement for air-carrier spraying.

Figure 23.6 Spray swath measurement for strip-spraying weeds.

the obligation of the application team—all individuals involved in pest management, including growers, applicators, manager, and pest control advisers—to ensure that spray drift is managed so that there is a reasonable certainty of no harm to nontarget organisms. The proper equipment should be set up and maintained to maximize efficient targeting of the crop. A safety system should be in place that ensures that best management practices (BMPs) for sprayer operations are followed before and during a spray application. The BMPs described below are similar for both weed sprayers (spray directed down) and air-blast sprayers (spray directed upward or horizontally). Every orchard location is unique, so BMPs for a certain block may not be exactly the same as another, but the process of developing those practices should be the same.

Best management practices

Plan ahead. Know the label requirements of each material planned for use. Monitor pest populations so a timely application can be made as needed when risk of drift or runoff is low, as better control is often obtained when applications are properly timed. Avoid spraying when winds are above 10 miles per hour (too windy) or below 3 miles per hour (higher inversion risk) or just before a significant rain event or irrigation (higher risk of runoff).

Everyone involved in the spray application should be aware of sensitive areas such as roads, homes, adjacent crop fields, open water, and so on that could be harmed by drift. Special precautions should be taken to avoid spraying at certain times, or spray practices should be changed to protect sensitive areas, for example, spraying near a road very early in the morning with the help of a spotter, spraying into the outside one or two rows of an orchard that is adjacent to a sensitive area, spraying next to a sensitive area when the wind is blowing away from that location, or using a low-pressure hand application adjacent to the sensitive area. (Additional suggestions to protect sensitive areas are found later in this section.)

Equipment should be set up for the most efficient application possible. This includes using sensor-guided orchard and weed sprayers, placing properly sized nozzles in the correct positions on the spray boom, and adjusting airflow and directional vanes to match the air delivery to the target.

Manage the volume of airflow. The amount of air used to help carry the pesticide should be managed to minimize movement of the spray beyond the target trees. During fall, dormant, and bloom sprays, reducing the volume of airflow can save money in fuel costs while reducing drift. Using the low gear on the sprayer, throttling down the engine on an engine-drive sprayer, or using a "Cornell donut" air intake restrictor on PTO sprayers with no gearbox reduces airflow to match the target orchard. The goal is to use only enough air to carry the pesticide to the opposite edge of the target tree or just a little beyond.

Manage the droplet size. Droplet size is the single most manageable factor that affects spray drift (see table 23.1). Larger droplets are heavier and so drop faster than do small droplets. Select nozzles that deliver the largest droplets possible without compromising spray efficacy. Information helpful in nozzle selection may come from pesticide manufacturers, PCAs, university researchers, farm advisors, and others. Consider using different nozzles in the same orchard for different spray jobs, especially if the orchard is close to a sensitive area (road, edge of property, other crop, open water, etc.)

Often, larger droplets do not provide the same degree of spray coverage as do smaller droplets, so make a compromise between coverage and droplet

Fill in as per your spray operation

A. Tree row spacing, in feet = ________

B. Gallons of spray desired per acre (gpa) = ________

C. Speed of travel selected in orchard (mph) = ________

Calculate (use below formulas)

A. Gallons per minute (gpm) necessary per side

$$GPM = \frac{\text{gpa x mph x tree row spacing}}{1{,}000} = \_\_\_\_\_\_\_\_$$

B. ⅔ of above gpm (for top ½ of manifold) = ________

C. ⅓ of above gpm (for bottom ½ of manifold) = ________

Check (with a pressure gauge)

Your sprayer pressure at the manifold = ________

Select nozzles

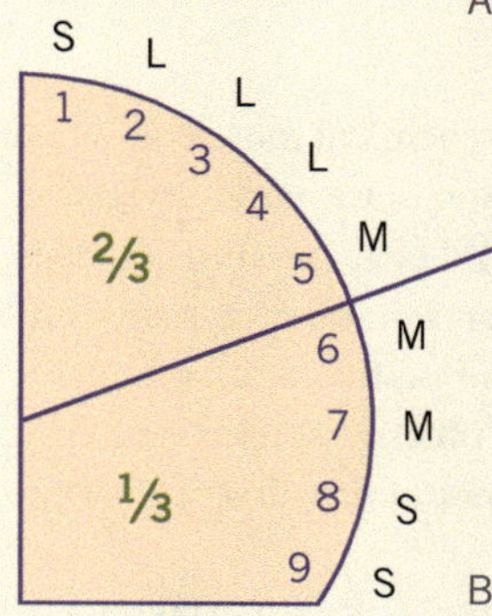

A. Obtain a manufacturer's nozzle chart for your type of nozzle tips (orifice sizes) from the column on the chart having the same pressure (psi) as the pressure at which your sprayer operates. Working with only one side of your sprayer, first select tips for the top ½ of the manifold that total approximately ⅔ of the gpm discharge for one side. The balance of gpm discharge (⅓) should then come from the nozzles on the bottom ½ of the manifold. Preferably these nozzle tips should not all be the same size but in general arranged so that those nozzles in the 2, 3 and 4 positions are the largest; 5, 6 and 7 the next to largest; and 1, 8 and 9 the smallest. This is done to conform to the tree's height, width, the distance the spray must travel, and the density it must penetrate. If your sprayer has more or fewer nozzles than the example, adjust the gpm/nozzle accordingly but maintain the ⅔ and ⅓ relationship. In each case, the total gpm discharge of all nozzles should equal the amount calculated.

B. Now place the same nozzle arrangement on the other side of your sprayer, and the proper nuzzling steps have been completed.

Speed of travel

Check your desired speed in the orchard by traveling the required number of spaces in 1 minute.

$$\text{Tree spaces/min} = \frac{\text{mph x 88}}{\text{tree spacing}}$$

Speed check

A. Spray out an acre or part of an acre of trees with plain water in order to check the pattern as well as the total gallons per acre (gpa) output. If the spray pattern needs adjusting in regard to the observed penetration and deposit, rearrange the nozzles accordingly. For example, the larger nozzles may need to be moved up or down on the manifold to better conform to tall or low bushy trees.

B. If the gallonage per acre is off, check whether your original pressure (psi) reading was wrong, whether your nozzle size selections were incorrect, or whether your speed of travel was too slow or fast.

Figure 23.7 Calibration worksheet for spraying trees with an air-blast sprayer.

size. Systemic pesticides do not require complete coverage to be effective, so larger droplets are more compatible with these materials than, for example, protectant fungicides that require excellent coverage for effective pest control. Venturi-type nozzles (sold under names such as Air Induction, TurboDrop, ID, etc.) that deliver larger droplets than do standard nozzles are very effective in reducing drift when spraying weeds with systemic herbicides (glyphosate, etc.), but tests with these nozzles in airblast sprayers have yielded inconsistent results.

Drift control adjuvants can increase droplet size and may help reduce spray drift. Many drift control adjuvants are on the market; some are effective, some are not. Carefully consider whether to use a drift control adjuvant, as the performance of the adjuvant can affect deposition characteristics (coverage) as well as droplet size.

Growers must balance the spray location, the pesticide being applied, the pest pressure, spray conditions, and distance from the orchard to sensitive areas when choosing nozzles, spray pressure, or drift reducing adjuvant to influence droplet size for a particular application.

Manage the spray pressure. Lowering the spray pressure within the nozzle manufacturer's recommended pressure range can reduce the number of small droplets produced by conventional nozzles

GALLONS PER MINUTE = (gpm, for one side of sprayer)	$\frac{\text{gpa} \times \text{mph} \times \text{row spacing (ft)}}{1{,}000}$
GALLONS PER ACRE = (gpa)	$\frac{\text{gpm (for one side)} \times 1{,}000}{\text{row spacing (ft)} \times \text{MPH}}$
MILES PER HOUR = (mph)	$\frac{\text{gpm (for 1 side)} \times 1{,}000}{\text{gpa} \times \text{row spacing (ft)}}$
ACRES PER HOUR = (ac/hr)	$\frac{12 \times \text{mph} \times \text{row spacing (ft)}}{100}$
SPEED OF TRAVEL = (expressed as tree spaces passed per minute)	$\frac{\text{mph (desired)} \times 88}{\text{tree spacing (ft)}}$
NUMBER OF TREES = PER ACRE	$\frac{43{,}560 \text{ (square feet per acre)}}{\text{row spacing} \times \text{tree spacing}}$
For example:	$\frac{43{,}530}{30 \times 30} = 48$ trees per acre

Figure 23.8 Formulas useful for calibrating air-carrier sprayers. Replace the "row spacing" value with the sprayed width in feet to use many of these formulas for weed spraying when less than the entire orchard floor is to be sprayed.

without reducing efficacy. High spray pressure increases the risk of equipment failure (hoses, seals, etc.). Follow the sprayer manufacturer's recommendations for the sprayer pressure range.

Avoid low-humidity or high-temperature conditions. Spraying in low-humidity or high-temperature conditions, common during much of the summer daylight hours in the prune-producing regions of California, produces increasingly smaller droplet sizes as the spray moves away from the sprayer due to evaporation. The risk of drift increases with decreasing humidity and rising temperatures. Avoid spraying when the difference in wet and dry bulb temperatures is 8°C or larger. This is referred to as Delta (Δ) 8. A psychrometer is needed to determine wet bulb temperature. Combinations of dry bulb thermometer readings (°F) and % relative humidity (RH) that equal Δ8 include: 86°F and 50%; 77°F and 43%; and 68°F and 37%.

Reduce the distance from the nozzle to the target. The shorter the distance the droplet has to travel, the better the efficiency of the spray, since there is less chance for drift from wind, evaporation, and misdirected application. Herbicide sprayers can reduce the distance by lowering the boom, while orchard air-blast sprayers can use a tower to get the nozzle closer to the target.

Manage spray volume. Optimize the volume of material sprayed to balance the chemical concentration in spray droplets with the economic efficiency of application. High-volume sprays deliver droplets containing lower chemical concentrations than do low-volume sprays when applying the same rate of chemical per acre. High-volume sprays also often deposit more of the chemical on the orchard floor. They may be considered when spraying close to a sensitive area, so that less chemical moves off-site in a sudden change in the direction of the wind. Irrigation timing and rain forecasts should be taken into account when considering using higher-volume sprays adjacent to sensitive sites, due to the risk of pesticide run off.

Use protective shields. Protective shields on herbicide sprayers help reduce movement of fine droplets away from the target.

Select materials less prone to drift. Material selection can be a key part of drift management, as certain materials, such as 2,4-D, are more prone to drift via volatilization (changing from a liquid to a gas). In addition, using a lower-toxicity pesticide, even if it costs more, when spraying close to sensitive areas will help reduce harm from any potential drift that may occur.

Turn off sprayer when exiting any target row. This basic practice can reduce drift in any spray application, whether it be weed or canopy spray. If the applicator shuts off the sprayer outside the tree row regardless of the location in the orchard, the practice will more likely be followed when it is most necessary. "If you always do it, you always do it" is a good strategy when managing drift.

References

Dexter, A. G. 1993. Herbicide spray drift. North Dakota State University Extension Publication A-657. NDSU website, http://www.ag.ndsu.edu/pubs/plantsci/weeds/a657w.htm.

Giles, D. K., P. Klassen, F. J. A. Niederholzer, and D. Downey. 2011. Smart sprayer technology provides environmental and economic benefits in California orchards. California Agriculture 65(2): 85–89.

O'Connor-Marer, P. J. 2000. The safe and effective use of pesticides. 2nd ed. Oakland: University of California Agriculture and Natural Resources Publication 3324.

Part 5 Prune Harvest and Postharvest Management

24 Fruit Maturity and Harvest Management

• G. Steven Sibbett and James T. Yeager

Profitable prune production requires a consistent yield of high-quality prunes. Although considerable attention must be paid to horticultural inputs during the growing season for high yield and quality, improper harvest management can negate in-season efforts. This chapter reviews the role of prune fruit maturity and predehydration harvest management in achieving the highest dry yield of best-quality fruit to maximize profit.

Prune Fruit Maturity

Prune fruits begin maturation 20 to 30 days prior to harvest maturity. The first, most obvious development is skin color as it starts to change from green to purple. After the initiation of skin color changes, the flesh changes from green to yellow. Coincident with these changes, fruits soften and starches within the fruit are converted to sugars, increasing sweetness. These latter two physiological changes are easily measurable and are the most important indices of prune maturity for scheduling harvest.

Prune Fruit Maturity Indices

Flesh firmness

Prune fruits soften as they ripen. They are considered to be at horticultural maturity and at their best quality when the external pressure required to penetrate the flesh with a 5⁄16-inch-diameter tip declines to 3 to 4 pounds-force (lbf) for the French and Muir Beauty cultivars, a maturity index referred to as flesh firmness. At this point, both maximum sugar content and best potential dry fruit size have been attained. It should be noted that although the Sutter cultivar is also at horticultural maturity at 3 to 4 lbf, its best dried fruit quality (i.e., without slabbing) will occur when harvested at 5 to 6 lbf. It is currently recommended that Sutter be harvested before fruits soften below 4 lbf for that reason.

As prunes begin to color from green to purple, it takes approximately 10 to 12 lbf pressure to penetrate the flesh. This pressure requirement declines approximately 1 to 2 lbf per week as ripening proceeds, depending on temperature; very hot temperatures (> 100°F) slow the process. Flesh firmness decline is a physiological process and is the most important prune maturity index because it takes place independently of crop load or tree condition.

Flesh firmness can be accurately measured with a penetrometer fitted with a 5⁄16-inch plunger tip (fig. 24.1) to measure pound-force. Measure flesh firmness on one randomly selected (exposed or unexposed) cheek of a fruit (not at the suture) by using a very sharp knife or potato peeler to remove a portion of the skin with a very thin portion of the flesh and then

Figure 24.1 Penetrometer, used to test prune flesh firmness. *Photo:* M. L. Poe.

applying enough pressure with the instrument to penetrate the remaining flesh (fig. 24.2). The flesh firmness will be displayed on the instrument's dial or gauge as pound-force. The cheek reading is then recorded for each individual fruit in the sample (see the section "Sampling to Determine Fruit Maturity," below).

It is important that only the skin and not flesh be removed when preparing fruits for flesh firmness measurements. Measuring flesh firmness deep within the flesh will give an erroneously low reading, as these cells soften first. The plunger may also come in contact with the pit, giving an invalid result.

Soluble solids content (SSC)

The soluble solids content is essentially a measure of a fruit's sugar content. As prune fruits ripen, soluble solids increase steadily until they reach horticultural maturity, that is, 3 to 4 lbf flesh firmness. The ultimate soluble solids level that prune fruits from a given orchard can achieve depends on crop load and in-season culture. Once horticultural maturity is reached, a measurable increase in percentage of soluble solids will still occur as fruits dehydrate and lose weight (water loss), but actual soluble solids content does not increase beyond horticultural maturity (3 to 4 lbf flesh firmness).

Soluble solids content is measured as a percentage with a refractometer (fig 24.3.). The technique requires that juice be extracted from each fruit in a given sample and placed on the instrument to determine the juice's percentage of soluble solids content. Readings are most accurate when a seedless half of each fruit in the sample (see "Sampling to Determine Fruit Maturity," below) is placed in a blender and ground, and the resulting juice strained through cheesecloth (or similar media) onto the instrument. With this method, all cells of the fruit sample are ruptured and an accurate representation of the total soluble solids contents can be obtained. An alternative method requires squeezing, combining, and mixing a standard number of drops of juice from each half fruit in the sample and placing a portion of that mix on the instrument. This technique is biased toward a higher reading because the riper cells with a higher soluble solids content rupture first, often precluding the juice of firmer cells with a lower sugar content from being included in the sample.

Soluble solids content is an important component of profit. It directly influences both yield (weight) and fruit size (quality).

Figure 24.2 The correct procedure for preparing a prune for testing flesh firmness. *Photo:* M. L. Poe.

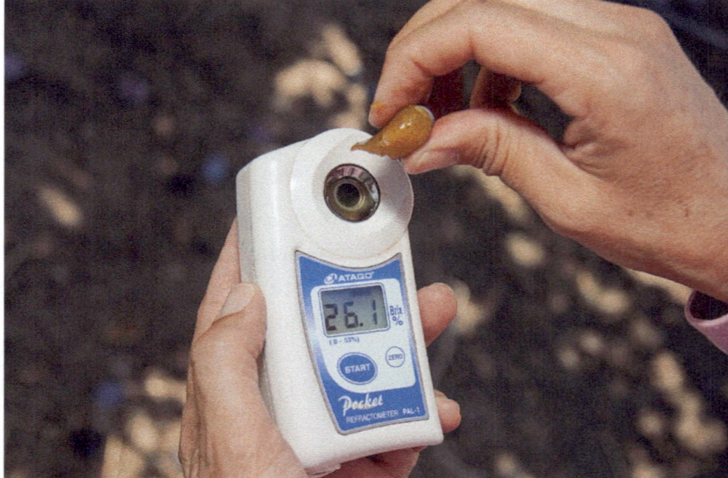

Figure 24.3 Refractometer, used to measure soluble solids. *Photos:* M. L. Poe.

Yield

Harvesting when prunes reach maximum sugar content (i.e., at horticultural maturity of 3 to 4 lbf) translates directly into the best potential dry yield. For example, 1 ton of fruit with a percentage soluble solids content that has a drying ratio of 3 fresh to 1 dry ton yields 666 pounds of dried fruit. That same ton, harvested earlier at a lower percentage of soluble solids content that had a drying ratio of 3.5 fresh to 1 dry, yields 571 pounds of dried fruit, 95 pounds (16%) less dried fruit per ton. Of course, harvesting after horticultural maturity, when some on-tree dehydration has already occurred, has an even lower drying ratio because there is less water to remove. But the ton that is obtained when dried to a given percentage of moisture content will be the same; it simply required less moisture removal. Harvest delays can have significant economic disadvantages (see below).

Fruit size

Soluble solids content directly influences dry fruit size, the "value component" of yield. For example, 22 fresh prunes per pound containing 18% soluble solids has a drying ratio of 3.5:1, giving 77 dried fruit per pound, whereas the same 22 prunes at 22% soluble solids content has a drying ratio of 3:1, giving 66 fruit per pound, an approximate 16% increase in fruit size. Harvesting at maximum soluble solids content, as determined by flesh firmness measurements (true horticultural maturity), is essential for best yield and quality.

Sampling to Determine Fruit Maturity

Begin sampling the orchard when prunes begin to color. At this time, flesh firmness will approximate 10 to 12 lbf. Use the following procedure:

- A sample consists of 25 fruits from each 10-acre block. Take 5 fruits at shoulder height or above from five trees (flag the sample trees), making sure to sample proportionally from the trees' inside and outside fruit. The sample trees should be randomly selected within the orchard. Do not select edge trees or trees atypical of the block.
- Sample each block each week. Make sure to sample the same trees each time.
- Record the date of sampling, average the percentage of soluble solids concentration and flesh firmness from each fruit in the sample on a data sheet (fig. 24.4).

Note that uneven fruit maturity exists within the tree. Fruits in the lower inside portion of the tree mature later and have lower sugar content than fruits on the well-lighted periphery of the tree. Fruits that mature late and have a low sugar content are unprofitable and detract from the crop's overall quality. During pruning, efforts should be made to eliminate fruit growing in low, shaded portions of the tree and allow light to those fruit remaining.

Harvest Management

Harvest management involves harvest timing, in-field sizing of fresh prunes, and cold-storing fresh prunes prior to dehydration.

Harvest Timing

It is important that harvest be timed as close to horticultural fruit maturity, 3 to 4 lbf (5 to 6 lbf for Sutter, above horticultural maturity), as possible to maximize profit. As a practical matter, however, it is usually impossible to harvest the entire crop at horticultural maturity due to the size of the acreage, dryer prorate, or harvest equipment limitations. Attempting to harvest most fruit at horticultural maturity usually requires harvesting a small portion before horticultural maturity and some of the crop overripe.

To time harvest as close as possible to horticultural maturity and for maximum profit, consider the following.

Crop size. Harvest light crops before heavy ones. Light crops mature sooner than heavy ones. Their fruits will have a higher sugar content at any given state of maturity and dry down to a profitable size even if maximum sugar content has not been attained. Heavy crops, on the other hand, should be harvested as close to horticultural maturity as possible so fruits gain their highest percentage of sugar and dry to their best potential size.

Fruit drop. During crop maturation, some fruit drop because they are overripe; these are lost from the crop. Harvesting as close to horticultural maturity as possible usually means that some fruit will have fallen while the bulk of the crop reaches maturity. Harvesting before any fruit drop will yield largely immature fruit that have an excessive drying ratio, less-than-optimal fruit size and value, poor skin color, and less-sweet flavor. On the other hand, delaying harvest past horticultural maturity will yield both excessive crop loss due to drop and poor-quality dried fruit. Slabbing, slip-skins,

Preharvest Data Sheet
Orchard Block Number ____________________

Fruit sample number	Sample dates				
	Flesh firmness readings (lbf)				
1					
2					
3					
4					
5					
6					
7					
8					
9					
10					
11					
12					
13					
14					
15					
16					
17					
18					
19					
20					
21					
22					
23					
24					
25					
Average firmness reading (lbf)					
Percent soluable solids					

Figure 24.4. Sample preharvest data sheet.

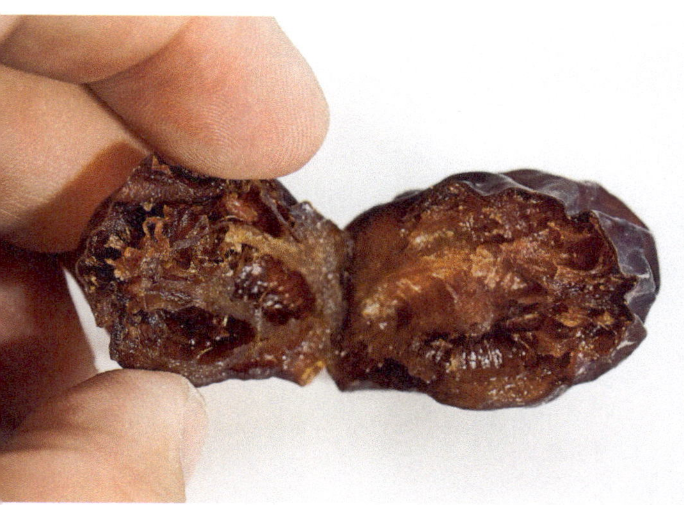

Figure 24.5 Internal flesh discoloration after drying overripe fruit. *Photo:* M. L. Poe.

Figure 24.6 A chain and bar sizer attached to a harvester. *Photo:* M. L. Poe.

internal flesh browning, and gas pockets occur in overly mature prunes (fig. 24.5).

Dehydrator prorate. Most prune dehydrators, especially in heavy crop years, prorate the amount of fruit growers can deliver during the peak season. To obtain the highest return, time and coordinate harvest with the dryer to account for potential prorating.

In-Field Sizing of Fresh Prunes

The most important component of prune value is the size of dry fruit. The value of small dried prunes varies with the crop year, but in most years excessively small fruit have less value than the cost to harvest, haul, and dehydrate them.

In-field fresh fruit sizing is a common practice in California prune orchards to eliminate small prunes and avoid the cost of their processing. Either chain or, more commonly, adjustable bar sizers are attached to the harvester to eliminate small fruit as the crop is conveyed from the harvester into bins (figs. 24.6 and 24.7).

This process must be managed carefully to avoid discarding valuable fruit. For example, with a given sizer setting early in the harvest period when the soluble solids concentration is relatively low and fruit are firm, undersized fruit can reliably be removed. Later in the harvest, at the same sizer setting, when soluble solids concentration accumulation has peaked and fruit are softer and larger, more valuable fruit can easily be discarded. Careful, continuous monitoring of fruit being discarded and adjusting sizers accordingly is required if in-field sizing is to be profitable.

Note that crop load and in-season management influence dried fruit size. Small fruit are caused by overcropping, inadequate nutrition, and poor water management. See chapters 15, 16, and 12, respectively, for more information about these components of crop management.

Cold-Storing Fresh Prunes

At peak harvest, the ability to harvest often exceeds the capacity to dehydrate the fruit. In this situation, dehydrators prorate the amount of fruit growers can deliver each day. In addition to the inability to harvest for a full day, prorates extend the harvest period, which increases the potential for loss of crop to natural fruit drop and more overripe fruit.

Research has shown that French prunes can be safely stored at 32° to 36°F for up to 3 weeks before dehydration provided the fruit are rapidly cooled; vented bins and forced-air cooling are required. Harvesting early in the day while fruits are cool minimizes cooling time. If proper bins and storage

Figure 24.7 Small fruit eliminated in the field. *Photo:* M. L. Poe.

facilities are available, cold storage can allow fruit to be harvested in a timely manner and dehydrated later. Because each grower's situation is different, the cost of cold storage and additional handling must be weighed against the economic benefit of a timely harvest.

Summary

Early harvest, before horticultural maturity is reached, reduces total tonnage and fruit size. Conversely, late harvest causes direct crop loss to fruit drop and quality loss. Harvest management that times harvest as close to horticultural maturity as possible provides an important monetary return to the grower.

25 Growing Prunes for the Fresh Market

• Kevin R. Day and Richard P. Buchner

Traditionally, virtually all of California's prunes were dried after harvest. Prunes have a very high sugar content that enables them to be dried without fermenting at the pit. Prunes in the dry form are very convenient for long-term storage. Depending on industry demand, dried prunes are rehydrated and processed into marketable product. Over the years, enterprising prune growers have developed a small market for selling prunes as a fresh product similar to plums. Large fruit is hand harvested and usually picked with higher flesh pressure and lower sugar content than prunes bound for the dehydrator. The fresh market is not a large industry, but it does represent a significant use for prunes.

Industry Development

Fresh market prunes have been a recent addition to the California plum industry. Records from the California Tree Fruit Agreement indicate that in 1985 there were 10,000 boxes of prunes shipped, less than 0.1% of the total plum production, marking it as the first season in which more than 1,000 boxes of prunes had ever been shipped (fig. 25.1). The following year, that figure exploded to 250,000 boxes, or about 2.3% of total plum production. Production grew steadily and peaked in 1995, when 625,000 boxes of prunes were shipped, representing about 7.1% of the total plum crop. (Note that reliable figures are not available from 1991 to 1993 due to the temporary cessation of the California Plum Marketing Order.) Recently, the demand for fresh market prunes has softened, and production has decreased: fresh market prunes currently represent about 2 to 3% of the total plum crop.

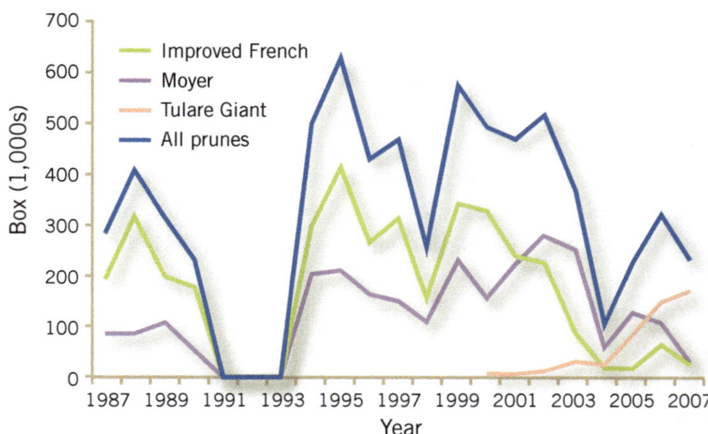

Figure 25.1 California fresh market prune production 1985–2007.

Varieties

Traditionally, Improved French and Moyer were the major varieties of prunes packed and shipped fresh in large quantities. Improved French is characterized by superior soluble solids content and smaller size than Moyer. While Moyer is preferred over French for its superior size, it lacks the higher soluble solids concentration of French. These disparate attributes allowed the presence of both varieties in the market for some time to meet consumer preferences. In the year 2000, the Regents of the University of California released the variety Tulare Giant. A product of the UC Davis Prune Breeding Program, it was intended exclusively for fresh market production. Production of Tulare Giant has risen steadily since its introduction, and it has now supplanted both Improved French and Moyer in terms of production and marketing importance, currently comprising more than 70% of the total fresh market plum production. Tulare Giant is characterized by an earlier harvest date, excellent quality, and much larger potential size than Improved French or Moyer. Since it is not self-fertile, it requires cross-pollination to set a sufficient crop. The currently recommended pollenizer is the UC variety Muir Beauty. Both

bloom 7 to 10 days earlier than traditional prune varieties.

As with other fresh fruits, fruit size is of vital importance to the grower and consumer. Large fruit is in greater demand, is worth more money, and is easier to harvest, pack, and handle. Consequently, anything that can be done to improve fruit size is of benefit to the grower. Fresh fruit sizes are expressed as the number of fruit needed to fill a 10-pound sample. Thus, a 120-count sample contains larger fruit on the average than does a 150-count sample. More than 80% of the French prunes shipped fresh are unsized or are in the 140 and 150 count categories. About 50 to 70% of the Moyer fruit fall into the 80 to 100 size categories, and more than 50% of Tulare Giant fruit are commonly in the 60- to 80-count range and larger. These differences represent the genetic fruit size potential of the varieties. Beyond this, growers must implement cultural practices that further benefit fruit size.

Cultural Practices

Pruning

Production of high-quality fruit begins with proper pruning. Because fruit size is so important, fresh market prune trees are pruned significantly more than trees managed for the dry market. Pruning is similar to that performed on Japanese plums, with the center of the tree opened to allow light penetration for improved fruit coloration, tree height reduced to allow for easier harvest, removal of long whips, and heading back of spur-bearing units to improve vigor.

All of these actions reduce flower number and crop load. This in turn improves fruit size, color, and soluble solids. Hand thinning most varieties of prunes is rarely a viable option because it is so expensive (see "Thinning," below). Consequently, it is important to remove enough wood during pruning so that fruit set is not excessive. The proper level of pruning will vary between orchards, and it may take several years to establish what is best for a particular location. A very rough starting guideline would be to remove about 30% of the spurs on a tree. Using this formula, spurs would be replaced about every 3 years, helping to ensure young, strong spurs.

Unlike prunes intended for drying, the fruit surface color is very important in the fresh market. Buyers and consumers tend to equate color with quality, and highly colored fruits are often considered more valuable than those that are green, despite similar fruit soluble solids concentrations. Maturing fruit need sufficient light exposure for proper color to develop. Trees should be opened up to sunlight, but care must still be taken to ensure that interior branches do not sunburn. Again, shortening the length of fruit-bearing shoots and removing whips usually achieve this. Also, because fruit from the interior of the tree is generally of poorer color and quality, there should be no premium placed on trying to retain those fruits, and most of this interior fruitwood should be eliminated. Additionally, some secondary or tertiary branches can be removed to allow for large holes in the tree to allow light into the canopy.

Tree height is usually limited to 12 to 13 feet. Heights greater than this make it too difficult and dangerous to harvest with conventional 10-foot and 11-foot ladders. Because the largest, sweetest, and best-colored fruit are at the top of the tree, it is necessary to have a tree structure that allows this fruit to be easily accessed. Care should be taken to avoid routine heading or topping of all shoots in a tree, as this may cause excessive shoot growth at the top of the tree that will increase shading.

Summer pruning is rarely practiced on fresh market prunes, but it can be performed if trees are excessively shaded. Summer pruning usually involves the removal of vigorous interior and exterior shoots that are not needed for fruit production in subsequent years. Where sunburn potential exists, especially on the interior scaffolds exposed to afternoon sun, head shoots to a length of 4 to 8 inches rather than removing them. This will provide some shade to the scaffolds but still allow for an improvement in light penetration.

Like other prune orchards, young trees should usually be long pruned. This will help bring the trees into production earlier. Summer pruning and light heading can be used on young trees to help improve spur development on exterior fruiting shoots. Branch bending can also reduce vigor and induce fruitfulness of shoots, but care should be taken to avoid bending branches so flat as to cause sunburn.

Moyer prunes are notoriously difficult to fruit when the trees are young, but they make excellent orchards from a fruit size standpoint throughout their lives. Young Moyer trees should always be minimally pruned, and dormant heading cuts should especially be avoided to help bring the trees into production as soon as possible.

Tulare Giant prunes, on the other hand, are very precocious and require dormant heading of the main leaders to prevent them from losing vigor and bending over. For the same purpose, it also is common to shorten fruiting shoots.

Thinning

French prunes intended for the fresh market are rarely, if ever, thinned by hand, as are Japanese plums. Nor are they machine thinned. Fruit thinning is accomplished by dormant pruning. In general, if fruit set on these trees is so heavy as to need hand thinning to bring fruit up to fresh market size ranges, the tree can be considered too heavily cropped to economically divert to the fresh market. Such trees should instead be managed for the dry market.

Moyer prunes are sometimes hand thinned. This typically amounts to quickly breaking up fruit clusters. Again, because of the small size of the fruit, it is difficult to perform extensive thinning economically. The bulk of the effort of limiting crop and improving fruit size should be put into pruning.

Because of the tree structure and bearing habit, it is necessary to hand-thin Tulare Giant trees. This provides improved fruit size and greater sugar concentrations, prevents tree and limb breakage, and ensures sufficient crop in subsequent years.

Tree Nutrition

Since fruit color and harvest date can be very important in successful fresh market operation, care should be taken not to overfertilize trees. Excessive nitrogen fertilization can cause significant delays in fruit maturity and can also increase shading and further reduce fruit color. Trees should be monitored annually by tissue sampling, and nitrogen should be in the lower range of the established critical levels. In general, for fresh market prunes, it is better to have nitrogen concentrations that are too low rather than too high.

Prune tree potassium needs are high and must be met. Leaf scorch tends to be less on fresh market trees because crops are significantly lighter than on trees intended for drying. However, because of the important role that potassium plays in fruit color, it is best to continue on a good potassium fertilization program in all orchards.

Irrigation

Because of the great importance of fruit size, fresh market prunes are a little more sensitive to irrigation stress than are drying prunes. Additionally, care should be taken not to overwater trees so as to cause excessive vigor that might reduce fruit color and delay maturity. Many orchards intended exclusively for the fresh market have low-volume irrigation systems that allow for much greater irrigation precision. Applying excessive amounts of water should be avoided prior to harvest. This will not pump up the size of the fruit but can reduce soluble solids concentrations and delay maturity. As long as trees are receiving 100% of their evapotranspiration needs, fruit growth, quality, and orchard yield will be maximized.

Fruit Maturity

The current maturity standards for prunes include both a surface color and soluble solids concentration requirement. For prunes of all types, the fruit surface must be at least 50% mottled red, and the balance of the surface a light green color. Unlike any other fresh market stone fruit, there is a soluble solids standard as well. Moyer prunes must average 16% soluble solids concentration, Tulare Giant 16.5%, and French 19%.

Harvesting

Fruit are harvested by hand into normal fruit bins that contain about 1,000 pounds of fruit. Harvesting is often done on a tree-by-tree basis to avoid trees or portions of orchards that are overcropped. Those trees or areas with poor-sized fruit are generally skipped completely. Because of the expense in harvesting an orchard, it is rare to harvest more than once. Under the best of circumstances, an orchard may be harvested twice, but this is unusual. After the fresh market harvest, the remaining fruit are then used for drying, although this option is usually limited only to Improved French orchards.

Harvest costs for fresh prunes are usually two to three times higher than those for Japanese plums. This is primarily a function of fruit size, although tree cropping patterns can play a role as well. Because of this, growers must constantly monitor orchards, harvest crews, costs, and markets to determine whether the harvest activity is profitable or should be discontinued.

Other Cultural Practices

Few, if any, other cultural practices are significantly different than those used for drying trees. Care should always be taken to ensure that trees are not vigorous enough to reduce fruit color and delay maturity.

Packing and Marketing

After harvest, fruit are transported to the packing shed, where they are sorted for grade and size. Fruit are generally packed into volume fill containers that hold 28 pounds of fruit. Fruit are then cooled. Because of their high soluble solids content, they can be held in storage for several weeks without noticeable deterioration in quality.

Most of the prunes that are packed for the fresh market are exported to Asia. People in that region tend to place a very high value on fruits with high soluble solids content and refer to prunes as sugar plums. Currently, Taiwan and Hong Kong are the largest markets for California fresh prunes. Significant volumes of fruit are also sent to Singapore and Malaysia.

Major Diseases and Pests

Russet Scab

The external appearance of fresh market fruit is very important, and trees and orchards intended for the fresh market should be protected against russet scab. Essentially, the tolerances for fruit surface injuries are very small, and affected fruit cannot be packed. Annual bloom sprays to reduce russet scab are necessary. For further information on this disorder, see Chapter 18, "Diseases and Physiological Disorders."

San Jose Scale

For cosmetic reasons similar to those described above, San Jose scale can also become a significant pest of fresh prunes. While drying prunes can tolerate small levels of infestation without much of a problem, fresh prunes must be free of damage in order to be packed. All orchards should receive a dormant spray and must also be monitored for scale. The presence of fruit damaged by scale generally means that additional sprays, usually in-season, are warranted. For further information on monitoring and controlling San Jose scale, see Chapter 17, "Insect and Mite Pests."

Mites

Because of the limited potential for fresh market trees to withstand reductions in fruit growth, mites must be kept to a minimum. Population treatment threshold levels are lower than those that can be tolerated in drying prunes. Again, a good monitoring and scouting program is essential.

Worm Pests

Both codling moth and oriental fruit moth can be significant pests of fresh prunes. This is often because fresh prunes are grown near other stone fruit that serve to also attract these pests. As with other insect pests, the amount of damage that a fresh market orchard can withstand is very low. Accurate monitoring is essential.

References

California Tree Fruit Agreement. Annual report: 1987, 1997, and 2007. Sacramento: The Agreement.

26 Dehydration and Storage

• James F. Thompson and Judy Johnson

California prunes are dried with heated-air dehydrators. Freshly harvested prunes are loaded about two layers deep on trays and stacked on wheeled carts. The carts move in a concurrent flow direction through a horizontal tunnel with air heated to 185° to 195°F. Sixteen to 20 hours of drying bring the fruit to 18 to 22% moisture content. Minor changes to and proper management of dehydrators can minimize the use of natural gas. Insect damage to the dried prunes can be minimized by good storage sanitation, keeping the prunes at recommended storage moisture content, and the occasional use of fumigation based on insect trapping.

Dehydration

Until the early 1920s, California prunes were dried in the open sun. The prunes were dipped in a hot solution of water and lye to produce small cracks in the skin that hastened drying, then spread on wooden trays and set out in the sun. When the prunes were two-thirds dry (in 5 to 10 days), the trays were stacked and left to complete the drying process. The entire process usually took from 10 to 14 days. In 1925, about 30% of the 145,000 dry tons of prunes produced that year were dried in heated, forced-air dehydrators (Mrak and Perry 1948). Today all prunes are dried in dehydrators heated with direct-fired natural gas or propane burners (fig. 26.1). Dehydrator configurations have varied in the past, but most tunnels used a counter-current design with the dry fruit exiting the hot (165°F) end of the tunnel (figs. 26.2 and 26.3). Fruits for the dehydrator were originally placed on wooden trays similar to those used for sun drying, but the trays were stacked on low-wheeled trucks (usually called cars) and dried as a unit. In the 1960s, research showed that prunes could be dried in a tunnel with a concurrent (parallel) flow design in about 30% less time than with countercurrent designs. Since that time, most of the countercurrent flow tunnels have been converted to concurrent flow, and

Figure 26.1 Typical commercial dehydrator. *Photo:* M. L. Poe.

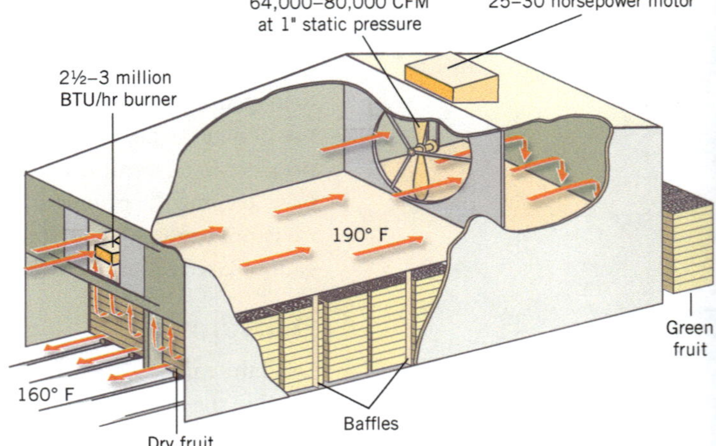

Figure 26.3 Another early style of prune dehydrator. *Source:* Perry et al. 1946.

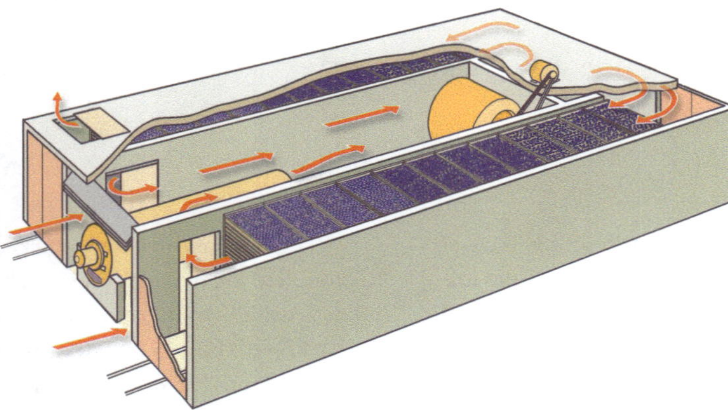

Figure 26.2 Early style of prune dehydrator. *Source:* Perry et al. 1946.

new tunnels are built with a concurrent flow design (figs. 26.4 and 26.5).

Since the early 1970s, the cost of natural gas, the primary fuel used for drying, has increased dramatically, and there does not appear to be a cheaper source of heat for existing tunnels. However, Thompson et al. (1981) demonstrated that improved tunnel management and minor changes in tunnel design can reduce energy consumption by about 30% (table 26.1). Most dehydrators have been modified to increase air recirculation and tray loading has been maximized, producing natural gas consumption in the range of 22 to 26 therms (100,000 Btu) per ton of undried prunes. Thompson and Studer (1981) demonstrated that improved methods of fruit handling may make sun drying an economic alternative to heated forced-air dehydrators.

Figure 26.4 Schematic of a typical nine-truck concurrent flow (parallel flow) dehydrator.

Prune Dryer Operation

Prunes typically arrive at the dehydration facility by truck in bulk bins (1,000 lb of fresh fruit per 4-ft by 4-ft by 2-ft bin). The bins are emptied into a cleaning unit that separates dirt, leaves, and twigs from the prunes and washes them with water. The prunes are loaded onto wooden (or sometimes metal) trays 3 feet wide by 6 feet long, weighing at least 3.5 pounds per square foot. The trays are stacked on a low four-wheeled car that is loaded into the hot end of the dryer. Every 105 to 135 minutes, a car of dried prunes is pulled from the cold end of the tunnel and a car with fresh prunes is loaded into the hot end. Sixteen to 20 hours are required to dry a car of prunes.

Instructions for Operating a Parallel Flow Dehydrator

Starting

1. Load tunnel with cars filled with empty trays.
2. Start fan and burner to heat tunnel to about 185°F.
3. Put in first loaded car after tunnel is heated.
4. Add loaded cars at intervals of 1½ to 2 hours.
5. After 13½ hours, the first car to enter the tunnel should be in position 9 and ready to be removed. Check moisture 20 minutes prior to removal.

Table 26.1. Summary of methods to reduce energy consumption in prune dehydrators

Method	Energy savings (%)
Install passive air-to-air heat exchanger on exhaust air to preheat fresh air.	15
Seal air leaks, especially leaks around doors and motor well.	5–10
Increase recirculation of drying air, which usually requires adding doors to exhaust end of tunnel and installing an air duct to supply fresh air to the burner.	5–15
Fully load trays; a 3-by-6-ft tray should hold at least 60 lb of fresh fruit.	0–25
Insulate roof and place radiant heat shield above burner.	3–5
Properly install and maintain burner.	0–7
Avoid overdrying fruit.	0–10

6. Set the thermostat at the hot end of the tunnel to 185° to 195°F for normal operating conditions. Small prunes or prunes that have been partially dried in the field may require slightly lower temperatures.
7. Make slight adjustments of the cycle time and hot end temperature to obtain proper drying.

Running

1. Once dryer is operating, the loading time interval and temperature should be changed only slightly to compensate for prune size and time of harvest. Parallel flow is a timed operation. In fact, each tunnel can be set on a timer and warning horn to tell the operator when to add another car.
2. Periodically check the moisture content of the dried prunes.
3. Although the cold end temperature will approach 165°F when it is time to remove a car, the temperature should never exceed 165°F or prunes will caramelize (burn). Adjust the fresh air intake to maintain the highest possible wet bulb temperature (122° to 125°F) that will not significantly increase drying time. Closing the fresh air intake increases recirculation and causes the wet bulb temperature to rise. Opening the doors decreases the wet bulb temperature.
4. If the prunes in the ninth car have not completely dried when its normal removal time has arrived, the car can be moved into the tenth car position for extra drying.

Stopping

1. Remove the ninth car at the normal time and add a car of empty trays at the hot end.
2. At the next normal time interval, remove the next two cars in line and add cars of empty trays at the hot end.
3. Lower the hot end temperature 5° to 10°F to make sure that the cold end temperature does not exceed 165°F.
4. Continue removing cars at normal 1½-hour intervals, lowering hot end temperatures by about 5°F each time a car is removed until all loaded cars are removed and replaced with cars of empty trays.

Storage

After drying, the prunes are scraped from the wooden trays and loaded into pallet bins (2,000 lb of dried fruit per 4 by 4 by 4-ft bin). The moisture content of the prunes is typically in the range of 18 to 22% wet basis moisture after drying; they must continue drying to a safe, long-term storage moisture content of 16 to 18%. Prunes near the long-term storage moisture content are transferred into a new bin (turned) to facilitate final drying. Turning also facilitates moisture equalization between overdried small fruit and underdried large fruit. Bins of the wettest fruit are left in the open sun and turned several times to finish drying.

After final drying, the bins are placed in an unrefrigerated building for storage until processing. Inadequate drying or accidental rewetting will encourage the growth of fungi on the prunes. The actual safe storage moisture is somewhat variable and depends on the prunes' sugar content and temperature. A better measure of storage stability is water activity, abbreviated a_w (fig. 26.6). Water activity is the decimal equivalent of the relative humidity of the air in equilibrium with the fruit; it accounts for variation in sugar content and other compositional characteristics of the prunes. Water activity below 0.60 in prunes prevents fungal growth, although safe water

Figure 26.5 Prune dehydrator. *Photo:* M. L. Poe.

activity increases slightly as the temperature of the prunes decreases below 70°F.

Storage Insects and Their Control

Although the drying process exposes prunes to temperatures sufficient to kill any insect, prunes are immediately susceptible to infestation by several insect species. Control of these pests is important because there are no tolerances for live insects, and infestations can reduce consumer confidence in the product. The most commonly found insects in stored prunes are Indianmeal moth, driedfruit beetle, and sawtoothed grain beetle.

The most serious pest of stored prunes is the Indianmeal moth (fig. 26.7). Indianmeal moth larvae cause damage by reducing quality, both by feeding directly on the prunes and by contaminating them with excrement, cast skins, cocoons, silk, and living or dead insects. Under ideal conditions, Indianmeal moth can complete its life cycle in less than 30 days. During the summer months, eggs will hatch in 2 to 4 days, and larvae can develop to maturity in 3 weeks. At cooler temperatures, development may take several months. Indianmeal moth survives the winter as mature, diapausing larvae, usually in cocoons formed in protected locations. Larvae normally enter diapause in late October or early November. As temperatures warm in the spring, larvae pupate and emerge as adults 4 to 9 days later. Adults are normally first observed in April or May.

The sawtoothed grain beetle is primarily a pest of stored grain products, but it is also the beetle most often found on stored dried fruits (fig. 26.8). The narrow, flat, dark brown beetles are quite small, about ⅛ inch long, with six "teeth" on either side of the thorax (just behind the head). Larvae pupate in or near a prune, producing a cocoon by cementing together fine particles of food. Adult sawtoothed grain beetles may live as long as 2 years. They begin laying eggs 1 or 2 weeks after emergence, laying 3 to 10 eggs each day for several months. Female beetles produce an average of 280 eggs, with some individuals producing more than

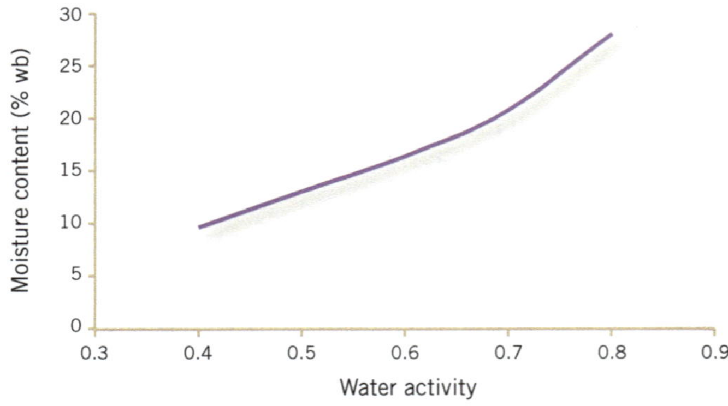

Figure 26.6 Water activity of dried prunes.

400 in their lifetime. The developmental time from egg to adult is about 27 days in the summer, but it may last for several months at lower temperatures. Both larvae and adults cause damage by feeding and soiling prunes with feces. The sawtoothed grain beetle prefers dry foods and is more of a problem in prunes that have been in storage for some time.

The driedfruit beetle (*Carpophilus hemipterus* (L.)) is more commonly a pest of field-dried raisins or figs, but it can be attracted to improperly dried or stored prunes. Adults are about ⅛ inch long and are dark brown with two light brown spots on their backs. Adults and larvae cause damage by feeding directly on prunes and by soiling them with feces and cast skins. Adults may live for several months, and females may produce up to 1,000 eggs. Development is rapid during the summer; eggs hatch in 2 days, larvae develop in 11 days, and the pupal stage lasts about 8 days. Larvae pupate within cells formed in the soil or directly within prunes; no cocoons are formed. Adults, larvae, and pupae are all capable of overwintering. Adults are strong daytime fliers; flight occurs when temperatures are above 64°F. Larvae require food of high moisture content, usually over 30%, with some yeast growth, and adults are attracted to fermenting fruit. Properly stored prunes should have no damage from driedfruit beetles.

Management Guidelines

Recently dehydrated prunes should be free of insects; management practices that prevent infestation by storage insects are most useful. Dehydrators and storage facilities should be kept as clean as possible to minimize populations of Indianmeal moth and sawtoothed grain beetle. Spilled and waste prunes should be removed quickly, and empty storage bins should be cleaned before new prunes are added. Maintaining prunes at the proper moisture content levels and removing rotting prunes prevent driedfruit beetle infestations. Storing prunes below 50°F usually prevents the development of damaging pest populations.

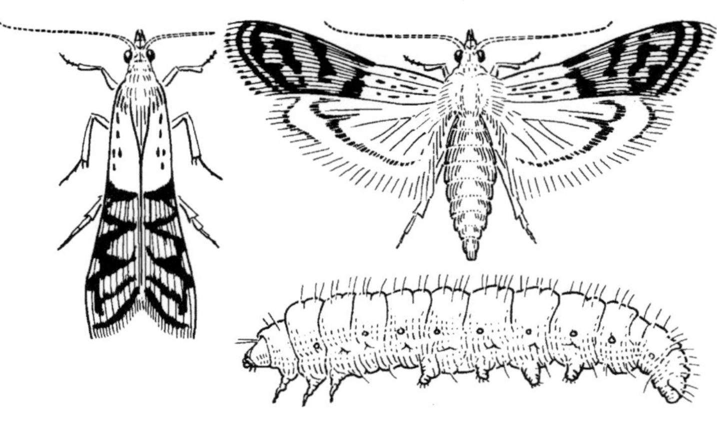

Figure 26.7 Indianmeal moth.

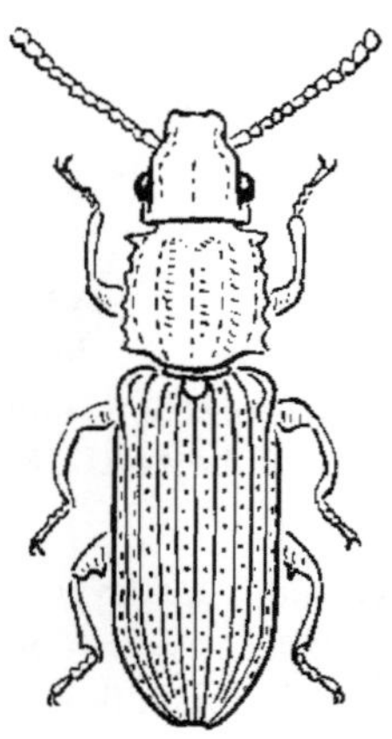

Figure 26.8 Sawtoothed grain beetle.

As insect populations develop slowly on dried prunes, processors often rely on a single fumigation of the entire facility after all prunes have been placed in storage, usually in late November. Occasionally, additional fumigations may be needed in the late spring or summer. The need for additional treatments can be determined with commercially available traps baited with sex pheromones or food lures, or by visual examination of the facility. Automated dispensing systems of synthetic insecticides can be used on a periodic basis to control exposed insects.

Historically, the most commonly used fumigant for dried fruits has been methyl bromide. Due to environmental restrictions, this fumigant is no longer available except for quarantine treatments or through critical use exemptions. Phosphine may be used in some cases, but it is less suitable for treating entire storage facilities. Other fumigants such as sulfuryl fluoride may prove useful as replacements for methyl bromide. Currently, nonchemical treatment methods under consideration include use of subfreezing temperatures, low oxygen atmospheres, and methods to exclude pests. These methods are commodity treatments only and are not suitable for treating entire facilities, but they may be coupled with improved sanitation procedures to provide an adequate level of control.

For more information, see the UC IPM Prune Pest Management Guideline at the IPM website, http://www.ipm.ucdavis.edu.

References

Adams, R. L., and J. F. Thompson. 1985. Improving drying uniformity in concurrent flow tunnel dehydrators. Transactions of the ASAE 28(3): 890–892.

Gentry, J. P., M. W. Miller, and L. L. Claypool. 1965. Engineering and fruit quality aspects of prune dehydration in parallel and counter flow tunnels. Food Technology 19(9): 121–125.

Mrak, E. M., and R. L. Perry. 1948. Dehydrating prunes. California Agricultural Experiment Station Circular 383.

Perry, R. L., et al. 1946. Fruit dehydration. 1. Principles and equipment. Berkeley: California Agricultural Experiment Station Bulletin 698.

Simmons, P., and H. D. Nelson. 1975. Insects on dried fruits. USDA Agricultural Handbook 464.

Thompson, J. F., and H. E. Studer. 1981. Sun drying of prunes on continuous trays. American Society of Agricultural Engineers Paper No. 81-3051.

Thompson, J. F., M. S. Chinnan, M. W. Miller, and G. D. Knutson. 1981. Energy conservation in drying of fruits in tunnel dehydrators. Journal of Food Processing Engineering 4(3): 155–169.

27 Grade Standards, Inspection, and Processing

• Greg Thompson, George Sousa Sr., Patrick Ferreira, Franz J. A. Niederholzer, Mark Krause, and Ken Peterson

The California prune industry is highly competitive in the world market because the industry works continually to meet consumer preferences and improve quality and food safety. These keys to successful marketing are supported by industry grade standards and fruit inspection, which help growers and packers maintain quality and match market demands. P-1 grade sheets report inspection results to growers so they can adjust cultural and harvest practices to improve quality. Maximizing fruit sugar while maintaining fruit integrity are vital to good quality and efficient processing of dried prunes.

The Importance of Quality

It has been said many times that California agriculture leads the world in terms of quality, food safety, and environmental protection. The California prune industry is certainly part of that proud heritage. Yet leaders cannot rest on their laurels and expect to remain leaders. So the industry must continue to improve quality and meet the needs of consumers. With prunes, quality, food safety, and environmental protection begin with the grower and the cultural care provided to the crop in the orchard. An understanding of quality factors and market trends helps growers and the California prune industry not only to remain competitive but also to continue the fine tradition of leading the world in production of high-quality prunes.

It is important to understand what we mean by "quality" when it comes to prunes. Quality factors include appearance (size, color, gloss, freedom from defects and skin damage), texture, and flavor. Industry standards for cosmetic factors allow for a small percentage of substandard fruit that does not meet grade. It goes without saying, however, that there is no allowance for fruit that fails to meet safety factors such as pesticide residues above allowable limits or presence of human pathogens or foreign materials. The industry actively monitors fruit production and processing and educates its members to guarantee the safety of prunes sold to consumers.

The consumer ultimately defines what constitutes a good-quality prune. Beyond food safety, beauty is in the eye of the beholder. So as consumers' needs change in the future, the term "quality" may mean quite different things than what we might have in mind today. Nonetheless, certain basic standards are not likely to change. Most of these were worked out years ago and are reflected in grade standards still in use today. For example, most consumers don't want to see defects, so prunes with cracks, splits, and scab are considered substandard.

Fruit size is also a major part of what constitutes the basic quality of prunes. Larger prunes are meatier, more tender, and sweeter. So larger size will likely always remain a mark of better quality. Flavor is part of the quality equation as well. Good cultural practices and proper harvest timing produce a better-tasting prune. The care that California growers and dehydrators put into growing and drying the crop enhances the reputation of California-grown prunes for consistency and flavor.

In today's markets, the trade demands quality and size, and consumers expect convenience and ease of use. Pitted prunes now account for well over 80% of prune sales in the domestic market (excluding prunes sold for prune juice). With such a high demand for pitted prunes, most packers are looking for crop deliveries with average fruit size in the 60-count per pound range. The days are over when a grower could grow 80-count prunes and expect to survive. Small prunes don't pit well, and juice and concentrate manufacturers are moving away

from using smaller prunes for juice because of their higher acid content.

It is useful to consider that the grower really has two customers, the packer or processor and the buyers who buy from the processor. The packer needs a certain pack and size of fruit to match the market outlets. Again, most packers, like most buyers, want to see clean fruit of good size. While the grower may not always be able to deliver exactly what the packer needs because of the variability in production from year to year, growers and packers benefit when they strive to grow and process what best fits the market.

It would be nice if the story ended there. However, in today's market environment growers are facing new challenges, and what constitutes doing a quality job of growing prunes can seem like a moving target. For example, in 1990, California became the first state to require full reporting of agricultural pesticide use—due in part to consumers' concerns about dietary exposure to pesticide residues. The record keeping and reporting burden fell on growers, but the real challenge came as "quality" in the consumer's eye came to mean not only large, good-tasting, and free of defects, but also grown with few or no chemical inputs.

For a number of reasons, the trend in food quality is toward increased food safety, reduced use of pesticides, and more natural products. While the issues vary by market and consumer needs, one trend is very clear: buyers are demanding that fewer pesticides be used. Safe pesticide use and reporting is critical. The California Dried Plum Board tests prunes for residues, so the industry has documentation to back up the safety of its product. Growers have become more sophisticated and are better able to pinpoint whether sprays are needed, and they have been reducing pesticide usage through integrated pest management programs. These efforts are important to show that growers are using pesticides responsibly and only when necessary.

Quality is also closely tied to the industry's ability to compete in world markets. In order to maintain our advantage, we must have the best quality. Costs are higher in California, so we produce a premium prune that demands a higher price. In today's market, we have the advantage that South American producers tend to grow smaller prunes, and much of that production is sun dried and not well suited for pitting. France grows very large prunes for the whole prune market, but the fruit has large pits and is not as well suited for pitting as are prunes from California. While today we have an advantage in producing prunes that are well suited for pitting, things can change. For example, some growers in South America are adopting California growing and drying practices. In sum, growing global competition increases the need for growers and processors to keep an eye on market trends and to be able to adapt new practices, since quality is relative and our competitors are constantly trying to close the gap between the quality of their prunes and ours.

A Brief History of Prune Inspection

In the old days, prunes were inspected by the packer at the door of the packing facility. Hence the name "door test." Prunes were delivered in burlap bags on wagons or trucks from the grower's barn to the packer. A probe was used to cut the sacks and sample the prunes; so another name for the original way of sampling field deliveries of prunes was the "sack test."

With the advent of the federal marketing order in 1949, with its quality standards and volume regulation, a neutral third party was needed to inspect and certify both incoming and outgoing prunes. The Dried Fruit Association (today known as DFA of California, or DFA) was named as the inspection agent, and they worked closely with the marketing order to provide inspection certificates to growers and packers and also provide statistical data regarding volume and quality for the Prune Marketing Committee.

In the 1950s, with the advent of increased mechanization and the resulting change from field lugs to harvest bins, new methods of sampling were needed. Prunes were no longer gathered in piles in barns and sacked for delivery. By the 1960s, it was common practice for DFA inspectors to go out into the field to the dehydrators to sample growers' prunes. The inspectors would sample the prunes from bins, taking ten 1-pound samples to determine the average size count. They would also collect a 100-ounce sample and analyze the defects on the spot.

Eventually, samples were collected and taken to a central location for sizing and grade analysis. Sizing was done in a series of circular drums. The sample was moved through a set of drums with progressively larger holes. The overs were the prunes left in the last drum. This procedure was used to determine the percentage of a given size range in each load. The DFA continued to use the 100-ounce sample to determine the percentage of defects.

In due course, an in-line grader was developed that saved the labor of moving the prunes from one drum to the next. This modified grader was adopted as the standard in 1974. For some years, samples from Sunsweet growers were collected in lug boxes and delivered to the DFA. Samples from independent growers were collected in bags.

In 1975, the Technical Grading Subcommittee was established. This was an important innovation to make sure that prune defects were being scored appropriately. The meetings provided oversight of the scoring of defects and an open forum for growers to voice any concerns they might have. The whole process of prune inspection is very important because the resulting grade sheets determine the value per ton that growers receive for their production.

Incoming Inspection of Prunes Today

Prune grade standards were regulated under a federal marketing order from 1949 through the 2004 season. In 2005, the federal marketing order and its grade standards were suspended indefinitely, and industry-wide mandatory inspections came to an end.

In 2005, the Prune Bargaining Association incorporated the grade standards from the federal order (table 27.1) into their field price agreement. This was done since inspection standards were no longer required under the federal order and no standards were adopted by the industry under the state marketing order.

With suspension of the federal order, the DFA provides private inspection services under contract to packers and reports their findings to the packer. If the DFA collects the sample, a certificate of inspection is issued for the delivery. If the packer pulls the sample, the DFA issues a statement of findings for the sample provided by the packer. Grower identities are kept confidential, so there is no fear of favoritism. Whether a grower has 20 bins or 400, the grading is done consistently.

Most packers sample incoming fruit as it is turned onto the size grader at their processing plants, but samples may also be drawn as prunes are turned from one bin to the next or by an automatic sampler. Maximum lot size is 30 "ton boxes," or 60,000 pounds if prunes are delivered in other sizes of containers. Samples are retained for a minimum of 30 days, should a question arise about size or grade.

Customarily, a DFA inspector draws a representative sample of about 40 pounds by scooping fruit as it is dumped from bins. The inspector seals the sample in a bag along with a P-4 form, which identifies the sample by grower name, weight certificate number, and an assigned P-1 number. Samples are consolidated into harvest bins and transported to the DFA grading station in Yuba City.

Once the sample is received at the DFA grading station, it is run over a series of screens that resemble a scaled-down version of a commercial size grader. This grader has become known as the modified grader since it is a modified version of a commercial grader.

The modified grader consists of five separate screen sizes:

- a trash screen of ¼-inch slots
- an undersize screen of $^{23}/_{32}$-inch- or $^{24}/_{32}$-inch-diameter holes
- a D screen of $^{26}/_{32}$-inch-diameter holes
- a C screen of $^{28}/_{32}$-inch-diameter holes
- a B screen of $^{30}/_{32}$-inch-diameter holes

Category A prunes, or overs, are prunes that do not fall through any of the screens and go over the end of the grader. Each sizing section consists of three 2-foot-square screens so that prunes run over a 6-foot length of each screen size.

A sample box for each size category collects the prunes as they fall through the screens or go over the end of the grader. After all prunes have cleared the grader, the material from each box is weighed, and all prunes are counted by a mechanical counter. Undersize prunes are not included in the calculations of size counts.

As the prunes are run through the mechanical counter, a subsample is taken for defect analysis. Depending on the weight in the size category, 100, 200, 300, or 400 prunes are counted out and placed in a paper bag. Each subsample is kept apart so that each size category is scored for defects separately. Inspectors visually examine each prune in these subsamples and note any scoreable defects. A P-2 worksheet is used to record the weight, number of prunes, and scoreable defects in each size category.

The raw data on the P-2 is keypunched, and a computer program generates the P-1 grade results. Results of the P-2 grade counts are converted from a count basis to a weight basis. The door test average count and defect percentage are computed from the weighted average of the four size categories. The door test size count is the total number of prunes in the A, B, C, and D categories (not including undersize or trash) divided by the total weight of those same prunes.

Once the raw data have been keypunched and processed by the DFA, the results are placed on computer disk and sent to packers or transmitted electronically.

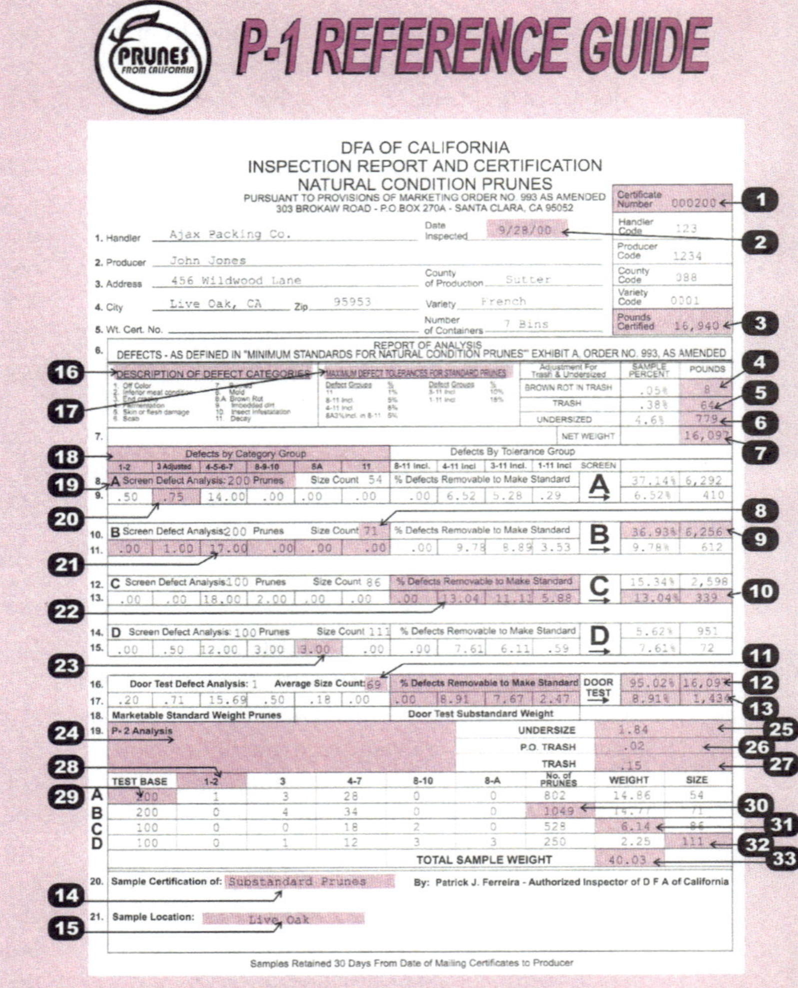

P-1 REFERENCE GUIDE

PRUNES FROM CALIFORNIA

DFA OF CALIFORNIA
INSPECTION REPORT AND CERTIFICATION
NATURAL CONDITION PRUNES
PURSUANT TO PROVISIONS OF MARKETING ORDER NO. 993 AS AMENDED
303 BROKAW ROAD - P.O BOX 270A - SANTA CLARA, CA 95052

Certificate Number 000200

1. Handler Ajax Packing Co. — Date Inspected 9/28/00 — Handler Code 123
2. Producer John Jones — Producer Code 1234
3. Address 456 Wildwood Lane — County of Production Sutter — County Code 088
4. City Live Oak, CA Zip 95953 — Variety French — Variety Code 0001
5. Wt. Cert. No. — Number of Containers 7 Bins — Pounds Certified 16,940

6. REPORT OF ANALYSIS
DEFECTS - AS DEFINED IN "MINIMUM STANDARDS FOR NATURAL CONDITION PRUNES" EXHIBIT A, ORDER NO. 993, AS AMENDED

DESCRIPTION OF DEFECT CATEGORIES: 1. Off Color; 2. Inferior meat condition; 3. End cracks; 4. Fermentation; 5. Skin or flesh damage; 6. Scab; 7. ; 8. Mold; 8A. Brown Rot; 9. Imbedded dirt; 10. Insect Infestation; 11. Decay

MAXIMUM DEFECT TOLERANCES FOR STANDARD PRUNES:

Defect Groups	%	Defect Groups	%
11	1%	3-11 Incl	10%
8-11 Incl.	5%	1-11 Incl	15%
4-11 Incl	8%		
8A3%Incl. in 8-11	5%		

Adjustment For Trash & Undersized	SAMPLE PERCENT	POUNDS
BROWN ROT IN TRASH	.05%	8
TRASH	.38%	64
UNDERSIZED	4.6%	779
7. NET WEIGHT		16,097

Line	1-2	3 Adjusted	4-5-6-7	8-9-10	8A	11	8-11 Incl.	4-11 Incl	3-11 Incl.	1-11 Incl	SCREEN		
8.	A Screen Defect Analysis: 200 Prunes				Size Count 54		% Defects Removable to Make Standard				A	37.14%	6,292
9.	.50	.75	14.00	.00	.00	.00	.00	6.52	5.28	.29	→	6.52%	410
10.	B Screen Defect Analysis: 200 Prunes				Size Count 71		% Defects Removable to Make Standard				B	36.93%	6,256
11.	.00	1.00	17.00	.00	.00	.00	.00	9.78	8.89	3.53	→	9.78%	612
12.	C Screen Defect Analysis: 100 Prunes				Size Count 86		% Defects Removable to Make Standard				C	15.34%	2,598
13.	.00	.00	18.00	2.00	.00	.00	.00	13.04	11.11	5.88	→	13.04%	339
14.	D Screen Defect Analysis: 100 Prunes				Size Count 111		% Defects Removable to Make Standard				D	5.62%	951
15.	.00	.50	12.00	3.00	3.00	.00	.00	7.61	6.11	.59	→	7.61%	72
16.	Door Test Defect Analysis: 1			Average Size Count: 69			% Defects Removable to Make Standard				DOOR TEST	95.02%	16,097
17.	.20	.71	15.69	.50	.18	.00	.00	8.91	7.67	2.47	→	8.91%	1,434
18.	Marketable Standard Weight Prunes						Door Test Substandard Weight						

19. P-2 Analysis

UNDERSIZE	1.84
P.O. TRASH	.02
TRASH	.15

	TEST BASE	1-2	3	4-7	8-10	8-A	No. of PRUNES	WEIGHT	SIZE
A	200	1	3	28	0	0	802	14.86	54
B	200	0	4	34	0	0	1049	14.77	71
C	100	0	0	18	2	0	528	6.14	86
D	100	0	1	12	3	3	250	2.25	111
							TOTAL SAMPLE WEIGHT	40.03	

20. Sample Certification of: Substandard Prunes — By: Patrick J. Ferreira - Authorized Inspector of D F A of California

21. Sample Location: Live Oak

Samples Retained 30 Days From Date of Mailing Certificates to Producer

Figure 27.1 P-1 grade sheet. Note: The black reference numbers 1–33 are not used in the text.

How to Read a P-1 Grade Sheet

The P-1 grade sheet has baffled many people over the years. The myriad rows and columns of data can be intimidating at first glance (see fig. 27.1). The key to reading the P-1 is to take it slow and look at one piece of information at a time. While there are over 130 different numbers on the P-1, only about 5 or 6 numbers are of interest to most growers. Yet identifying and understanding these few numbers is critical if growers are to connect the impact of cultural and harvest practices with the resulting fruit size and quality.

One of the first things to look at is the Pounds on line 5. This number should match the net weight from a weight receipt. While prunes are generally shipped in 20-bin lots, packers often combine bins from one truckload with another to make up a 30-bin lot—the maximum lot size—in order to reduce inspection fees, since fees are charged on a per-lot basis.

Next, to get an idea of the overall quality findings for the lot, look at lines 16 and 17, the Door Test Defect Analysis. The average size count, found in the middle of line 16, ideally should be around 65-count or less. Higher numbers indicate smaller prunes; lower numbers mean better size. Generally, an average size in the low- to mid-60-count range means that the crop load has been managed through pruning and thinning to achieve a size mix that will optimize returns and provide a marketable size mix for the processor.

An average count in the mid-70s or higher usually means that the crop load was excessive, the dry-away and percentage of undersize fruit will be high, and the orchard may have suffered from limb breakage and potassium dieback. The stress to the orchard from an excessive crop load often produces a light crop the following year, setting up a pattern of alternate bearing whereby the grower either has crops that are too heavy or too light to optimize profitability.

After checking the overall size count on line 16, drop down one line to line 17 and go almost all the way to the right side. The second number from the right is the overall percentage of off-grade, or to be technically correct, the % Defects Removable to Make Standard. This is the percentage of the P-1 net weight (line 7) that must be removed in order to make the remaining prunes standard (within off-grade allowances).

On average, over the years, this figure usually amounts to around 3% for the state as a whole. Ideally this number should be zero, but if it is higher than a few percentage points check to see what the major type of off-grade is by looking at the numbers in the left columns of line 17 (these are the columns labeled Defects by Category Group). These numbers are percentages, and the off-grade categories for each column are labeled just above line 8.

The most common types of off-grade fall into

the 4 to 7 category. This category covers a range of defects, so if this number is high, the only way to determine the cause is to call the DFA and request to see the sample. This should be done since there is no other way to know whether the cause was climatic conditions, cultural or harvest practices, or dehydration and handling without looking at the sample (see table 27.2 for a list of defects and possible causes).

Next look at the grade results for each of the screens A, B, C, and D. In the second column from the right is the percentage of the total delivered weight on each screen and the percentage of off-grade on each screen. Look at the percentage on line 8. This figure represents the percentage of the sample weight that falls on the A screen. Normally, if the grade sheet

Table 27.1. Grade standards in effect in 2004

993.97 Exhibit A; minimum standards.

I. Minimum standards for natural condition prunes:
A. Defects. Defects are: (1) Off-color; (2) inferior meat condition; (3) end cracks; (4) fermentation; (5) skin or flesh damage; (6) scab; (7) burned; (8) mold; (9) imbedded dirt; (10) insect infestation; (11) decay.
B. Explanation of terms.
(1) Off-color means a dull color or skin differing noticeably in appearance from that which is characteristic of mature, properly handled fruit of a given variety or type.
(2) Inferior meat condition means flesh which is fibrous, woody, or otherwise inferior due to immaturity to the extent that the characteristic texture of the meat is substantially affected.
(3) End cracks means callous growth cracks, at the blossom end of prunes, aggregating more than three-eighths of one inch (3/8 inch) but not more than one-half of one inch (½ inch) in length.
(4) Fermentation means damage to the flesh by fermentation to the extent that the characteristic appearance or flavor is substantially affected.
(5) Skin or flesh damage means growth cracks, splits, breaks in skin or flesh of the following descriptions:
(a) Callous growth cracks, except end cracks as defined in this section, aggregating more than three-eighths of one inch (3/8 inch) in length;
(b) Splits or skin breaks exposing flesh and affecting materially the normal appearance of the prunes;
(c) Any cracks, splits, or breaks open to the pit;
(d) Healed or unhealed surface or flesh blemishes caused by insect injury and which materially affect appearance, edibility, or keeping quality;
(e) Skin damage caused by rain or over dipping to the extent that the prunes cannot be processed normally without material sloughing of the skin.
(6) Scab means tough or thick scab exceeding in the aggregate the area of a circle three-eighths of one inch (⅜ inch) in diameter or by unsightly scab of another character exceeding in the aggregate the area of a circle three-fourths of one inch (¾ inch) in diameter.
(7) Burned means injury by sunburn or excessive heat in dehydration to the extent that the characteristic appearance, flavor, or edibility of the fruit is noticeably affected.
(8) Mold means a characteristic fungus growth and is self-explanatory.
(9) Imbedded dirt means the presence of dirt or other extraneous material so imbedded in, or adhering to, the prune that it cannot be removed in normal processing.
(10) Insect infestation means the presence of insects, insect fragments, or insect remains.
(Editor's note: Decay (#11) is not defined in the marketing order; thus it is not scored by DFA)
C. Maximum tolerances.
(1) The tolerance allowance for decay shall not exceed one percent (1%).
(2) The combined tolerance allowance for mold, imbedded dirt, insect infestation, and decay shall not exceed five percent (5%).
(3) The combined tolerance allowance for mold, brown rot, imbedded dirt, insect infestation, and decay shall not exceed five percent (5%), and, within such tolerance, brown rot shall not exceed three percent (3%).
(4) The combined tolerance allowance for end cracks, fermentation, skin or flesh damage, scab, burned, mold, imbedded dirt, insect infestation, and decay shall not exceed ten percent (10%), except that the first eight percent (8%) of end cracks shall be given one-half value and any additional percentage of end cracks shall be given full value.
(5) The combined tolerance allowance for off-color, inferior meat condition, end cracks, fermentation, skin or flesh damage, scab, burned, mold, imbedded dirt, insect infestation, and decay shall not exceed fifteen percent (15%), except that the first eight percent (8%) of end cracks shall be given one-half value and any additional percentage of end cracks shall be given full value.
(6) Prunes showing obvious live insect infestation shall be fumigated prior to acceptance.
D. Natural condition prunes must be properly dried and cured in original natural condition, without the addition of water, and free from active infestation, so that they are capable of being received, stored, and packed without deterioration or spoilage.

Source: Federal Marketing Order for Prunes 993.

Table 27.2. Prune defects, possible causes, and possible solutions

Defect	Possible causes	What to consider
1. Off-color	Immature fruit at harvest or sunburn.	Time harvest for proper maturity using fruit pressure gauge. Reduce amount of low, inside fruitwood during dormant pruning. (see below for sunburn)
2. Inferior meat condition	Overripe fruit may lead to gas pockets.	Time harvest using fruit pressure gauge. Is it possible to increase harvest volume at a more optimum maturity?
3. End cracks	Climatic conditions leading to water stress, followed by rainfall or irrigation and rapid fruit growth.	Avoid tree water stress April–July.
4. Fermentation	Improper drying or handling of dried fruit.	Not common; was fruit dried and handled properly?
5. Skin or flesh damage		
a) Callous growth cracks	Climatic conditions, diurnal contraction and expansion of fruit, rapid fruit growth.	Avoid irrigation during rapid fruit growth in late June, early July.
b) Splits or skin breaks	Excessive fruit pressure at harvest can cause fruit to split when it hits branches or hard objects; dried prunes may be damaged during tray scraping after drying.	Avoid harvesting when fruit pressure is above 4.5 lbf, pick later in day, pad equipment, keep drying trays clean.
c) Breaks open to pit	Climatic conditions, rapid growth of flesh pulls pit apart resulting in split pits.	Avoid tree water stress April–July.
d) Insect injury	Peach twig borer, obliquebanded leafroller, and codling moth are most common causes.	Review/improve insect management program.
e) Box rot	Breakdown of fruit skin caused by excessive temperature and length of time in harvest bins after picking.	Was fresh fruit held too long in bins after picking? Was harvest quota ignored? Can coordination between harvester and dehydrator be improved?
6. Scab	Cool and wet conditions at bloom, rare in dry years.	Was orchard treated at full bloom with fungicides such as captan or chlorothanil (Bravo or Echo) to help reduce scab development?
7. Burned: sunburned or dehydrator burn	High daytime orchard floor temperatures or excessive heat or time in dryer tunnels damages flesh, causing skin to adhere to the pit.	Would improving fertilizer program or irrigation practices improve shading by tree canopy? Would lower heat settings or shorter pull times reduce dryer burn?
8. Mold	Breaks to skin or insect damage allow mold to grow on fresh fruit; improper drying or storage of dried fruit result in moldy fruit in dry bins.	Can in-season damage to fruit skin be reduced? Was fruit handled properly after picking? Are drying and storage facilities adequate? Are proper drying and storage procedures followed?
a) Brown rot	Warm, wet conditions at bloom, rain in-season.	Review brown rot management program.
9. Imbedded dirt	Overripe fruit, improper harvesting and handling procedures, excessive amounts of leaves and twigs mixed in with fruit at harvest, foreign material in bins.	Are the harvest machines adequate for the job? Are crews going too fast for the orchard conditions in order to allow for adequate trash removal?
10. Insect infestation	Insect damage that leaves frass or droppings on fruit, including dried fruit beetle and Indianmeal moth. Poor sanitation around dehydrator, rotting or fermenting discards left in area.	Is the in-season pest management adequate? Are proper drying, handling, and storage procedures followed? Is sanitation around the dehydrator adequate?

has an overall average size count (door test, line 16) in the mid-60-count range, the A screen will be 50% or more of the total weight, and the B, C, and D screens will be roughly 22%, 15%, and 5%, respectively. With good-size fruit, the A and B screens together will represent roughly 75% or more of the total delivered weight.

The percentage figure on line 9 in the second column from the right shows the amount of off-grade on the A screen. The figure in the first column on the right is the weight of substandard prunes that must be removed so that the remaining prunes are within grade tolerances. This is the off-grade or substandard weight. Subtract this figure from the one just above (the figure on line 8) to determine the standard prune weight for the A screen.

The undersize percentage and weight is just above the net weight on line 7, and above that are the trash and brown rot figures. Ideally the trash and brown rot percentages will be close to zero, and the undersize will be 3% or less for a 23/32-inch undersize screen, or 5% for a 24/32-inch undersize screen. If the undersize is more than 5% and the D screen is more than 10%, the crop load may have been too heavy, or not enough fruitwood was pruned out of the lower inside portion of the tree, or sugar was poor or fruit was picked too early, or tree nutrition was inadequate. It might also be helpful to consider screening the smaller prunes at harvest since these have low value and the drying ratio can be very high.

On the bottom third of the P-1, starting at line 19, is the actual sample weight on each screen and the count of off-grade prunes. The numbers above line 19 are computed from this data (which were originally recorded on the P-2 worksheet at the time of grade analysis). The data are presented here for growers who like to see the sample weights and counts. The test base in the first column is the number of prunes (actually a subsample of the total number of prunes on each screen) that was analyzed for defects.

It cannot be emphasized enough the importance of reviewing each P-1 to see whether fruit size and quality goals are being met and to determine whether adjustments to orchard management are needed. Also, review each grade sheet, and if necessary make arrangements to view the sample, to determine whether harvest or drying practices were adequate, or if not, what changes are needed for the future.

Grading Prunes to Size

Prune grading is a very simple process. The goal is to end up with clean prunes of uniform size in the same bin. If there is a high percentage of trash or small prunes in the delivery, it slows down the grading process and makes it difficult to get uniform sizing. Achieving uniform sizing is very important in today's market because such a high percentage of the crop is plunger-pitted. If sizing is not uniform, the pitting efficiency and yield drop off significantly.

Growers do their part by keeping the trash (i.e., leaves, twigs, brown rot, etc.) to a minimum in each load delivered. Sometimes if the crop is light and the trees are old, growers cannot do much other than keep the leaf blowers working at top efficiency and keep a worker on the back of the harvester to get out the worst of the trash. Using a size grader on the back of the harvester also allows more trash to drop out and eliminates the smallest prunes that are not worth dehydrating.

Most growers understand that trash in a load hurts the grower's pocketbook as well as the processor's. Trash not only incurs a trash sorting charge, it also increases the drying costs because of the additional weight delivered, and sticks and twigs and dirt can damage the fruit and lower the overall grade. For the packer, debris slows down grading and decreases the accuracy of grader sizing and electronic sorting. Leaves that stick to prunes are nearly impossible for an electronic sorter to distinguish from scab or cracks.

In some years, brown rot is a major issue and causes headaches for grading and sorting. Brown rot clumps break apart and the pieces often stick to the good prunes, creating another defective prune. Brown rot is an inedible mold, so every piece of it must be removed from each lot, even for juice or cattle feed. In a bad year, brown rot creates a huge economic loss for growers and packers because so much fruit and time is lost and because of the increased costs of sorting.

Paying attention to cultural practices that help produce clean fruit, such as bloom sprays and irrigation timing, and practices that produce good size such as pruning and thinning really shows up at the grader. For pitting, we need prunes in the 40- to 60-count range. Too much small fruit makes it difficult to get the separation of sizes on the grader.

Besides good cultural practices to help produce clean fruit, harvest timing and handling is critical to producing prunes that grade well. Prunes that are sticky are more difficult to grade, and trash sticks more easily to them. Getting the crop picked at the proper maturity, and then dipped and on the drying trays in a rapid manner, is critical to producing high-quality fruit.

Dehydrators do everything they can at the dryer and grader to eliminate the trash before grading. Besides the trash screen on the dipper, they have personnel on the dipper to rake out the leaves and sticks, and they run vacuum machines on the dry side as the prunes come off the trays. Later, they use the vacuums on the box turners as prunes are sampled or as bins are dumped for grading.

Drying prunes to the right moisture is important. If prunes are overdried, they are more difficult to process and nearly impossible to pit. Too much moisture prevents them from storing well and causes them to break down in the bins. Dehydrators aim to dry the prunes to 21% moisture. This is beneficial for the grower and processing, but early in the season the newly dried prunes are soft and must be handled carefully. Any debris increases the chance of skin damage to the fruit.

Prune Defects

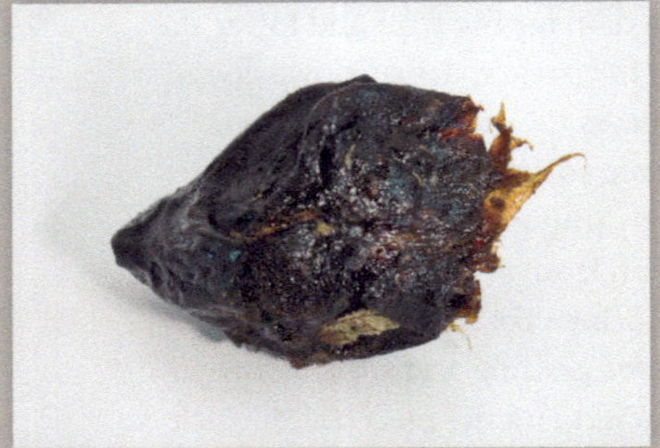
Box rot

Brown rot

Burnt and sunburned

Callous growth cracks

Embedded dirt

End crack

Hard scab

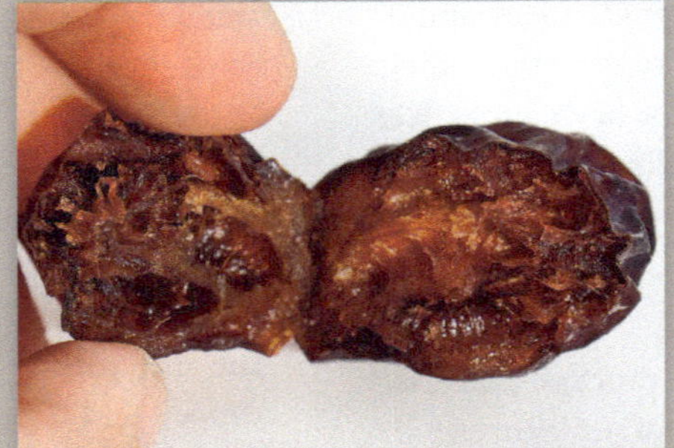
Inferior meat condition

Insect injury

Mold

Open to the pit

Skin or flesh damage

Soft scab

Prune Production Calendar of Operations in the Sacramento Valley

Crops are complex biological systems, which makes it difficult to accurately predict every management practice. The bars below represent a reasonable probability that the cultural practice would occur at the time indicated. Not all of these management practices are implemented in every field every year. Site-specific conditions will determine whether a specific management practice is necessary.

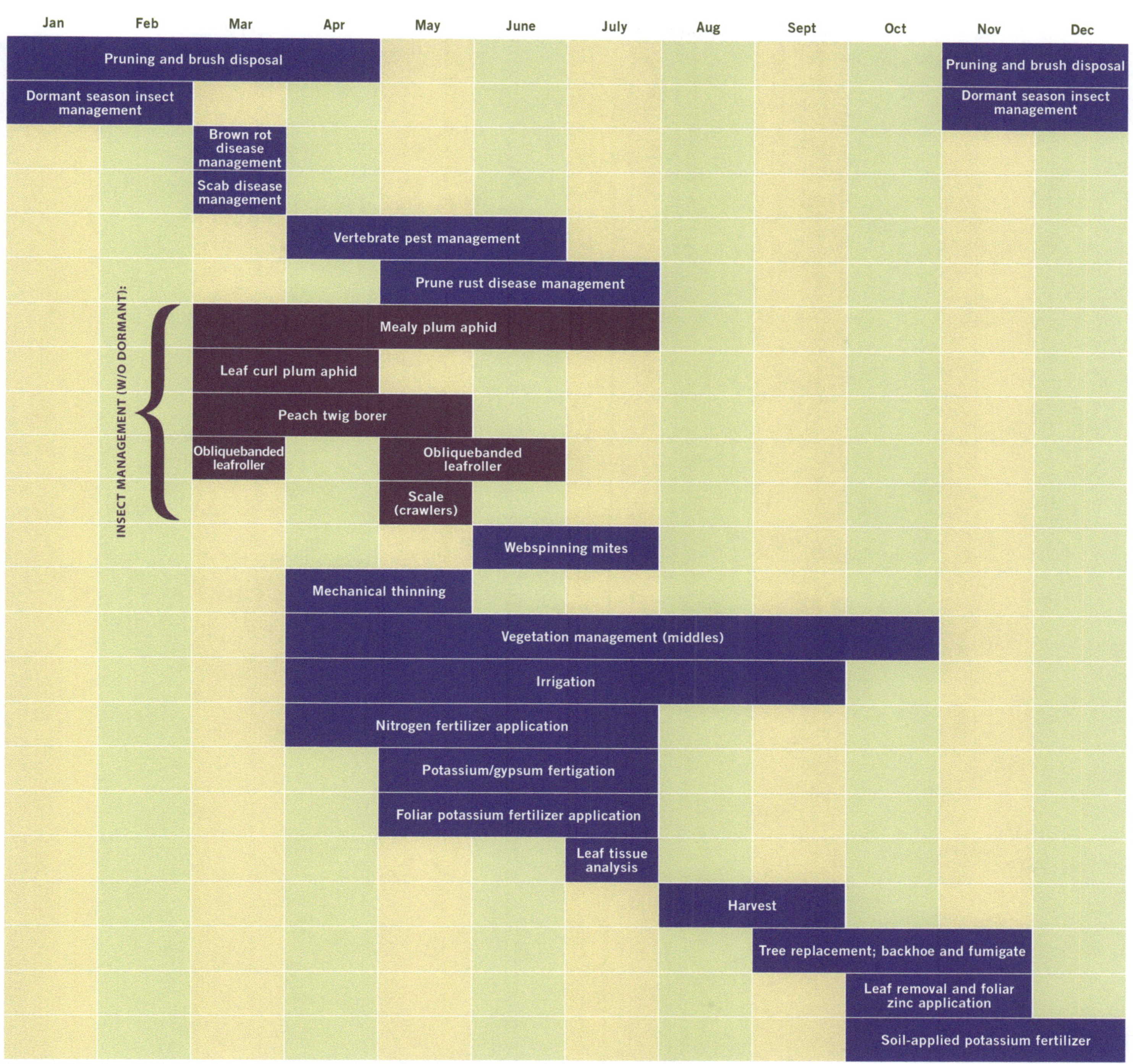

Source: R. P. Buchner, W. H. Olson, F. J. A. Niederholzer, W. O. Reil, W. H. Krueger, and A. E. Fulton, University of California Cooperative Extension.

Measurement Conversion Table

U.S. Customary	Conversion factor for U.S. Customary to Metric	Conversion factor for Metric to U.S. Customary	Metric
Length			
inch (in)	2.54	0.394	centimeter (cm)
foot (ft)	0.3048	3.28	meter (m)
mile (mi)	1.61	0.62	kilometer (km)
Area			
acre (ac)	0.4047	2.47	hectare (ha)
square foot (ft^2)	0.0929	10.764	square meter (m^2)
square mile (mi^2)	2.59	0.386	square kilometer (km^2)
Volume			
ounce, fluid (fl oz)	29.57	0.034	milliliter (ml)
pint, liquid (pt)	0.473	2.11	liter (l)
pint, dry (pt)	0.55	1.82	liter (l)
quart, liquid (qt)	0.946	1.056	liter (l)
quart, dry (qt)	1.1	0.91	liter (l)
bushel (bu)	0.035	28.37	cubic meter (m^3)
gallon (gal)	3.785	0.26	liter (l)
acre-inch (ac-in)	102.8	0.0097	cubic meter (m^3)
acre-foot (ac-ft)	1,233	0.000811	cubic meter (m^3)
cubic foot (ft^3)	28.317	0.353	liter (l)
gallon per acre (gal/ac)	9.36	0.106	liter per hectare (l/ha)
Mass			
ounce (oz)	28.35	0.035	gram (g)
pound (lb)	0.454	2.205	kilogram (kg)
ton (T)	0.907	1.1	metric ton (t)
pound per acre (lb/ac)	1.12	0.89	kilogram per hectare (kg/ha)
ton per acre (T/ac)	2.24	0.446	metric ton per hectare (t/ha)
Pressure			
pound per square inch (psi)	6.89	0.145	kilopascal (kPa)
Energy			
BTU	1.055	0.947	kilojule (kJ)
Temperature			
Fahrenheit (°F)	°C = (°F – 32) ÷ 1.8	°F = (°C x 1.8) + 32	Celsius (°C)

Glossary

absorption. To take in and make part of.
acaricide. A pesticide that kills mites; a miticide.
active ingredient. The compound in a pesticide that kills a pest.
adjuvant. Any solid or liquid added to a substance such as a fluid pesticide or fertilizer to increase its effectiveness.
ADP. Adenosine diphosphate, a compound which upon phosphorylation forms high-energy bonds as ATP.
adsorption. Adhesion of ions or molecules to a solid body.
advection frost. A frost that occurs when a mass of cold air displaces a mass of warmer air at the earth's surface.
aerobic. Living or active only in the presence of oxygen.
aestivation (estivation). A state of inactivity during summer months.
air-carrier sprayer. Sprayer using air velocity to carry pesticide, plant growth regulator, or nutrient to a target.
alkaline. Soils that have a pH above 7.0, containing sodium salts sufficient to damage plants.
alluvial soil. A fine-grained fertile soil deposited by water flowing over floodplains or in riverbeds.
alternate bearing. Condition of bearing fruit every other season.
amendments. Materials added to change the composition of soil or irrigation water.
amino acid. The chief component of proteins.
anaerobic. An environment or condition that lacks oxygen.
anchorage. How well a rootstock remains upright and structurally stable in the soil.
anions. An atom or group of atoms that carries a negative electrical charge.
annual. A plant or organism that completes its life cycle in 1 year.
anther. Pollen-bearing part of a flower's stamen.
anticoagulant. A substance that hinders blood clotting.
antifoaming agent. A liquid substance that minimizes foaming in a spray tank.
antioxidant. Organic compound that prevents or neutralizes an oxidizing agent.
apex. Tip or distal end.
apical. At the tip of a root or shoot.
apical dome. The cells of the growing point at the tip of a bud that divide to form the new shoot or flowers.
apical dominance. Tendency for the apical vegetative buds at or near the shoot apex to grow and develop while suppressing the growth of lower buds.
Ascomycete. A group of fungi producing their sexual spores, ascospores, with asci.
ascospores. A sexually produced spore borne in an ascus.
ascus. The saclike cell of a hypha of an Ascomycete fungus within which ascospores are formed.
aseptic. Free of the living agents of disease, fermentation, or putrefaction.
asexual spores. Minute fragments of the mycelium that separate into spores without any nuclear change.
assimilation. Process by which carbohydrates become incorporated into the structural part of the plant.
atomization. Reduced to fine particles or a fine spray.
ATP. Adenosine triphosphate, a compound formed by phosphorylation of ADP, which stores and releases energy for cell functions.
auxin. Plant hormone that promotes cell growth and regulates other growth processes.
axial flow fan. Fan whose blades force air to move parallel to the shaft.
axil. The angle between a petiole and the stem it is attached to.
bacteria. Large, widely distributed group of typically one-celled organisms, chiefly parasitic or saprophytic.
bare root. Plant root system with no soil attached; usually observed when trees are removed from the nursery.
basidiospore. A sexually produced spore borne on a basidium.
basidium. A club-shaped structure on which basidiospores are borne.
bicarbonate. A naturally occurring alkaline substance that influences pH, salt content, and buffering capacity of soils and water; HCO_3^-.
biennial. A plant that completes its life cycle in 2 years and does not flower until the second year.
biological control. Pest control by natural means.
biomass. Living matter, as in a unit area or volume of habitat.
blight. Result of a disease characterized by general and rapid killing of leaves, flowers, and branches.
bract. A specialized leaf or leaflike part, usually at the base of a flower or inflorescence.
Brix. Unit of measure (in degrees) for soluble solids, usually used to measure sugar in fruits or berries.
bud. A plant structure at the base of the leaf axil or the tip of a shoot from which a shoot, flower, or flower cluster develops.

budding. Grafting by inserting a single bud of a desired variety onto a stem or rootstock.

bud stick. A stick of wood from which buds are excised for propagation.

callus. Soft tissue that forms over a cut or damaged plant surface.

calorie. A unit used to express the energy producing value of food.

calyx. Base of the flower enclosing the ovary.

cambium. Plant tissue between the xylem and phloem of most vascular plants that produces new cells and is responsible for secondary growth.

canopy. The periphery or upper structure of a tree.

carbohydrates. Organic compounds produced by plants that contain chemically bonded energy composed of carbon, hydrogen, and oxygen.

carbon to nitrogen ratio. The ratio obtained by dividing the percentage of organic carbon by the percentage of nitrogen.

carrier. An inactive accessory substance combined with an active ingredient in a spray material formulation.

cation. An atom or group of atoms that carry a positive electrical charge.

cation exchange. The capacity of a soil to hold or bind cations.

cell. The structural and functional unit of a plant.

chilling requirement. The number of hours of cold temperature a plant requires, during the dormant season, to resume normal growth and fruit development the following spring.

chlamydospore. A thick-walled spore formed from the cell of a fungal hypha.

chlorophyll. Green pigments mainly in leaves, which, under the stimulus of light, manufacture carbohydrates used in plant growth and fruit development.

chloroplast. A plastid, usually located in a leaf cell or green plant tissue, containing chlorophyll.

chlorosis. Yellowing of leaves.

chlorotic. The yellowing or whitening of normally green plant tissue because of a decreased amount of chlorophyll, usually a result of disease or nutrient deficiency.

chromosome. An organized structure of DNA and protein found in cells.

cleistothecia. Spherical closed structures of certain Ascomycete fungi where one or more asci and ascospores are formed.

clonal. Having reproduced or propagated asexually.

clone. A group of plants originating as parts of the same individual usually from buds, cutting, division, or tissue culture.

conidia. Asexual fungal spores formed by fragmentation or budding at the tip of a specialized hypha.

conidiophores. Special hyphae in some fungi that produce conidia.

cork cambia. The layer of cambium on the outside of the phloem that gives rise to the outer bark.

cornell donut. A metal attachment on a spray machine that restricts air intake.

cover crop. A second crop grown to improve the production of a primary crop, e.g., grasses or legumes grown in orchards to improve soil conditions.

cultivar. A taxonomic group of plants that have unique characteristics within a species; in common horticultural usage, synonymous with variety.

cytokinins. Naturally occurring plant hormones that stimulate cell division and delay senescence.

cytoplasm. The living substance of a cell outside the nucleus.

cytoplasmic membrane. The membrane that encloses or surrounds the cytoplasm, separating it from the environment.

dehydrator. Facility that dries prunes.

dichotomous. Divided or dividing into two parts or classifications.

dieback. Progressive death of branches or shoots beginning at the tips, characteristic of unhealthy prune trees.

diffusion. The process whereby particles of liquids, gases, or solids intermingle as the result of thermal agitation and move from a region of higher to lower concentration.

diluent. A material that reduces the concentration of a solution.

diploid. An individual with double the number of chromosomes per cell.

dipper. A tank through which freshly harvested prunes are conveyed to separate leaf and stem material prior to dehydration.

diurnal. Having a daily cycle.

DNA. Deoxyribonucleic acid, a nucleic acid that contains the genetic instructions used in the development and functioning of all known living organisms and some viruses.

doliform. Barrel shaped.

dormancy. A physical or physiological state in which a plant is not actively growing.

drosophilids. Insects in the order Diptera (flies), family Drosophilidae, including pomice flies and small fruit flies.

dry bulb temperature. Reading from an ordinary thermometer alongside a wet bulb thermometer that is to measure relative humidity.

drying ratio. Ratio of green fruit weight to dry fruit weight; sometimes referred to as dryaway.

electrostatic attraction. Attraction of a negatively charged body to a positively charged body.

embryo. Rudimentary plant found in the seed.

emulsifiable concentrate. A liquid concentrate that is added to water to create a pesticide solution.

endocarp. Stony part of the pit enclosing the seed of a stone fruit.

endoconidia. A conidia formed inside a hypha.

endogenous. Developing or originating within an organism or part of an organism; having no apparent external cause.

endoparasitic nematode. A nematode that lives inside the host.

endosperm. Tissue surrounding the embryo in seeds.

enzyme. Any of a class of complex organic substances that accelerate chemical reactions within a plant.

epidermis. The thin layer of cells forming the outer covering of a plant.

evapotranspiration (ET). The loss of soil moisture due to evaporation from the soil surface plus transpiration through plant tissue.

exocarp. The outer layer of fruits derived from the ovary, as with the skin of a prune.

explants. Plant material used to initiate tissue cultures, usually taken from field- or greenhouse-grown plants.

exuvia. The cast skins of previous growth stages found on armored scale insects such as San Jose scale.

Federal Marketing Order. A regulation of an executive agency that sets prices and other conditions for the sale of certain fruits, nuts, or vegetables.

feeder root. Small and relatively short-lived lateral roots arising from the main root system that take up water and nutrients.

fermentation. The biochemical process of converting sugar into carbon dioxide and alcohol.

filament. The stalk on which the anther is attached.

fixed-wing aircraft. A heavier-than-air aircraft capable of flight whose lift is generated not by wing motion relative to the aircraft, but by forward motion through the air.

floral bud. A bud containing a flower or cluster of flowers.

flowable. Pesticide formulation in which the pesticide is a liquid toxicant uniformly dispersed throughout a stable emulsified system, allowing for long periods of storage and redispersal with mild agitation.

frass. A mixture of feces and food fragments produced by an insect from feeding.

fructose. A simple sugar found in fruits.

fumigation. Gas-based chemical treatments to kill or reduce soilborne pathogens or insects in fruit storage facilities.

fungicide. A pesticide that kills fungi.

fungus (pl., fungi). A multicellular lower organism, such as mold, mildew, rust, or smut, whose body normally consists of mycelia.

gametangium. A cell containing gametes or nuclei that act as gametes.

gamete. Specialized haploid cells of each sex that must fuse during fertilization to form a diploid zygote and embryo.

genotype. The genetic makeup, expressed and latent, of an organism.

germination. To begin to grow or develop, as a plant from a seed.

germplasm. Genetic material that provides the physical basis of heredity; a collection of genotypes of an organism.

gibberellin. Naturally occurring growth-regulating plant hormone that promotes shoot vigor and root growth.

girdle. To remove a thin strip of bark around the trunk or scaffold limbs to increase fruit size or yield.

glucose. A six-carbon sugar that plays an important role in cellular metabolism.

grafting. A form of asexual reproduction in which a scion of one variety is placed in contact with the cambium layer of another variety or rootstock.

green tip. Stage of spring flower bud swell in which expanding sepals appear as green tips at the terminal end of the bud; also known as green bud.

growth regulator. A naturally occurring or synthetic chemical that is transported within a plant to a location where it evokes a response that regulates plant growth; a growth hormone.

guard cells. Specialized epidermal cells that regulate the opening and closing of stomata.

habitat. An ecological or environmental area that is inhabited by a particular animal or plant species.

haploid. A cell or an organism whose nuclei have a single complete set of chromosomes.

hardwood cuttings. Cuttings taken from dormant shoots usually for propagation of rootstock selections or for rooting scion selections to grow on their own roots.

heading cut. Removing a terminal portion of a limb during pruning.

heel in. To temporarily cover tree roots to prevent desiccation.

herbicide. A chemical substance that kills plants.

herbivorous. Feeding on plants; plant eating.

hermaphroditic. Condition in which both male and female reproductive structures are present.

hibernation. Passing the winter in a torpid or dormant state in which the body temperature and metabolic rate drop to very low levels.

hormone. A substance that is produced in minute amounts in one part of a plant and transported to another part, where it evokes a response.

horticulture. The science of growing fruits, vegetables, and ornamental plants.

hyaline. Colorless; transparent.

hybrid. The offspring of two individuals differing in one or more heritable characteristics.

hypha (pl., hyphae). The branching elements of a mycelium.

IBA. Indole-3-butyric acid, an auxin commonly used to induce rooting.

in vitro. In a culture; in a controlled environment outside the host.

incubation. Time between penetration of a host by a pathogen and the first appearance of symptoms.

infiltration. The physical process of water entering the soil.

ingestion. The act of taking a pesticide into the body by mouth (eating).

inoculant. A bacterial substance topically introduced onto seed or soil to ensure atmospheric nitrogen fixation.

inoculum. Any part or stage of a pathogen, such as spores or virus particles, that can infect a host.

inorganic. Relating to compounds that do not contain carbon.

instar. Form assumed by insects between successive molts.

interveinal. Area of a leaf between the veins.

invertebrate. An animal having no internal skeleton.

ion. An atom or group of atoms that carries a negative (anion) or positive (cation) electrical charge.

iron precipitate. Formation of iron in a solution during a chemical reaction.

juvenile. Growth stage of a seedling plant in which floral initiation is not possible.

larva (pl., larvae). The immature form of an insect that hatches from an egg, feeds, and enters a pupal stage.

lateral bud. A bud on the side of a shoot, spur, or branch.

leach. To remove nutritive or harmful elements from soil by the downward movement of water.

legume. A plant of the family Leguminosae (Fabaceae) capable of fixing atmospheric nitrogen.

lesion. A localized area of diseased tissue, such as a canker or leaf spot.

lichen. A fungus, usually of the class Ascomycetes, that grows symbiotically with algae, resulting in a composite organism that characteristically forms a crustlike or branching growth on rocks or tree trunks.

lime precipitate. Formation of lime in a solution during a chemical reaction.

locule. The cell or cavity of an ovary or anther.

macronurtients. Chemical elements used in large quantities by plants for growth and development.

meiosis. The first two divisions of a zygote that produce the gametes, or haploid individuals.

meristem. Actively dividing cells of undifferentiated tissue, as cambium or growing tips of shoots and roots.

mesocarp. Flower tissue that forms the fleshy portion of a fruit.

metabolism. The biochemical changes in living cells by which energy is provided for plant processes and activities.

metabolite. Any substance produced by metabolism or by a metabolic process.

metamorphosis. Change in form during development of an insect.

micronutrients. Chemical elements used in small quantities by plants that are essential for growth and development.

microprogagation. Propagating plant tissues, primarily shoot tips and meristems, by using in vitro techniques.

micropyle. The minute opening in the integuments of an ovule through which the pollen tube penetrates to the embryo sac.

mineralization. The conversion of organic tissues to an inorganic state as a result of decomposition by soil microorganisms.

miticide. A pesticide that kills mites; acaricide.

morphological origin. The origin of terminal or lateral vegetative buds on a prune tree and how they perform.

mummy. Dried, shriveled fruit remaining on a tree; often the dried remains of a prune decayed by brown rot.

mutagenesis. Developing new varieties from bud or tissue mutations.

mutation. Spontaneous alteration or change in genetic material.

mycelium (pl., mycelia). Mass of hyphae that make up a fungus.

mycophagous. Fungus eating.

mycorrhiza. A symbiotic, beneficial association of a soilborne fungus with plant roots.

necrotic. Dead or dying tissue of a plant part.

nectar. A sweet liquid secreted from flower nectaries.

nectary. A glandlike organ, located outside or within a flower, that secretes nectar.

nematicide. Chemical compound or physical agent that kills or inhibits nematodes.

nematode. Nonsegmented microscopic roundworms that live in soil and parasitize roots.

nitidulids. Insects in the order Coleoptera (beetles), family Nitidulidae, including sap-feeding beetles.

nitrogen fixation. The conversion of free nitrogen to nitrogen compounds; usually the conversion of atmospheric nitrogen to plant-available nitrogen.

noctuids. Insects in the order Lepidoptera (butterflies and moths), family Noctuidae, including the citrus cutworm and green fruitworm.

node. The point on a stem from which buds and leaves arise.

nonbearing trees. Young trees that have not yet become reproductive.

no-till. Leaving orchard floors uncultivated throughout the growing season.

nucellus. The tissue of the ovule in which the embryo sac develops.

nucleus (pl., nuclei). The dense protoplasmic body found in all cellular organisms that is essential for all synthetic and developmental activities of a cell.

nurse limb. A temporary limb, often kept when topworking trees to provide nutrients to the root system.

open pollination. Pollination by pollen not from the same flower or tree.

ovary. Lower portion of the pistil enclosing the ovules or young seeds in a flower.

ovoid. Egg shaped.

ovule. Outgrowth of the ovary that develops into a seed after fertilization.

oxidation. Energy releasing biochemical process involving removal of electrons from a substance.

parasite. An organism that spends all or some of its life cycle in or on the body of a larger living organism (its host) from which it derives food without killing the host directly.

pathogen. Any disease-producing organism.

pectin. Substances that bind cell walls in plants, yielding a gel.

pectinolytic enzyme. An enzyme that breaks down pectin.

pedigree. In horticulture, a list of ancestors for a plant variety.

perennial. A plant that lives 3 or more years and flowers at least twice.

pericarp. The ripened and variously modified walls of the ovary, usually exhibiting three distinct layers: endocarp, mesocarp, and epicarp.

perithecia. Globular or flask-shaped structures produced by certain Ascomycete fungi, within which asci and ascospores are formed.

pesticide. A synthetic, natural, or biological material used to kill pests.

petal. Showy structures around the reproductive organs of a flower.

petiole. Slender stalk that attaches to the stem of a plant and supports the leaf blade.

pH. Hydrogen ion concentration or activity of a soil or water; lower pH is acidic, higher pH is alkaline.

phenotype. The external visible appearance of an organism.

phloem. Conducting tissues that transport sugars and other essential elements within the plant.

photosynthates. Carbohydrate products of photosynthesis.

photosynthesis. The biochemical process by which energy from the sun is captured in green pigments (chloroplasts), converted into chemical energy, and used by the plant to convert carbon dioxide and water into sugar.

phytotoxic. Causing injury to plants.

pistil. Female part of the flower consisting of a stigma, style, and ovary.

plant hormones. Substances produced by plants that control or regulate germination, growth, metabolism, or other physiological activities.

plastid. Small bodies of specialized protoplasm located in the cytoplasm of plant cells.

pollen. Spore-like grains or particles originating in the anther of a flower that contain the male gamete.

pollen tube. Pollen growth extension from the stigma to the ovule following germination of the pollen grain.

popcorn bloom. The stage of bloom in which most prune flowers are about to open; the white unopened flowers resemble popcorn.

postemergent herbicide. An herbicide designed to kill weeds after they have emerged.

precocious. Bearing fruit or nuts early in the life of a plant or tree.

predator. An organism that attacks and feeds on other organisms (prey), usually consuming all of the prey and consuming many prey during its lifetime.

preemergent herbicide. An herbicide designed to kill weeds before or as they germinate, prior to emergence.

prepupae. A quiescent instar between the end of the larval period and the pupal period.

primordia. The first recognizable but undifferentiated stage of a developing organ.

primordial leaves. The first recognizable leaves.

proliferating mass. A dividing cluster of cells, such as callus.

propagation. To generate or multiply plants by sexual or asexual techniques.

protoplast. The organized living unit of a single cell; everything inside the cytoplasmic membrane.

prune. Variety of European plum (*Prunus domestica* L.) that is high in sugar and can be satisfactorily dried whole without fermenting at the pit.

psychrometer. An instrument that uses the difference in readings between two thermometers, one having a wet bulb and the other having a dry bulb, to measure the moisture content or relative humidity of air.

PTO. Power take off accessory on a tractor.

pupa. The nonfeeding inactive stage between larva and adult in insects with complete metamorphosis.

pycnidia. Small spherical or flask-shaped structures formed by certain types of fungi, inside which asexual spores are produced.

radiated. Sent out in the form of rays or waves.

radiation frost. Frost that occurs on cold nights when the air is clear and dry and heat is radiated from the earth's surface into the atmosphere.

raptor. A carnivorous predatory bird.

reference date. The physiological date in the development of a prune when 80 to 90% of the seed show the presence of an endosperm.

replant. Replacement tree in an established orchard.

residual herbicide. Plant-killing chemical applied before weed seeds germinate or emerge that remains active in the soil after application.

residue. Material remaining after completion of a chemical or physical process, such as environmental degradation or evaporation.

respiration. The orderly breakdown of photosynthates.

rhizome. An underground horizontal stem.

rhizomorph. A rootlike structure produced by the fungus *Armillaria mellea* that grows from the root of an infected host plant to the root of an uninfected host plant.

rootstock. A plant having certain desirable root characteristics to which other varieties are grafted or budded to produce a commercially acceptable compound tree.

rotary-wing. An aircraft that uses a moving wing to generate lift, such as a helicopter.

rugose. Rough and wrinkled.

russetting. Brownish, roughened areas on the skin of fruit.

saline. Soil or water containing excessive salts that can reduce crop productivity.

salinity. The level of soluble salts in water or soil.

saprophyte. An organism that lives off the dead body or nonliving products of another plant or animal.

scaffold. Larger branch arising from the central portion or trunk of a tree; the primary branching structure of a tree.

scion. A branch, shoot, or bud removed from one plant and grafted onto another plant.

self-compatible. A flower that can be pollinated and fertilized with its own pollen or pollen from another flower on the same tree.

self-incompatible. A flower that must be pollinated and fertilized with pollen from flowers on a different cultivar.

sepal. One of the outermost flower structures which usually encloses the outer flower parts. The individual lobes of the calyx.

septate. Divided by a septum, a thin partition or membrane that divides two cavities or soft masses of tissue in an organism.

shoot meristems. Actively dividing growing tips of shoots.

sod culture. Managing an orchard floor by mowing (not cultivating) native weeds or planted vegetation.

soil amendment. A material added to soil to improve plant growth and health.

soil texture. A description of a soil's property based on the relative proportion of different grain sizes of mineral particles in a soil.

solubility. The degree to which a substance can be dissolved in water.

soluble solids content. A measure of fruit sugar content.

sorbitol. Six-carbon plant sugar.

sporangia. A structure containing asexual spores.

sporulation. Formation of spores.

spot-treat. To apply pesticide to a single target, such as a clump of weeds.

spreader-sticker. An adjuvant that enables pesticides to flow and stick more readily, increasing effectiveness.

spur. Small stubby branch on which fruit is borne.

stamen. Male part of a flower, consisting of an anther on a filament, which produces pollen.

starch. The principal polysaccharide storage product of vascular plants.

stigma. Top of pistil where pollen grains germinate.

stipule. Outgrowths borne on either side of the base of a leaf stalk (petiole), which may appear as glands, scales, hairs, spines, or laminar (leaflike) structures.

stolon. A stem that grows horizontally along the surface of the ground.

stomata. Small openings in leaves through which gas exchange takes place.

stratification. The process of exposing seed to cool, moist conditions, causing the seed to break dormancy.

stratified soil. Soil composed of layers of differing materials.

stylar. The area of a fruit opposite the stem end.

style. Region of the pistil connecting ovary and stigma.

stylet. Insect mouthpart used for penetrating tissue for feeding.

suberization. Process in which plants deposit suberin, a waxy substance created in plant cells that repels water.
subglobose. Almost globe shaped.
subhyaline. Pale yellow.
subsample. A sample of a sample.
subsoil. The layer of soil under the topsoil; generally, the layer of soil below the plow layer.
subtending wood. Wood underneath the bark when a bud is removed for budding.
sucrose. A double sugar composed of a unit of glucose and a unit of fructose.
surfactant. Materials that lower the surface tension of a liquid, allowing improved spreading.
suspended solids. Organic or inorganic particles dispersed in and carried by water.
symbiosis. The living together of two or more organisms in which each benefits from the association.
synthesis. Biochemical process of building compounds.
tanked. Irrigation by pouring water in a basin around the plant.
thinning. Removing excess, nonproductive, or older fruitwood within a tree by cutting branches at their origin.
topsoil. The uppermost layer of soil, usually the top 2 to 8 inches.
topworking. Using grafting techniques to change varieties.
tortricids. Insects in the order Lepidoptera (butterflies and moths), family Tortricidae, including the fruittree leafroller.
translocate. To move water and other dissolved substances through the vascular system of a plant.
transpiration. The loss of water vapor from plant surfaces.
tuber. An enlarged, short, fleshy underground stem.
turgid. Swollen or plump as a result of internal water pressure.
turgor. The normal distention of plant cells resulting from internal pressure exerted against cell walls, as when water is absorbed.
vacuole. The cavity within the cell containing the cell sap.
variety. A taxonomic group of plants that have unique characteristics within a species; in common horticultural usage, synonymous with cultivar.
vascular. Relating to system of plant tissues that conducts water, mineral nutrients, and photosynthates through a plant.
vascular cambium. A cylinder of meristematic tissue that lies between the wood and the bark.
vector. Organism that carries and transmits disease-causing microorganisms.
vegetative bud. A bud that produces only a shoot with leaves.
vegetative propagation. Reproduction of a variety or clone by asexual reproduction using tissue from the desired plant.
venturi. A short tube with a tapering constriction at the center that increases fluid velocity and decreases fluid pressure.
vertebrate. Organisms in the phylum Chordata, having backbones or spinal columns.
virus. Submicroscopic infectious agent consisting of nucletic acid and a protein coat that can reproduce only within the living cells of a host.
volatilization. Conversion of a chemical substance from a liquid or solid state to a gaseous or vapor state.
volute. Metal or plastic covering over the fan of a spray machine that channels air to increase velocity and direct spray placement.
water amendment. A chemical added to water to improve quality.
water potential. Measure of the tendency of water to move from one location to another within a plant.
water-soluble concentrate. Concentrated pesticide that can be dissolved in water.
water table. The soil level below which the ground is saturated with water.
weed. A plant that is not valued where it is growing.
wet bulb thermometer. Thermometer whose bulb is covered by a moist muslin bag; used with a dry bulb thermometer to measure relative humidity.
wettable powder. Pesticide material that will not dissolve in water but remains suspended in it.
wetting agent. A chemical substance that increases the spreading and penetrating power of a liquid.
woody perennial. A tree or shrub that lives 3 or more years and flowers at least twice.
xylem. Woody portion of a tree located inside the cambium containing nonliving vascular tissues that conduct water and nutrients from the roots to the leaves.
zoospore. A swimming asexual spore.
zygote. The diploid cell formed by the union of the male nucleus and the egg.

Index

Note: Page numbers in **bold** type indicate major discussions. *Italic* type indicates tables (*t*) and figures (*f*).

Q

R